Biocompatibility
of Co-Cr-Ni Alloys

NATO ASI Series

Advanced Science Institutes Series

A series presenting the results of activities sponsored by the NATO Science Committee, which aims at the dissemination of advanced scientific and technological knowledge, with a view to strengthening links between scientific communities.

The series is published by an international board of publishers in conjunction with the NATO Scientific Affairs Division

A	**Life Sciences**	Plenum Publishing Corporation
B	**Physics**	New York and London
C	**Mathematical**	Kluwer Academic Publishers
	and Physical Sciences	Dordrecht, Boston, and London
D	**Behavioral and Social Sciences**	
E	**Applied Sciences**	
F	**Computer and Systems Sciences**	Springer-Verlag
G	**Ecological Sciences**	Berlin, Heidelberg, New York, London,
H	**Cell Biology**	Paris, and Tokyo

Recent Volumes in this Series

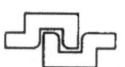

Series A: Life Sciences

Biocompatibility of Co-Cr-Ni Alloys

Edited by

Hartmut F. Hildebrand

Institute of Occupational Medicine
Lille, France

and

Maxime Champy

University Hospital Center of Strasbourg
Strasbourg, France

Springer Science+Business Media, LLC

Proceedings of a NATO Advanced Research Workshop on
Biological Incidences of Co-Cr-Ni Alloys Used in
Orthopaedic Surgery and Stomatology,
held September 30–October 4, 1985,
in Bischenberg Obernay, France

Library of Congress Cataloging in Publication Data

NATO Advanced Research Workshop on Biological Incidences of Co-Cr-Ni Alloys
Used in Orthopedic Surgery and Stomatology (1985: Bischenberg, France)
    Biocompatibility of Co-Cr-Ni alloys / edited by Hartmut F. Hildebrand and
Maxime Champy.
        p.   cm.—(NATO ASI series. Series A, Life sciences; vol. 158)
    "Proceedings of a NATO Advanced Research Workshop of Biological In-
cidences of Co-Cr-Ni Alloys Used in Orthopedic Surgery and Stomatology, held
September 30–October 4, 1985, in Bischenberg Obernay, France"—T.p. verso.
    "Published in cooperation with NATO Scientific Affairs Division."
    Includes bibliographies and index.

    ISBN 978-1-4612-8067-5   ISBN 978-1-4613-0757-0 (eBook)
    DOI 10.1007/978-1-4613-0757-0

    1. Cobalt-chromium-nichel alloys—Biocompatibility—Congresses. 2. Im-
plants, Artificial—Congresses. I. Hildebrand, Hartmut F. II. Champy, Maxime. III.
North Atlantic Treaty Organization. Scientific Affairs. IV. Title. V. Series: NATO
ASI series. Series A, Life sciences; v. 158.
    [DNLM: 1. Biocompatible Materials—congresses. 2. Chromium
Alloys—adverse effects—congresses. 3. Dental Implantation—congresses. 4.
Implants, Artificial—congresses. QT 34 N2788b 1985]
R857.C63N37   1985
610'.28—dc19
DNLM/DLC   88-38501
for Library of Congress   CIP

PREFACE

For several years now scientific and medical staff have recognised the risks of toxicity of certain metals contained in alloys used in the manufacture of biomaterials : protheses, implants, and artificial organs. A number of scientific and industrial research centres have focussed their investigations in this direction and international societies and commissions have organised meetings with specialists from complementary disciplines in attendance in attempts to guage the importance of biological risks and to determine the toxicity of certain metals, with the aim of establishing preventive measures and guidelines.

In the last century great efforts have been made to reduce unwanted biological effects caused by orthopaedic implants. The problems of pain and infection were overcome and the development of modern technology has resulted in a convincing decrease in corrosion problems and mechanical failure, such that ostosynthesis and endoprosthesis have rapidly progressed beyond the level of tentative investingation. However, a number of problems still remain to be solved, such as the influence of the material type on the healing process and its relative speed.

The increasing use of cobalt-,chromium- and nickel-containing alloys in surgical and dental implants has raised various questions concerning the biological consequences of chronic internal release of these elements in the human body. A total of 55 delegates representing 16 countries heard presentations of fundamental aspects, local and remote tissue response, immunopathology, clinical aspects, and manufacturing quality control issues.

The discussion was energetic and heated at times, with informal conversations stretching far into the evenings. A workshop held on the final day drew several widely-supported conclusions. It was generally agreed that there was considerable evidence for unfavourable biological reaction to these alloys, reactions associated primarily with the nickel content of such alloys. The most common such responses are of an immunological nature, sometimes requiring device removal in order to eliminate symptoms. Nevertheless, it was felt that the use of these alloys should continue in those situations where they remain the best choice, based on their critical properties and in the absence of any reasonable alternatives.

It was agreed that comprehensive studies on clinical epidemiology and device retrieval and analysis are needed to better identify the magnitude of clinical problems and to try to distinguish between the effects of the various alloys in use. There is a great need for improved data-gathering systems and clinical tests, especially in immunolgy where is considerable doubt as to the safety and accuracy of the conventional skin (patch) test. Finally, the delegates were pleased by the evidence of the growing number of well-structured, interdisciplinary research groups in many countries which are currently addressing the problems of biological responses to implants.

H. F. HILDEBRAND and M. CHAMPY

ACKNOWLEDGEMENTS

The authors and editors gratefully acknowledge
the World Gold Council for the financial
support allowing the printing of the colour
plates.

CONTENTS

FUNDAMENTAL ASPECTS

TISSUE RESPONSE

IMMUNOPATHOLOGY

CLINICAL ASPECTS

QUALITY CONTROL

CONCLUSIONS AND RECOMMENDATIONS

THE INTERNATIONAL STANDARDS ORGANIZATION

TECHNICAL COMMITTEE ON SURGICAL IMPLANTS

Bernard Bloch

Chairman ISO/TC 150

Central Secretariat Geneva

INTRODUCTION

Implants have been defined as "objects or devices which are surgically placed in the body either temporarily or permanently for diagnostic or therapeutic purposes". This is indeed a broad definition. Practically every surgical procedure thus involves the use of an implant.

There is need for surgical implant standards and assurance of their quality control. The achievements of ISO's technical committee (TC150) on Implants for Surgery, are described.

SURGICAL IMPLANTS

Recent innovations in almost every branch of surgery have been accompanied by, and often were dependent upon, remarkable new materials and composites derived from the fields of metallurgy, chemistry, textiles and bio-ceramics. Processed materials of biological origin must also be included. The surgeon, who is untrained in the materials and engineering sciences, is justifiably bewildered by the confusion, particularly in the polymer field, between chemical and generic names. He is unaware of the different properties engendered by variations in chemical formulation, molecular weight and methods of fabrication and sterilization. He may be unaware that "nylon" is a trade name for polyamide and that there are many variations in type. Or that a cobalt-chromium or titanium alloy may have excellent qualities in tension or bending but is less satisfactory in tension, torsion and bending--the typical forces that exist when inserting a bone screw. He may not know, for instance, that polyamides (such as nylon) lose tensile strength in the tissues, and incorrect grades will degrade. PTFE (Teflon) is hystiocytic when abraded. Dissimilar metals, if both implanted, produce an electromotive force and reaction. If incorrect catheter materials are selected for intravascular and intracardiac use, the result may be embolization due to fragments of material, or cardiac rupture. Often he is unversed in the formulation and composition, details of fabrication, effects of sterilization and of in-service conditions on these materials, and in other important technological details. He thus accepts implants in blind faith.

Implants vary in the service expected of them, and this also must be taken into consideration. For instance a bone plate and screws have served their purpose once the fracture has united, whereas the components of a hip replacement or vascular graft must last the life-span of the patient. Should infection occur then the whole problem of implant compatibility and performance is altered, as is the success of the exercise.

Surgical implants form only a proportion of expenditure compared to the 1984 world-wide assessment of $90 billion turnover in the pharmaceutical industry, but this proportion is not minor. Often manufacturers of implants also make instruments as a further profitable outlet. Ancillary items, such as disposable drapes, syringes, drains, catheters, bandages, transfusion sets, telemetric devices and many others, add to commercial potential.

One could quote figures such as over 500 000 total joint replacements per year worldwide or 250 000 cardiac pace-makers, or that an annual market for internally applied surgical staples in Western countries could be $220 million, in 46 million operations and external skin staples $200 million. Or that basic components and cements in an average total hip joint replacement may vary between $250 to $1000. Or that country X spends 5.6% of its gross national product on health care, and country Y only 2%. Trade figures are not always accurate in detail. Country Z may import $50 million of surgical goods, but these could include $10 million of bedpans and $30 million of bandages. Detailed breakdowns of imports and exports are difficult to obtain, but the potential volume of the implant market is enormous, particularly with the increasing industrialization of developing countries.

MATERIALS

The use of stainless-steel, chromium-cobalt and titanium alloys for orthopaedic implants is a well-established major industry. Composition and mechanical properties continue to improve in quality. These contain varying amounts of Ni and Cr as dictated by properties required. Ceramics and carbon composites also have application.

The four dominant medical polymers are polyvinyl chloride, polycarbonates, polyesters and methyl methacrylate, but growing use of fluoro polymers, cellulosic materials and silicone rubbers is increasing, as is the use of composites. Ultra high molecular weight polyethylene has proved an essential component in joint prostheses.

Unless the source is from an identical twin, biological materials or organ transplants still have limited function, even if rendered immunogen-free. Most such materials are so processed that they become inorganic. Porcine heart valves or artificial skin are examples.

THE MANUFACTURER

The surgeon is indeed indebted to the manufacturer, who produces low-volume, high-cost items with all the worries of biological application. Surgical progress would have been impossible without the many manufacturers of repute.

The engineering sciences have become increasingly oriented towards biological problems, and surgeons have welcomed this. The materials scientist, mechanical engineer, metallurgist, polymer scientist---these and many others form the nucleus of an interdisciplinary approach to the problem. The increasing liaison between surgeon, scientist, and manufacturer is essential.

It used to be customary for the quality of a product to be accepted on the good name of a particular manufacturer. Indeed, this may still be the case, but in the last decade, the number of manufacturing concerns in the "medical" field has greatly increased. Many of these are no longer organizations with particular expertise, but offshoots of large combines covering a wide range of merchandise. This, in such a specialist field, can be a potentially hazardous situation. It is also one that is further complicated as the full impact of international trade agreements and "local" economic communities take effect.

It should be borne in mind that the vast majority of manufacturers in this field are guided by codes of high ethical behaviour and are very community-oriented in their thinking. However, the quality of health care achieved in any one country is a reflection not only of the manufacturers' contributions but also of the calibre and training of clinicians, enlightened policies and administration.

STANDARDS

Standards are an all-pervasive part of life today and affect everything from moral attitudes to systems of measurement. In industry, standards aim at establishing a safe working milieu and exchange of goods, services and information, and are based on the consolidated results of science and technology and of experience.

Herein lies the crux of the problem as it applies to surgical implants. The surgeon, in his desire to alleviate suffering, has often to perform reconstructive procedures in the absence of the full patho-physiological, biochemical and biomechanical data that would allow scientific certainty in the fabrication, design and performance of the intended implant.

A high level of safety of implants is mandatory. The economic and clinical significance of failure of surgical implant procedures can be enormous: for example failure of a total hip replacement can increase costs by a factor of 10. The risk of further complications is increased. In the extreme, the patient may become severely handicapped.

Over decades national standards have been produced in the USA, UK, Australia, Germany F.R. and other countries. The range of these has varied, mostly limited to orthopaedic implants. The exception is the USA where most aspects of the implant field have been covered.

TECHNICAL COMMITTEE 150 "SURGICAL IMPLANTS"

The need for an ISO technical committee on surgical implants was apparent. The author has been the chairman of ISO/TC 150 since its inception in 1972. Australia was the member body acting as the secretariat until late 1982, since replaced by Germany F.R.

The format of TC 150 activities is outlined in Table 1. Participating (P) and observer (O) member bodies are indicated as well as liaising organizations.

National delegations reflect the technological and scientific capabilities of their countries. National or personal influences are minimized by the democratic and loose-knit structure. The calibre and capability of the delegations ensure that at sub-committee level no one body can enforce its wishes. The full technical committee endorses the final recommendations by majority consensus. Delegates may be surgeons, scientists, manufacturers, administrators or others representing the inter-disciplinary interests required.

A topic under study progresses by conference and correspondence. When the technical committee brings draft proposals to a draft international standard for a particular item, this is circulated to all ISO member bodies for comment. Each country's member body is free to consult with any of its national authorities. Finally, all comments are collated by the technical committee secretariat and if there is consensus then the draft standard is sent to the Central Secretariat of ISO in Geneva for publication as an international standard. Such a standard can be updated as required.

Where problems defy immediate technical solution a code of practice may result.

As a general rule, only the permanent officials of ISO member bodies are funded, other delegates have to make their own arrangements for financing the time and travel involved. The use of correspondence, backed by efficient technical committee management, and improving telecommunications facilities will increase the production of international standards.

TABLE 1
ISO/TC150 IMPLANTS FOR SURGERY
Chairman: Dr. B. Bloch
Secretariat: Australia 1972-82 Germany FR 1982-

TC Working Groups (WG's):

WG1-Terminology WG2-Marking/Packaging W3-Certification
WG4-Biocompatibility.

TC Subcommittees (SC) and their Working Groups (WG):

SC1 Orthopaedic: WG1 Materials WG2 Osteosynthesis

SC2 Cardiovascular: WG1 Valves WG2 Pacemakers
                     WG3 Vascular prostheses
                     WG4 Blood oxygenators WG5 Haemodialysis

SC3 Neurosurgical

SC4 Bone and joint replacement: WG1 Fatigue testing

P-members: Australia, Belgium, Canada, France, Germany, F.R.
India, Ireland, Italy, Mexico, Netherlands, Poland, Spain,
Switzerland, Thailand, United Kingdom, USA, USSR

O-members: Austria, Brazil, Chile, China, Colombia,
Czechoslovakia, Denmark, Finland, Hungary, Israel, Japan,
Korea Rep., Malaysia, Norway, Portugal, Romania,
Rep. South Africa, Sweden, Tunisia, Yugoslavia

Liaisons: IFSC. WHO. CCE. ISPO. OECD. SICOT.

TC/150 ACHIEVEMENTS

To date more than 30 standards have been published or
are in the process of finalization. These cover materials
such as stainless steel, cobalt chrome, titanium alloys,
acrylic resin cements, UHMW polyethylene, and ceramics.

Devices in osteosynthesis include a wide range of bone
screws, plates, nails, pins and wires. Cardiovascular
implants refer to heart valves and pacemakers, and in
neurosurgery to shunt assemblies. Hip and knee joint
replacements to date have standards on classification,
dimensions and bearing surfaces. A standard on marking and
packaging of implants has been finalized.

Work proceeds on many items. Examples include
osteosynthesis of the spine, new alloys such as high nitrogen
steel, niobium and vascular prostheses of biological origin.
Also on oxygenators, chemicals and materials used in
haemodialysis, methods of test and evaluation of joint
prostheses. Retrieval analysis of implants is under
consideration.

STATISTICS AND DATA

Although the World Health Organization has published international compendia for coding of diseases and procedures since 1978, this is as yet not universally applied. Prior coding and recording of procedures have varied from being almost non-existent to being detailed and accurate. Nonetheless, accurate statistics on the appraisal of surgical implants and procedures are unavailable at national or even international level. Epidemiological studies in this field are sometimes inaccurate. Some clinics have produced valuable figures, but at best these are guides only and they do not reflect overall solutions at national or international level.

Both surgeons and manufacturers are understandably reticent on the subject of implant failure. Clinical records and thorough scientific evaluation of failures vary enormously in accuracy. Litigation, or fear of it, adds to such reticence. Nevertheless, analysis of failed implants has become of increasing interest. Retrieval analysis is costly and time consuming. It requires close co-operation between the surgeon and the scientists concerned. Histopathology, tissue analysis, metallurgical or materials scientist investigation, physical, chemical and mechanical tests--all form a protocol of such practice. A United States costing gives a figure of $600 for a retrieval analysis of an implant. In Australia, the figure could be halved.

There is intensive multidisciplinary scientific endeavour in the study of biomaterials, compatibility, many laboratory investigations and animal experiments. Yet a vast amount of potential information that can be obtained from the "human experiments" is not investigated.

In some countries product liability and malpractice litigation proceed simultaneously preventing comprehensive tests on failed implants. This raises an ethical dilemma. Who "owns" the implant--the manufacturer, surgeon or patient? Hopefully as less partisan influences prevail the overall benefit of retrieval analysis will be internationally accepted. If not, national regulatory authorities, burdened by increasing health care costs, will surely legislate accordingly.

This important conference addresses itself to "biological incidences of Ni-Cr alloys". These metal ions (and many others) leach out of implants and complex with body proteins. The long term or metabolic effects of such products or others of degradation are undefined. The significance in "permanent" implants such as joint replacements or heart valves is obvious.

CERTIFICATION

At the insistence of the surgeon (and his patient) implants may be manufactured to comply with an accepted standard. Such compliance must be assured by regular surveillance. There are inherent difficulties.

The standard must include all necessary requirements ensuring total fitness for purpose. This may be possible with industrial items but it is not always so with surgical implants.

One can specify materials composition and other properties, to enhance implant performance and safety. However, although one knows what is bio-acceptable from proven clinical experience one cannot as yet specify biocompatibility. Fitness-for-purpose product standards are the ultimate aim, but to await such a "total" approach, seems unnecessary and harmful. A standard must reflect the highest level of current "state of the art" and be updated as required.

As yet no manufacturer of current implants or devices kite marks or makes same under licence in accordance with national or international requirements. To advertise that a product "conforms or complies" does not necessarily confirm compliance unless there is certainty of quality control and independant authoritative verification.

This is a complex problem involving legal trade and possibly political barriers. There are supplier assessment schemes in some countries ensuring good manufacturing practice, but not qualifying the product produced. There are some national "approved product" lists of implants, with similar limitations.

An implant may fail for many reasons, that of the device itself may not always be the key factor.

The high cost of health care and political influences of consumerism will demand quality control and product safety in implants. The activities of ISO/TC150 and other national member bodies are directed to this goal.

I appeal to the scientists at this conference to share their knowledge with ISO/TC150 through their respective national bodies.

Our mutual endeavours in the light of current progress should assist in improving the safety of surgical implants.

FUNDAMENTAL ASPECTS

CARCINOGENIC RISKS OF METAL IMPLANTS AND PROSTHESES

F. William Sunderman Jr.

Departments of Laboratory Medicine and Pharmacology
University of Connecticut Medical School
Farmington, Connecticut 06032, USA

ABSTRACT

Case reports of malignant tumors at sites of metal implants in humans and domestic dogs are reviewed; the results of carcinogenesis bioassays of implanted metallic nickel, chromium, and cobalt powders and alloys are summarized, and in vitro studies relevant to the carcinogenicity of metallic compounds and alloys are briefly discussed. These considerations suggest that local sarcomas may constitute a rare complication of orthopedic implants and prostheses fabricated of Ni, Cr, or Co alloys. The author recommends (a) that implant manufacturers and surgeons develop and use prostheses with minimal susceptibility to metal corrosion and wear; (b) that an international registry of implant-associated tumors be established; and (c) that epidemiological studies be undertaken to quantify cancer risks in patients with metal implants.

INTRODUCTION

In 1984, Hamblen and Carter (1) reviewed the occasional development of local sarcomas in patients following joint replacement with metal prostheses; they discussed the possibility that metal particles produced by wear or metal ions released by corrosion might be initiating factors for carcinogenesis in such cases. Metal pins, plates, and screws have been implanted by orthopedic surgeons for fifty years, and metal arthroplasties have been used to treat many thousands of patients during the past two decades. Ten patients and 24 domestic dogs with malignant tumors that developed at the sites of metal implants have been reported in the literature (2-11). There is current debate whether these sarcomas are coincidental occurrences, or whether they represent rare complications of the surgically implanted therapeutic devices.

The goals of this paper are to summarize the relevant clinical and experimental evidence and to appraise the carcinogenic hazards from implanted metal alloys that contain nickel, chromium, or cobalt. For background information on epidemiological evidence that certain occupational and environmental exposures to metal compounds are associated with excess cancer risks in humans and for comprehensive tabulations of carcinogenesis bioassays of metal compounds in experimental animals, readers should consult pertinent monographs

Table 1. Malignant Tumors at Sites of Metal Implants in Humans

Case	Authors	Age at implant	Site	Description of implant	Time-lapse	Tumor type	Comments
1	McDougall (2)	12 yr	Humerus	Stainless steel plate with screws of	30 yr	Ewing's sarcoma	External corrosion, attributed to 80 mV potential difference of plate and screws
2	Delgado (3)	37 yr	Tibia	Eggers plate with screws	3 yr	Osteo-chondro-sarcoma	Alloy not specified, probably stainless steel
3	Dube and Fisher (4)	54 yr	Tibia	Sherman stainless steel plates with screws of another alloy	26 yr	Hemangio-sarcoma	Corrosion around screw holes; Ni and Cr found in tumor tissue
4	Arden and Bywaters (5)	56 yr	Hip	McKee metal-to-metal total hip arthroplasty	2.5 yr	Fibro-sarcoma	Alloy not specified
5	Tayton (6)	11 yr	Femur	Sherman Vitallium plate with matching screws	7 yr	Ewing's sarcoma	Implant removed after one yr
6	McDonald (7)	31 yr	Tibia	Vitallium plate and screws	17 yr	Histiocytic lymphoma	Corrosion not recorded
7	Dodion et al (8)	49 yr	Femur	MacLaughlin plate, Strycker and Knowles screws (all Vitallium)	1 yr	Immunoblastic lymphoma	Osteosynthesis complicated by infection
8	Bago-Granell et al (9)	75 yr	Hip	Charnley-Muller hip arthroplasty (stainless steel and polyethylene)	2 yr	Fibrous histiocytoma	Polymethylmethacrylate cement was used
9	Swann (10)	63 yr	Hip	Vitallium McKee-Farrar hip arthroplasty (metal-on-metal)	3 yr	Fibrous histiocytoma	Corrosion and wear not recorded; no metal particles seen in tumor
10	Penman and Ring (11)	75 yr	Hip	Vitallium Ring-type hip arthroplasty (metal-on-metal)	5 yr	Osteogenic-sarcoma	Uncemented implant; corrosion and wear not recorded

published by the International Agency for Research on Cancer (IARC) (12-14) and recent review articles (15-18).

TUMORS AT SITES OF METAL IMPLANTS IN HUMANS

The reported patients with malignant tumors at sites of metal implants are listed in Table 1. In five of these ten patients (No. 1, No. 3, No. 5, No. 6, No. 10), the time-lapse between implantation of the metallic device and tumor detection ranged from five to thirty years, which is consistent with a possible etiologic relationship between the metal implant and subsequent tumor development. In two of these cases (No. 1 and No. 3), stainless steel plates were attached to bone with screws of another alloy, resulting in visible corrosion. In the other five patients (No. 2, No. 4, No. 7, No. 8, No. 9), the time-lapse between implantation and tumor detection ranged from one to three years. In these cases, the association of malignancy with metal implants was probably coincidental, since latent periods less than five years have rarely been associated with chemically induced tumors in humans. In patient No. 7, immunoblastic lymphoma was diagnosed in the hip at fourteen months after osteosynthesis of a subcapital fracture of the femoral neck. Infection at the osteosynthesis site may have acted as a localizing factor for a systemic lymphoma in this patient, since malignant lymphoblasts were present in sternal marrow aspirate obtained shortly after biopsy of the hip tumor.

TUMORS AT SITES OF METAL IMPLANTS IN DOMESTIC DOGS

Eight groups of veterinarians have reported a total of 24 sarcomas that developed in dogs around implanted orthopedic pins, nails, plates, and screws, mostly fabricated of stainless steel (Table 2). The tumor types included 21 osteosarcomas and 3 undifferentiated saromas; the elapsed times between the implants and tumor diagnosis ranged from 1 to 12 years, which is consistent with the latent periods for chemically induced tumors in domestic animals. Although delayed healing of fractures and chronic osteomyelitis may have been contributory factors, an etiologic relationship between the metal implants and sarcoma development seems plausible in these dogs.

CARCINOGENESIS BIOASSAYS IN RODENTS

Gaechter et al. (27) tested the carcinogenic activity of seven metallic alloys containing nickel, chromium, cobalt, and/or titanium, which are commonly used in orthopedic prostheses. Solid rods of each alloy were implanted in the gluteal muscles of Sprague-Dawley rats (22 to 34 rats/group). No local tumors, benign or malignant, developed within 24 months; no statistically significant differences in total tumor incidences were observed in the seven treated groups versus the two control groups. Powders of nickel, chromium, and cobalt metals have been tested for carcinogenicity in rodents following intramuscular injection or intramedullary administration into long bones (Table 3). The carcinogenesis bioassays of Ni and Co powders have generally been positive, yielding rhabdomyosarcomas or fibrosarcomas, whereas the carcinogenesis bioassays of Cr powder have been negative. Swanson et al. (37) observed 30% incidence of sarcomas when Co- and Cr-containing particles, prepared from in vitro wear tests of vitallium prostheses, were suspended in horse serum and administered to hooded rats by intramuscular injection. On the other hand, Meachin et al. (38) did not observe any tumors when similar wear particles from vitallium prostheses were implanted as a dry powder in the back muscles of hooded and Wistar rats. The discrepant outcomes of these two bioassays suggest a possible enhancing effect of the foreign proteins in horse serum. The positive

13

Table 2. Malignant Tumors at Sites of Metal Implants in Domestic Dogs

Authors	No. of cases	Sites	Types of implant	Time lapse	Tumor type	Comments
Banks et al (19)	2	radius, tibia	plate and screws, Jonas pin	3.5 and 6 yr	osteosarcomas	alloy not specified
Harrison et al (20)	2	humerus, tibia	Steinmann pin, plate and screws	11 and 5 yr	osteosarcomas	stainless-steel; corrosion in one case
Sinibaldi et al (21)	7	humerus, radius, femur	2 Steinmann pins, 5 Jonas pins, 1 plate and screws	1 to 6 yr	5 osteosarcomas, 2 undifferentiated sarcomas	corrosion around Jonas pins and cerclage wires
Madewell et al. (22)	1	femur	Jonas pin	11 yr	osteosarcoma	
Bennett et al. (23)	1	tibia	Venable plate	12 yr	osteosarcoma	
Brunnberg et al. (24)	3	humerus, tibia	plate and screws, Kutschner nail	4 to 7 yr	osteosarcomas	cerclage wires in one case
Van Bree et al (25)	1	humerus	Kutschner nail	5 yr	osteosarcoma	
Stevenson et al (26)	7	femur, tibia, radius, and ulna	plates and screws	3.5 to 8.5 yr	6 osteosarcomas, 1 undifferentiated sarcoma	stainless-steel, corrosion at contact of plates and screws

carcinogenesis bioassays of metallic Ni and Co, when administered as powders to rodents by parenteral injection (14,28-31,35), furnish indirect support for possible carcinogenic risks of Ni- and Co-containing wear particles and corrosion products derived from metal implants and prostheses.

IN VITRO STUDIES RELEVANT TO METAL CARCINOGENESIS

As summarized in recent reviews (15-18,38,39), positive results for certain compounds of Ni, Cr, and Co are obtained in genotoxicity and mutagenesis tests in bacteria and mammalian cell cultures. Karyotypic anomalies and morphologic transformation are produced by compounds of these metals in mammalian cells. DNA strand-breaks and cross-links, infidelity of DNA polymerase activity, and helical transition of B-DNA to Z-DNA are induced by Ni, Cr, and/or Co compounds in various experimental systems. In intact cells, soluble Cr[VI] compounds generally yield strongly positive results, whereas Cr[III] compounds are generally weakly reactive or negative. Cell membranes have low permeability to Cr[III], but are readily penetrated by Cr[VI], which can undergo intracellular reduction to Cr[III]. The outcome of in vitro mutagenesis tests of chromium compounds is consistent with epidemiological and bioassay evidence that carcinogenic activities are substantially greater for hexavalent chromium compounds than for trivalent chromium compounds (15). Metallic Ni, Cr, or Co powders have seldom been tested in in vitro systems; most of the relevant information is derived from tests of water-soluble compounds or certain highly carcinogenic but relatively insoluble compounds (e.g., crystalline nickel subsulfide, Ni_3S_2). Extrapolation of such in vitro results to the biological activities of wear particles and metal ions released by corrosion of orthopedic prostheses is therefore speculative. However, the frequently positive outcomes of in vitro genotoxicity and mutagenesis tests of Ni, Co, and Cr compounds are compatible with possible carcinogenic hazards from implants that contain these metals.

HYPOTHETICAL MECHANISMS OF METAL CARCINOGENESIS

Several hypothetical mechanisms of metal carcinogenesis have been discussed in detail elsewhere by the present author (15). [A] Metals may bind covalently to DNA, causing strand-breaks and excision of nucleotides, leading to frame-shift mutations during DNA repair. [B] Metals may induce DNA-protein cross-links, causing aberrant DNA replication or repair, and disturbing mitosis. [C] Metals may cause transition of B-DNA to Z-DNA, altering chromatin structure and derepressing oncogenes. [D] Metals may impair the fidelity of DNA replication by reacting with substrate-binding or catalytic sites of DNA polymerase, disturbing complementary base-pairing. [E] Metals may bind to histones, non-histone nuclear proteins, or RNA, influencing chromatin structure, and possibly modifying phosphorylation of regulatory proteins. Finally, [G] metals may induce DNA damage by indirect mechanisms mediated by intracellular production of oxygen free-radicals (40). The induction of lipid peroxidation by Ni and Co compounds (41-44) could be related to postulated free-radical mechanisms of metal carcinogenesis, since certain aldehydes produced by lipid peroxidation (e.g., malondialdehyde, 4-hydroxynonenal) are mutagenic and genotoxic in various in vitro test systems (45).

CONCLUSIONS AND RECOMMENDATIONS

Scattered reports of sarcomas at sites of metal implants in humans and domestic dogs have implicated carcinogenesis as a potential, albeit rare, complication of orthopedic prostheses fabricated of Ni, Cr, or Co

Table 3. Carcinogenesis Tests of Ni, Cr, and Co Powders by Intraosseous or Intramuscular Administration to Rodents

Authors	Material and vehicle	Species	Dose and route[a]	Local tumors and incidence
Hueper (28)	Ni powder in gelatin	Rabbit	31 mg, io	Fibrosarcoma (1/6)
		Wistar rat	5 mg, io	Fibro- and rhabdomyo-sarcomas (27/100)
Heath and Daniel (29)	Ni powder in fowl serum	Hooded rat	28 mg, im	Rhabdomyo-sarcomas (10/10)
Friedman and Bird (30)	Ni sponge	Sprague-Dawley rat	20 mg, im	Rhabdomyo-sarcomas (6/25)
Furst and Schlauder (31)	Ni powder in trioctanoin	Hamster	25 mg,[b] im	Sarcomas (2/50)
		Fischer rat	25 mg,[b] im	Fibrosarcomas (38/50)
Sunderman (14)	Ni powder in penicillin	Fischer rat	14 mg, im	Fibro- and rhabdomyo-sarcomas (13/20)
Hueper (32)	Cr powder in gelatin	Wistar rat	45 mg, io	No tumors (0/25)
	Cr powder in lanolin	Osborne-Mendel rat	45 mg, io	Fibroma (1/25)
Sunderman (33)	Cr powder in penicillin	Fischer rat	2 mg, im	No tumors (0/24)
Sunderman (34)	Cr powder in pencillin	Fischer rat	4 mg, im	No tumors (0/20)
Heath (35)	Co powder in fowl serum	Hooded rat	28 mg, im	Rhabdomyo- and fibrosarcomas (18/30)
Swanson et al (36)	Co-Cr powder[c] in horse serum	Hooded rat	28 mg, im	Fibro-, osteo-, and rhabdomyo-sarcomas (22/72)
Meachim et al (37)	Co-Cr powder[c]	Wistar and hooded rats; guinea pigs	20 mg, im	No tumors

[a] Intraosseous = io; intramuscular = im.
[b] 5 monthly injections of 5 mg.
[c] Prosthesis wear particles (65-68% Co, 27-28% Cr, 4-6% Mo).

alloys. Since physicians and veterinarians are not required to report such possible adverse reactions to prostheses, the actual incidences of sarcomas in association with metal implants cannot at present be reliably estimated, and the possibility that the apparent associations are coincidental cannot be excluded. Local sarcomas have been observed in rodents after parenteral injection of metallic Ni or Co powders, but not after injection of metallic Cr powder. In vitro genotoxicity and mutagenesis tests have yielded positive results with Ni, Co, and Cr[VI] compounds, consistent with the possibility that certain Ni, Co, and Cr compounds and alloys may be carcinogenic. In the light of these considerations, the author offers the following recommendations: [A] prosthesis manufacturers and surgeons should be cognizant of the potential problem and should develop and use prostheses with minimal propensity to liberate corrosion products and wear particles that contain Ni, Co, and Cr; [B] an international registry of implant-associated tumors should be established to compile cases and to assess their frequency; and [C] epidemiological studies should be undertaken to test the hypothesis that cancer risks are increased in patients with metal implants.

REFERENCES

1. Hamblen, D.L. and Carter, R.L., Sarcoma and joint replacement, J. Bone Joint Surg., 66B:625-626, 1984.
2. McDougall, A., Malignant tumor at site of bone plating, J. Bone Joint Surg., 38B:709-713, 1956.
3. Delgado, E.A., Sarcoma following a surgically treated fractured tibia, Clin. Orthopedics, 12:315-318, 1958.
4. Dube, V.E. and Fisher, D.E., Hemangioendothelioma of the leg following metallic fixation of the tibia, Cancer, 30:1260-1266, 1972.
5. Arden, G.P. and Bywaters, E.G.L., Tissue reaction, in: "Surgical Management of Juvenile Chronic Polyarthritis" (Arden, G.B. and Ansel, B.M., eds.), Academic Press, London, 1978, pp 269-270.
6. Tayton, K.J.J., Ewing's sarcoma at the site of a metal plate, Cancer, 45:413-415, 1980.
7. McDonald, I., Malignant lymphoma associated with internal fixation of a fractured tibia, Cancer, 48:1009-1011, 1981.
8. Dodion, P., Putz, P., Amiri-Lamraski, M.H., Efira, A., Martelaere, E. de, and Heimann, R., Immunoblastic lymphoma at the site of an infected vitallium bone plate, Histopathology, 6:807-813, 1983.
9. Bago-Granell, J., Aguirre-Canyadell, M., Nardi, J., and Tallada, N., Malignant fibrous histiocytoma of bone at the site of a total hip arthroplasty, J. Bone Joint Surg., 66B:38-40, 1984.
10. Swann, M., Malignant soft-tissue tumour at the site of a total hip replacement, J. Bone Joint Surg, 66B:629-631, 1984.
11. Penman, H.G. and Ring, P.A., Osteosarcoma in association with total hip replacement, J. Bone Joint Surg., 66B:632-634, 1984.
12. Anon., Nickel, in: "IARC Monographs on the Evaluation of Carcinogenic Risk of Chemicals to Man", Intern. Agency Res. Cancer, Lyon, 1976, pp 75-112.
13. Anon., Chromium and chromium compounds, in: "IARC Monographs on the Evaluation of the Carcinogenic Risk of Chemicals to Humans", Intern. Agency Res. Cancer, Lyon, 1980, pp 205-230.
14. Sunderman, F.W., Jr., (ed-in-chief), Nickel in the Human Environment, Intern. Agency Res. Cancer, Lyon, 1984, pp 3-266.
15. Sunderman, F.W., Jr., Recent advances in metal carcinogenesis, Ann. Clin. Lab. Sci., 14:93-122, 1984.
16. Kappus, H., Carcinogenic effects of metal compounds, Umwelt. Hyg., 14:83-95, 1982.

17. Helmes, C.T., Casey, S., Fung, V.A., Johnson, OH.H., McCaleb, K.E., Miller, A., Miller, J., Papa, P., Sigman, C., and Strauss, E., A study of metals for the selection of candidates for carcinogen bio-assay, J. Environ. Sci. Health, A18:203-295, 1983.

18. Belman, S. and Nordberg, G., (eds.) Symposium on metal carcino-genesis, Environ. Health Persp. 40: 1-245, 1981.

19. Banks, W.C.., Morris, E., Herron, M.R., and Green, R.W., Osteogenic sarcoma associated with internal fracture fixation in two dogs, J. Am. Vet. Med. Assoc., 167:166-167, 1975.

20. Harrison, J.W., McLain, D.L., Hohn, R.B., Wilson, G.P., Chalman, J.A., and MacGowan, K.N., Osteosarcoma associated with metallic implants, Clin. Orthop. Rel. Res., 116:253-255, 1976.

21. Sinibaldi, K., Rosen, H., Liu, S.K., and DeAngelis, M., Tumors associated with metallic implants in animals, Clin. Orthop. Rel. Res., 118:257-266, 1976.

22. Madewell, B.R., Pool, R.R., and Leighton, R.L., Osteogenic sarcoma at the site of a chronic nonunion fracture and internal fixation device in a dog, J. Am. Vet. Med. Assoc., 171:187-189, 1977.

23. Bennett, D., Campbell, J.R., and Brown, P., Osteosarcoma associated with healed fractures, J. Small Anim. Pract., 20:13-18, 1979.

24. Brunnberg, V.L., Gunsser, I., and Hanichen, T., Knochentumoren beim hund nach trauma und osteosyntheses, Kleintier. Prax., 25: 143-152, 1980.

25. Van Bree, H., Verschooten, F., and Hoorens, J., Internal fixation of a fractured humerus in a dog and late osteosarcoma development, Vet. Rec., 107:501-502, 1980.

26. Stevenson, S., Hohn, R.B., Pohler, O.E.M., Fetter, A.W., Olmstead, M.L., and Wind, A.P., Fracture-associated sarcoma in the dog, J. Am. Vet. Med. Assoc., 180:1189-1196, 1982.

27. Gaechter, A., Alroy, J., Andersson, G.B.J., Galante, J., Rostoker, W., and Schajowicz, F., Metal carcinogenesis: A study of the carcinogenic activity of solid metal alloys in rats, J. Bone Joint Surg, 59A:622-624, 1977.

28. Hueper, W.C., Experimental studies in metal cancerigenesis. IV. Cancer produced by parenterally introduced metallic nickel, J. Nat. Cancer Inst., 16:55-67, 1955.

29. Heath, J.C. and Daniel, M.R., The production of malignant tumors by nickel in the rat, Brit. J. Cancer., 18:261-264, 1964.

30. Friedmann, I. and Bird, E.S., Electron microscope investigation of experimental rhabdomyosarcoma, J. Pathol., 97:375-382, 1969.

31. Furst, A., and Schlauder, M.C., The hamster as a model for metal carcinogenesis, Proc. West. Pharmacol. Soc., 14:68-71, 1971.

32. Hueper, W.C., Experimental studies in metal cancerigenesis. VII. Tissue reactions to parenterally introduced powdered metallic chromium and chromite ore, J. Nat. Cancer Inst., 16:447-469, 1955.

33. Sunderman, F.W., Jr., Lau, T.J., and Cralley, L.J., Inhibitory effect of manganese upon muscle tumorigenesis by nickel sulfide, Cancer Res. 34:92-95, 1974.

34. Sunderman, F.W., Jr., McCully, K.S., Taubman, S.B., Allpass, P.R., Reid, M.C., and Rinehimer, L.A., Manganese inhibition of sarcoma induction by benzo(a)pyrene in rats, Carcinogenesis, 1:613-620, 1980.

35. Heath, J.C., The production of malignant tumours by cobalt in the rat Brit. J. Cancer, 10:668-673, 1956.

36. Swanson, S.A.V., Freeman, M.A.R., and Heath, J.C., Laboratory tests total joint replacement protheses, J. Bone Joint Surg., 55B:759-773, 1973.

37. Meachim, G., Pedley, R.B., and Williams, D.F., A study of carcinogenicity associated with Co-Cr-Mo particles implanted in animal muscle, J. Biomed. Mat. Res., 16:407-416, 1982.

38. Baker, R.S.U., Evaluation of metals in in vitro assays, interpretation of data and possible mechanisms of action, in: "Carcinogenic and Mutagenic Metal Compounds", (Merian, E., Frei, R.W., Hardi, W., and Schlatter, C., eds.), Gordon and Breach, New York, 1985, pp 185-206.

39. Hansen, K. and Stern, R.M., A survey of metal-induced mutagenicity in vitro and in vivo, in: "Carcinogenic and Mutagenic Metal Compounds", (Merian, E., Frei, R.W., Hardi, W., and Schlatter, W., eds.), Gordon and Breach, New York, 1985, pp 207-211.

40. Schlatter, C., Speculations on mechanisms of metal carcinogenesis, in: "Carcinogenic and Mutagenic Metal Compounds", (Merian E., Frei, R.W., Hardi, W., and Schlatter, C., eds.), Gordon and Breach, New York, 1985, pp 529-539.

41. Kuno, Y., Tochihara, N., and Koike, S., The effects of cobalt chloride on the formation of blood lipid peroxide related to glutathione peroxidase in the erythrocytes of rabbits, Japan J. Hyg., 35:665-669.

42. Morita, H., Kuno, Y., and Koike, S., Effects of cobalt on superoxide dismutase activity, methemoglobin formation, and lipid peroxide in rabbit erythrocytes, Japan J. Hyg., 37:597-600, 1982.

43. Sunderman, F.W., Jr., Marzouk, A., Hopfer, S.M., Zaharia, O., and Reid, M.C., Increased lipid peroxidation in tissues of nickel chloride-treated rats, Ann. Clin. Lab. Sci., 15:229-236, 1985.

44. Donskoy, E., Donskoy, M., Farouhar, F., Gillies, C.G., Marzouk, A.B., Reid, M.C., Zaharia, O., and Sunderman, F.W., Jr., Hepatic toxicity of nickel chloride in rats, Ann. Clin. Lab. Sci., 16:108-117, 1986.

45. Brambilla, G., Bassi, A.M., Faggin, P., Finollo, R., Martelli, A., Sciaba, L., and Mariani, U.M., Genotoxic effects of lipid peroxidation products, in: "Free Radicals in Liver Injury", (Poli, G., Cheesemen, K.H., Dianzani, M.V., and Slater, T.F., eds.), IRL Press, Oxford, 1985, pp 59-70.

THE TOXICITY OF CHROMIUM IN THE ZERO AND TRIVALENT FORM

U. Glaser

Fraunhofer Institute for Environmental
Chemistry and Ecotoxicology
D-5948 Schmallenberg-Grafschaft, F.R. Germany

Chromium is evaluated to contribute to a significant risk for the human health. But it is general accepted that the health effects of chromium for example irritation of skin and the upper respiratory tract, induction of contact hypersensitivity, of genotoxicity as well as the property to induce lung tumor and nephrotoxicity are mainly due to hexavalent chromium compounds exposed at very few occupational sites. In contrast trivalent chromium and chromium in the zero oxidation state are treated to be relatively 'inert' and not toxic to man and other species, because of their lower bioavailability (1-6). This paper will summarize the available data from literature on the acute and long-term exposure results of human and experimental studies.

Daily intake, metabolism and physicochemical properties:

Toxicological evaluations have to consider that chromium is an essential metal. The nutritional function of chromium depends on the trivalent form (1). In most humans the dietary Cr requirement appears to be satisfied by the ingestion of 50 - 200 micrograms of Cr per day as simple Cr(III) salts or less as specific bound Cr for example as the 'glucose tolerance factor' (7). There is evidence for an above average Cr nutritional requirement, especially in pregnant multiparous women, the elderly and the insulin-dependent diabetics.

From the data given by (8) it seems that there is a correspondence between the human requirement of chromium and the daily intake. The uptake of airborne chromium contributes only few percent (1-3 %) of the daily intake. However, depending on the chemical form and the particle size there is an estimated absorption rate of about 20 %. Cr(III) in food and drinks (90-120 µg/70 kg BW) is generally not well absorbed in the digestive tract. For simple Cr(III) salts less than 1 % absorption was the average literature value. But incorporated into plant-chromium complexes it was measured (9,10) that 20 to 30 % was bioavailable. In a contaminated environment, men and animals may be exposed to a variety of chemical forms of chromium. To assess toxicity, it is necessary to consider not only the total amount of chromium in the biosphere, but also its chemical forms. On the bioavailability of both the air-borne and ingested chromium further research is needed.

In simple solutions as well as in tissues trivalent chromium never exists as a free ion, but in a coordination form (1). The rate of ligant exchange of chromium(III) complexes and chelates is very low. Chromium(III) complexes are kinetically stable in solutions and in addition to their electrochemical stability these properties contribute to their 'relative inertness'. For simple Cr(III) salts at near neutral pH in the physiologic milieu hydroxyl ions replace the coordinated water from the metal and form bridges, linking chromium into large insoluble complexes. Coordination of Cr(III) to ligants is prerequisite for its biological function, and determine the low permeability of membrans for Cr(III) and thus the acute toxicity and other short-term effects of these compounds (1).

Acute toxicity effects

The NIOSH registry data (11) on the acute oral toxicity of Cr(III) compounds range from 1.8 to 3.3 g/kg BW in rats. It is obvious that in rats the oral LD 50 of the Cr(VI) compound sodium chromate (0.15-0.18 g/kg BW) (6) is about ten times or more lower than the LD 50's of the Cr(III) compounds. Their acute toxicities may mainly depend on their solubilities. But another result has also been published. In a recent study (12) the acute toxicities of chromium(III) chloride, nitrate, and sulfate were investigated. In comparison to potassium di-chromate and chromium(VI) oxide the Cr(III) compounds were administered to mice by intraperitoneal injection. The acute mortality within 3 days post-injection was much greater with the Cr(VI) compounds. However, 10 days post-injection the LD 50 doses averaged 18 mg/kg body weight, regardless of the oxidation state of chromium. The mice dosed once with Cr(III) also retained 7-fold more chromium at 3 weeks than the mice treated with chromium(VI). This interesting result, which has to be proved, shows a delayed bioavailability and toxicity of the chromium(III) compounds.

Chronic toxicity effects

In literature there were no observations indicating the long-term toxicity of <u>ingested</u> chromium in the zero and trivalent form, neigther in the men, nor in the experimental animals. In one experimental study (13) 5 ppm of chromium(III) as an acetate were added to the drinking water of male and female rats and mice throughout their lives. The authors did not find significant adverse effects. In another study (14) bread, contaminated with 1-5 % chromium(III) oxide was fed on 5 days a week to rats throughout their lives again without signifi-cant toxicity effects, except of mammary fibrosacromas.

However, with regard to the lung as the target tissue of chro-mium, several results suggest an inhalation hazard from chro-mium(III) in the chromium manufacturing industry. Pneumoconi-osis and reduced lung function seem to be rare lung effects not only in electroplating industry due to the irritive action of chromium(VI) (3), but also for working places, where dusts consisting of chromium(III) predominate. There is one report (15) on pneumoconiosis in chromite miners in South Africa,

where the responsible agents may be dusts that consisted of
chromium(III), but also chromium(VI) was questionable. Other
authors (16) described a chromite induced pneumoconiosis in
chromate workers. The lung chromite deposits contained
chromium(III), but again probably accompanied by chromium(VI).
There was also sufficient evidence to indicate an inhalation
hazard in the ferro-chrome industry, where again pneumoconiosis was diagnosed by (17) and (18). Another study (19) found
an excess of reduced lung function in workers of a ferro-
chromium and a ferrosilicon-producing plant. However, also in
these cases the responsible agents could not be identified.
Two of these studies (18,19) stated that besides of chro-
mium(III) the dust consisted mainly of silica.

The available experimental results do not clarify this situation. By two russian authors (20,21) was reported that long-
term exposure to chromite dust, chromium(III) oxide and chromium phosphate at exposure levels between 40 and 400 mg/m^3 can
induce bronchial irritations in rats. These few data need
further confirmations.

Dermal toxicity effects

The acute irritative dermatitis resulting from the effects of
chromium alone is attributed to the action of hexavalent
chromium (3) and never correlated with the exposure to zero
and trivalent chromium.

However, from their results several authors (22-24) believe
that trivalent chromium contributes to the risk to induce
allergic contact dermatitis. In contrast to nickel the chrome
allergic dermatitis is mainly associated with occupation (25).
There is no doubt that at several professions chromium(VI) is
the most important exzematogenic valency. But the incidence of
chrome dermatitis in workers associated with chromite ore, the
chromium(III) pigments, lubricating oils consisting of chro-
mium(III) and in workers associated with tanning materials
demonstrates that the contact to chromium(III) can also be
responsible (26).

Several investigators demonstrated cross reactions between
chromium(VI) and chromium(III). (25) described their results
of a well reproduced experiment, where in combination with
Freud's complete adjuvant guinea pigs can be sensitized whith
equal strength by intradermal injection of both chromium(VI)
and chromium(III) chloride given in equimolar doses. However,
the elicting tests (epicutaneous und intradermal tests) were
continually stronger with hexavalent chromium than with trivalent. In human, there is little published information regarding to the sensitizing capacity of both chromium valencies
(27). One study (28) using the maximization procedure, obtained sensitation rates of 48 and 56 % with chromium(III)
sulfate and chromium(VI) oxide in predictive tests on healthy
subjects.

As in guinea pigs chromium hypersensitive patients showed
different skin reactions either elicited by chromium(VI) or
chromium(III) compounds. It was indicated by (26), that hexavalent chromium was a strong elicitor, while probably in
dependence on solubility the reaction to trivalent chromium
alum, nitrate and acetate was poorer. These results confirm

the statement (29), that the sensitizing and eliciting action of chromium does not depend on the oxidation state alone, but also on the solubility of the compound. After penetration of the chromium(III) compounds or reduction of hexavalent chromium only chromium(III) binds to the proteins of the skin forming immunogenic hapten-protein conjugates (25,30).

The sensitizing activity resulting from metal prostheses is the subject of several other authors in this workshop that only few data should be given. It has been shown by several investigations that from long-term metal – to metal prostheses and in a lower extent from metal – to plastic prostheses of different materials several sensitizing metals can be liberated (31,32,33) and may induce hypersensitivity.

Thus, three retrospective studies (32,34,35) confirmed the result that namely patients with corrosions and or failures of long-term metal – to metal protheses showed positive skin reactions to several metals. Under the metals cobalt was the strongest sensitizer, while chromium was the poorest under the metals investigated. But it is well known that patients suffering from chromium-induced allergic dermatitis cross-react to other metals, in particular, to cobalt and nickel (36).

Systemic toxicity effects

Important systemic effects were not observed with respect to the exposure to Cr^0 and Cr^{3+}. In only few cases chromium(III) induced kidney diseases as a result from accidental dermal uptake of Cr(III) sulfate solubilized in acids (37). There are also no literature data indicating reproductive and teratogenic hazards for human subjects. However, in mice it was found (38) that chromium(III) chloride, intraperitoneally injected at doses of 9.8-24.4 mg Cr/kg bw., induced fetal retardations as well as malformations. This result was confirmed by another recent investigation (39) with mice, where it was indicative that high doses of chromium(III) injected into the dams induce trophic effects on the fetuses. However, probably the results of these high dosed experiments will be without relevance for human.

Cytotoxicity and genotoxicity effects

Cr(III) was found in RNA and DNA from diverse sources. Thus, it probably contributes to the stabilization of the structure of nucleic acids (40). Higher exogenic chromium(VI) as well as chromium(III) doses, however, results in cytoxicity studied by (41). Cultures of HeLa and rat embryonic cells were incubated with CrVI and various CrIII compounds for 3 days and the 50 % inhibitory doses (ID 50) for cell growth were determined. For HeLa cells the ID 50's varied over a wide range: oxalate: 4.0; acetate: 52.0; nitrate: 720.0; chloride: 1030.0; and bromide 1500.0 µg/ml, while with sodium chromate the ID 50 was 0.3 µg/ml. Similar results were found in the rat embryonic cell cultures. Thus, according to the solubility of the trivalent compounds the inhibiting effect on cell growth was ten to 500 times lower for the chromium(III) compounds compared to a soluble chromium(VI) compound.

A significant different result was obtained in a recent experiment (42) given as a representative for other cyto- and genotoxicity studies summarized by (43,44). In this study the cyto- and genotoxic effects of trivalent chromium consisted in tanning compounds and of the pigment chromite were compared with the action of analytical grade chromium(III) compounds and hexavalent chromium. By the use of water and hydrochloric acid it was also tried to solubilize the compounds to measure the liberated chromium(VI) by the coloured reaction with diphenyl-carbazid. Neigther cytotoxicity nor genotoxicity effects were found for chromium chloride and chromium nitrate purchased from Merck. But there was a cytotoxic level and also a positive clastogenic effect with chromium nitrate purchased from Riedel De Haën due to its contamination with 2 % soluble chromium(VI). Few parts of contaminating CrVI were also found in chrome alum and in the mostly unspecified tanning materials. Under these compounds only chrome alum was genotoxic probably due to its chromium(VI) content. But this compound was comparatively low cytotoxic, while the other tannings were cytotoxic, although containing negligible amounts of chromium(VI). The most surprising result in this study was obtained with the chromite, used. It may be not regular for chromite to contain 70 % of soluble chromium(VI), inducing similar cytotoxicity and genotoxicity as potassium chromate. However, this result demonstrates that chromium(III) compounds cannot be treated as inert substances. The mutagenicity and genotoxicity of uncontaminated trivalent chromium compounds is just the matter of discussion and need further investigations with respect to the prediction of the carcinogenic risk. But it has to be considered that besides the unnumerous positive and negative results obtained in vitro (44), there is up to now no result proving the in vivo genotoxicity of Cr^0 and Cr^{3+} to human subjects as well as in experimental animals.

Carcinogenic effects

For human subjects neigther clinical nor occupational results clearly indicate a carcinogenic risk from chromium in the zero and trivalent form (2,43,45). The suggestion that liberated chromium from long-term prostheses consisting of cobalt-chromium alloys is the causative agent for neoplasms in adjacent tissues may be another subject of this workshop. However, at present no conclusives have been presented a carcinogenic hazard for patients using protheses (46). The results from some animal experiments may describe a risk from several chromium compounds in the zero and trivalent oxidation state.

A carcinogenic potency was not shown for chromium metal. In a series of carcinogenicity studies (44) chromium metal was injected into mice, rats and rabbits in various doses and by several routes. They found only single tumors, all were of different origin. Two other studies (48,49), too, did not find increased incidences of specific neoplasms caused by chromium metal powders alone. But in one of these studies (49) intratracheally instilled chromium metal significantly increased the rate of squamous cell carcinomas in rat lungs induced by methylcholanthren. This result indicates the cocarcinogenic potency of chromium metal.

A real carcinogenic potency was shown for wear particles derived from cobalt-chromium alloys. These particles were intramuscularly injected into rats in a dose of 28 mg. The investigators (50) found injection-site sarcomas and other tumors in 7/74 treated rats.

Although it was shown that powders of chromite ore can be contaminated with chromium(VI) (42), a series of carcinogenicity studies by (47) was not indicative that unroasted chromite has a carcinogenic potency after intrapleural or intramedullar injection into rats or intravenious injection into rabbits. However, several other studies (51,53) showed low carcinogenic incidences of roasted chromite particles, intrapleurally or intramuscularly implanted into rats.

It was mentioned above that ingested chromium(III) compounds given to rats and mice for life-time did not induce adverse effects (13,14), except of higher incidences of mammary fibrosarcomas (7/180) in one of these studies (14). Injected chromium(III), however, was sometimes carcinogenic in the experimental animals, although there may be only a weak carcinogenic potency of these compounds. In one study (53), CrIII acetate was implanted intramuscularly and intrapleurally into rats. Only 1/34 implantation-site tumor was found for each method. In another study (54) the same result was observed after intramuscular injection of eight 25 mg portions of this compound. But no injection-site tumor resulted after intrapleural implantation of this dose. No implantation-site tumor was also found with an 20 mg CrIII oxid implant given to rats intrabronchially (55). This was a confirmed result (56), while in a series of russian studies (57) different doses of chromium(III) oxid injected intratracheally, intrapleurally or intraperioneally resulted in low carcinogenic incidences in rats. In mice several high intraperitoneal injections of CrIII sulfate did not increase the lung tumor incidence (58). Therefore there is may be a weak carcinogenic potency of both zero and trivalent chromium in experimental animals, but these data are inadequate to evaluate the carcinogenic risk for the human (2,4,45).

Conclusions

There are distinct differences between the results obtained from human studies and those from experimental animals. In human beings there was seen no acute and long-term toxicity of ingested chromium in the zero and trivalent form. There are some difficulties in the evaluation of inhaled dusts consisting of these compounds, probably due to other contaminating agents that caused pneumoconiosis and reduced lung function at few working places. It was seen no irritative dermal effects. But there may exist a risk from sensitation by trivalent chromium, especially by chromium liberated from corrosions or failures of metal - to metal prostheses. No genotoxicity and also no carcinogenicity was observed in human subjects. Summarizing these facts, it must be stated that till now there is no risk from environmental and occupational contamination by these compounds for men, except for patients with long-term metal prostheses.

Experimental animal research is necessary to evaluate the toxic potency of contaminating compounds. The chosen data are indicative that single-injected high doses of chromium(III) to mice may be as toxic as soluble hexavalent chromium to induce subacute lethality, whereas ingestion of chromium(III) compounds was not toxic. On the other hand there were chronic toxicity results from inhaled dusts consisting of this Cr form. These studies need further confirmations. Several authors improved that trivalent chromium in a soluble form has the potency to induce allergic contact dermatitis in experimental animals. Genotoxicity was not proved from experimental animal results. Genotoxicity was only obtained after in vitro incubation of pro- and eucaryotic cells. It is just the matter of discussion, whether trivalent or only contaminating hexavalent chromium is the causative agent. A weak carcinogenic potency was shown from both the zero and trivalent forms of chromium. Again these results need further confirmations. All the positive results from experimental animal research have to be confirmed, but already today it is necessary to state that chromium compounds of the zero and trivalent forms, although not very bioavailable, are not "inert" substances.

References

1. Mertz, W.: Physiol.Rev. 49, 163-239 (1969).

2. IARC (International Agency for Research on Cancer) monographs on the evaluation of carcinogenic risk of chemicals to man. Vol.2, IARC, Lyon (1973).

3. Langard, S. & T. Norseth: Chapter 22 In: Friberg, L., G. F. Nordberg & V.B. Vouk (Edit): Handbook on the toxicology of metals. Elsevier/North-Holland Biochemical Press, Amsterdam, New York, Oxford (1979).

4. Hayes, R.B. In: Lilienfeld, A.H. (Edit.): Reviews in Cancer Epidemiology Vol 1. Elsevier/North-Holland Biochemical Press, Amsterdam, New York, Oxford (1980).

5. Korallus, U.& N. Leonhoff: Arbeitsmed. Sozialmed. Präventivemedizin 12, 285-289 (1981) (In German).

6. Hertel, R.F.: Staub-Reinhalt.Luft 42, 135-137 (1982) (In German).

7. NAS/NRC: Recommended dietary allowances. 9th Edition. National Academy of Science, National Research Council, Washington, D.C. (1980).

8. Bonner, F.W. & D.V. Parke. Kapitel I 8b. In: Merian, E. (Edit): Metalle in der Umwelt. Verlag Chemie, Weinheim (1984) (In German).

9. Kieffer, F.: Sandoz Bulletin 52, 18-19 (1979).

10. Starich, G.H. & C. Blincoe: The Science of the Total Environment 28, 443-454 (1983).

11. NIOSH (National Institute of Occupational Safety & Health) Registry of toxic effects of chemical substances. Cincinnati, Ohio 45226 (1983).

12. Bryson, W.G. & C.M. Goodall: Carcinogenisis 4, 1535-1539 (1983).

13. Schroeder, H.A., J.J. Balassa & W.H. Vinton, Jr.: J. Nutrition 83, 239-250 (1964).

14. Ivankovic, S. & R. Preussmann: Food Cosmet. Toxicol. 13. 347-351 (1975).

15. Sluis-Cremer, G.K. & R.S. Du Troit: Brit. J. Medicine 25, 63-67 (1968).

16. Mancuso, T.F. & W.C. Hueper: Ind.Med.Surg. 20, 358-363 (1951).

17. Princi, F., L.H. Miller, A. Davies & J. Cholak: J. Occup. Med. 4, 301-310 (1962).

18. Davies, J.C.A.: Centr. Africa J. Med. 20, 140-143 (1974).

19. Langard, S.: Int. Arch. Occup. Environ. Health 46, 1-9 (1980).

20. Roschina, T.A.: Gig.tr.prof.zabol. 4, 28-32 (1959) (In Russian).

21. Blokin, V.S. & F.S. Trop: In: N.P. Dubinin (Edit.): Genetic effects of the pollution of the environment. Moscow, Nauka, pp. 173-176 (1977) (In Russian).

22. Kämmerer, H. & H. Michel: Allergische Diathese und allergische Erkrankungen. Bergmann, München, 3. Auflage (1956) (In German).

23. Morris, G.E.: A.H.A. Arch. Derm. (Chicago) 78, 612-618 (1958).

24. Fregert, S. & H. Rorsman: Arch. Derm 91, 233-234 (1965).

25. Polak, L., J.L. Turk & J.R. Frey: Prog. Allergy 17, 145-226 (1973).

26. Zelger, J.: Arch. Klin. Exp. Derm. 218, 499 (1964) (In German).

27. Marzulli, F.N. & H.J. Maibach: Chapter 13. In: Marzulli, F.N. & H.I. Maibach (Edit): Dermatotoxicology, Hemisphere Publishing Corporation, Washington, New York, London (1983).

28. Kligman, A.M.: J. Invest. Dermatol. 47, 393-409 (1966).

29. Bockendahl, H.: Derm. Wschr. 130, 987. (1954) (In German).

30. Samitz, M.H., S.A. Katz, D.M. Scheiner & P.R. Gross:
     Acta derm.-venereol. 49, 142-146 (1969).

31. Coleman, R.F., J. Harrington & F.T. Scales: Brit. Med.
     J. i, 527-529 (1973).

32. Evans, E.M., M.A.R. Freeman, A.J. Miller & B. Vernon-
     Roberts: J.Bone Jt. Surg. 56 B, 626-642 (1974).

33. Minski, M.J. & H.S. Dobbs: In Brätter, P. & P. Schramel
     (Edit): Trace element analytical chemistry in medicine
     and biology. De Gruyter & Co., Berlin, New York (1980).

34. Elves, M.W., J.N. Wilson, J.E. Seales & H.B.S. Kemp:
     Brit. Med. J. 4, 376-378 (1975).

35. Munro-Ashman, D. & A.J. Miller: Contact Dermatitis 2,
     65-67 (1976).

36. Clark, R. & G. Kunitsch: Berufsdermatosen 20, 222-238
     (1972) (In German).

37. Kelly, W.F., P. Ackrill, J.P. Day, M. Ottara, C.T. Tye,
     J. Burton, C. Orton & M. Harris: Brit.J.Ind.Med. 39,
     397-400 (1982).

38. Matsumoto, N., S. Iijima & H. Katsunuma: J. Toxicol,
     Sci. 2, 1-13 (1976)

39. Danielsson, B.R.G., E. Hassoun & L. Dencker:
     Arch.Toxicol. 51, 233-245 (1982).

40. Wacker, W.E.C. & B.L. Vallee: J.Biol.Chem. 234,
     3257-3262 (1959).

41. Susa, N., J. Furukawa & S. Tsubaki: Jap.J.Vet.Sci 39,
     325-336 (1977).

42. Vernier, P., A. Montaldi, F. Majone, V. Bianchi & A.G.
     Levis: Carcinogensis 3, 1331-1338 (1982).

43. Leonard, A. & R.R. Lauwerys: Mutation Research 76,
     227-239 (1980).

44. Baker, R.S.U.: Toxicol. & Environm. Chem. 7, 191-212
     (1984).

45. Langard, S.: Chapter 2 In: Burrows, D. (Edit.):
     Chromium: Metabolism and toxicity. CRC Press Inc., Boca
     Raton, Florida (1983).

46. Pedley, B., G. Meachim & D.F.Williams: In Willams, D.F.
     (Edit.): Fundamental Aspects of Biocompatibiliby. CRC
     Press Inc., Boca Raton, Florida (1981).

47. Hueper, W.C.: J. Nat. Cancer Institute 16, 447-462 (1955).

48. Sundermann, F.W.Jr., T.J. Lan & L.J. Cralley: Cancer
     Res. 34, 92-95 (1974).

49. Mukubo, K.: J. Nara.Med.Assoc. 29, 321-340 (1978).

50. Heath, J.C., M.A.R. Freeman and S.A.V. Swanson: Lancet 1, 564-566 (1971).

51. Payne, W.W.: Arch. Environ. Health 1, 20-26 (1960).

52. Hueper, W.C.: Arch. Ind. Hlth. 18, 284-291 (1958).

53. Hueper, W.C.: Cancer Res. 21, 842-857 (1961).

54. Hueper, W.C. & W.W. Payne: Arch. Environ Hlth 5, 445-462 (1962).

55. Kuschner, M. & S. Laskin: Amer. J. Pathol 64, 183-191 (1971).

56. Laskin, S., M. Kuschner & R.T. Drews: in Hauna et al. (Edit.): Inhalation Carcinogenesis, US Atomic Energy Commission, Div. Techn. Inform. Extension pp. 321-351 (1970).

57. Dvizhkov, P.P. & V.J. Fedorova: Vop. Onkol. 13, 57-62 (1967).

58. Stoner, G.D., M.B. Shimkin, M.C. Troxell, T.L. Thompson & L.S. Terry: Cancer Res. 36, 1744-1747 (1976).

MEASUREMENT OF THE UPTAKE OF CHROMIUM

Dennis Shapcott

Centre Hospitalier de l'Université de Sherbrooke

Sherbrooke Quebec J1H 5N4 Canada

SUMMARY

Chromium is held to be an essential trace element. It's measurement in biological media presents great difficulty due to the ubiquitous presence of the metal in the environment and it's very low concentration : The ideal technique to measure chromium combining sensitivity, precision and rapidity does not exist, but probably atomic absorption spectroscopy offers the best compromise. Hair, urine and serum are all analysed to determine the chromium status but none of these is directly related to tissue chromium level or to physiological activity in the individual. The site of activity of chromium at the molecular level has not yet been determined, but the physiological activity of chromium appears to be in conjunction with insulin. These are several body pools for chromium, the interaction between them remains to be defined. Improvement in glucose tolerance and a hypocholestremic effect have been attributed to chromium supplementation. Chromium intake, exercise, glucose tolerance and pregnancy have been shown to influence chromium values in serum and urine and must be taken into consideration when assessing chromium measurements. Chromium may be toxic both as hexavalent ion (industry) and trivalent ion (implants), the actuel mechanism may involve free radical toxicity and deformation of DNA. Much more information concerning tissue levels in healthy subjects and those at risk for chromium deficiency and excess, and also information concerning the biochemistry and physiology of chromium is mandatory before hard and fast conclusions can be drawn concerning the biological activity of this metal in implants.

INTRODUCTION

The measurement of ultra trace metals in biological materials by whatever technique imposes two absolute requirements, the first being technical competence in the actual analytical proceedure and the second, equally important, is the history of the sample and the physiological significance of the results. Ignorance of the latter has, in the past, resulted in the reporting by competent analysts of values which, even at the time, were patently grossly in error. In this review I will firstly discuss the advantages and disadvantages of the various analytical technique and results obtained thereby, but more point will be placed on the physiology of chromium and on what factors influence the actual analytical results.

Measurement of this element poses a number of difficulties : chromium is present at sub parts per million in living tissue which necessitates analysis by physical techniques, the near ubiquitous presence of Cr due to the use of stainless steel in the manufacture of chemicals and in sciencific apparatus, and also the possibility of different forms of Cr with different properties.

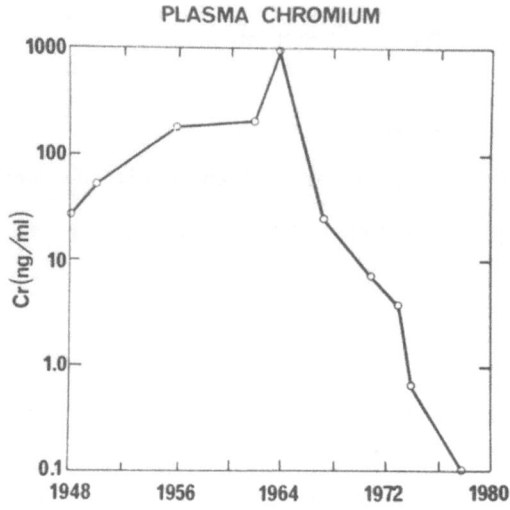

Figure 1 : Representative values for serum chromium taken from the literature.

Historically, the greatest problem with this analysis has been contamination at all levels from sample collection to the analytical proceedure itself, as is illustrated in figure 1 which gives representative values for serum or plasma chromium selected from the literature.

Regardless of the technique used to measure chromium in biological materials, the greatest source of error is still contamination with extraneous chromium during sample collection and manipulation, and meticulous care is mandatory for valid results. The lowest levels reported (and thus the least contamination) have been obtained using laminar flow hoods and clean rooms for sample preparation.

Table 1 : Techniques in current use for analysis of chromium

Technique	Type
Instrumental neutron activation (INAA)	Multielement
Chemical neutron activation (CNAA)	Multielement
Inductively coupled atomic Emission spectroscopy (ICP-AES)	Multielement
Proton-induced X-ray emission (PIXE)	Multielement
Atomic absorption spectroscopy (AAS)	Single and multielement
Isotope dilution analysis (IDA)	Single element

There are six techniques currently used for chromium analysis (Table 1) which can be grouped according to whether or not other elements can be measured simultaneously. Apart of the greater information which may be obtained,the outstanding advantage of multielement analysis is that the sample can be doped with a known quantity of a "foreign" element, and one can thereby assess sample loss during the proceedure.

In trace element analysis the ideal technique would involve no sample preparation and would be non-destructive so that analysis could be repeated, the closest technique available to this is neutron activation analysis. The two versions of this technique, instrumental neutron activation (INAA), and chemical neutron activation (CNAA) both offer advantages and disadvantages. In both techniques the sample size can be increased within reasonable limits to improve sensitivity, although with high volume samples such as dilute urine, some form of preconcentration is required. With INAA only a small fraction of a percent of the atoms are activated, the analysis can be repeated to obtain greater precision, and the specific radioactivity measured without manipulation of the sample. A major problem associated with INAA, is that the radioactive phosphorus which is formed at far greater concentration than radioactive chromium, by virtue of it's beta spectrum will completely obscure the gamma emission of the chromium. This means either leaving the sample for at least three months before counting thereby reducing the radiochromium content to about 10% or performing a chemical separation of the element of interest (CNAA). In practice the second alternative is taken even if the time factor and dilution is not of first importance. Thus, although CNAA overcomes the problems of INAA one loses the possibility of repeat analysis and a potential source of error – chemical separation, is introduced.

There are a number of factors which limit the utility of this technique ; the formation of gaseous products from radiation decomposition limit the size of the sample, the possibility of the interference from iron by the reaction ^{54}Fe n, α ^{51}Cr and above all the very low concentration of ^{51}Cr formed which necessitates prolonged counting times to arrive at statistically significant results. For these reasons and because of the necessity of a nuclear reactor and sophisticated gamma counting equipment, neutron activation is not practical for the routine measurement of chromium in a clinical setting, and its use has been restricted to small scale surveys and to its great utility as a means of calibrating materials as reference standards.

A technique which is increasingly used to calibrate reference materials is isotope dilution analysis by gas chromatography-mass spectrometry (IDA). This technique is ideal for the precise measurement of organic materials where an appropriate deuterated derivative is available and where environmental contamination by the product of interest is non-existant – such as cholesterol or glucose. In the case of chromium the situation is less ideal, there is a suitable isotope ^{50}Cr which is present at about 4.5% of natural chromium, the sample must be doped with a precisely measured amount of ^{50}Cr at the nanogram level and one must be capable of measuring the dilution of the isotope at high precision. The analytical requirements are more stringent than with NAA since the sample must be oxidised to eliminate organic matter, then a volatile derivative must be prepared, followed by chromatographic separation and high resolution mass spectrometry. Any contamination introduced before the actual preparation of the derivative will result in falsely high values. The necessity for high technical skill and expensive and sophisticated apparatus restricts IDA to a few specialised centers, and essentially to a role for the verification of the analytical precision of other techniques (1).

Atomic emission spectroscopy (AES) ; originally low-pressure carbon arc AES was used in earlier studies, more recently argon plasma AES has been used. The plasma is produced either by a DC arc, or more recently by induction (ICP-AES). The advantage of the technique is the capacity for simultaneous multielement analysis and a less expensive and simpler apparatus than NAA or IDA. There are two disadvantage, the sample must be nebulised for introduction into the argon stream resulting in loss of much of the sample, also the technique is very matrix sensitive, and so there is a marked loss of sensitivity when comparing aqueous solutions and urine. In practice this technique is restricted to the measurement of chromium in samples with high concentration and simple matrix such as hair.

Proton-induced x-ray emission (PIXE) is also used for the analysis of chromium (2), but the necessity for a source of protons (usually an accelerator) limits the use of the technique to a few centers. An advantage is the capacity for simultaneous multi-element analysis, the major disadvantages are the necessity for sample concentration and the somewhat lower sensitivity than NAA or AAS.

The most popular technique by far is atomic absorption spectrophometry (AAS), the low tissue level of chromium necessitates the use of electrothermal volatilization. The great weakness of AAS is the background signal produced by non-specific absorption due to molecules from the sample matrix, which is considerably greater than the absorbance from the chromium atoms, particularly in untreated plasma or urine (3). The non-specific absorption signal increases non-linearly with sample volume, this limits the volume of sample which can be analysed and also eliminates the possibility of increasing sensivity by concentrating the sample (4). Various means have been proposed to reduce this background signal including double beam correction with halogen lamp, separation of background and atomic signal by polarisation in magnetic field (Zeeman effect), the L'vov platform and Smith-Heftje background correction. Each of these has it's advocates, but in fact the latter three decrease the overall sensitivity and it is doubtful if the added expense and complication gives results superior to those obtained with halogen lamp correction and carefully optimised sample charring. A lower background signal can be obtained by pre-oxidation of the sample, although this has been criticised on the ground that it could lead to loss of "volatile" chromium, this was not found to be so with low-temperature oxygen plasma oxidation (5). A more cogent objection is that each additional sample manipulation increase the possibility of contamination. A major problem with AAS is the very small signal generated by samples with chromium level below 0.5 ppb, since although the signal may be amplified electronically the deviation is multiplied correspondingly. To overcome this problem standards addition is often used but if the chromium in the sample behaves differently from the inorganic standard i.e. more volatile on charring, then this will give a large negative bias to the results. In terms of sensitivity, it is doubtful if the lowest values for serum chromium reported using NAA can be obtained with adequate precision using AAS because one operates at the detection limit of the apparatus.

A recent modification of AAS-continous wavelength modulation AAS (CWM-AAS) overcomes to a great extent the two most important drawbacks of this technique, namely single element analysis and variable background signal (6). This is accomplished by the use of a continuum light source and scanning so that background may be measured at the spectral absorption line specific for the individual elements. However, the limitation of sentivity does not permit the direct measurement of serum chromium by this technique.

With refinements in sample collection and analysis to reduce contamination a minimum the problem of increasing significance, is that of sensitivity. As mentioned above, a potentially most useful technique CWM-AAS does not have the necessary sensitivity to enable direct measurement of plasma chromium. Using ID-MS it is possible to measure plasma glucose with a within sample deviation of less than 1%, this is not possible with chromium, indeed analytic variation in plasma chromium of 10% under practical working conditions would be considered most acceptable ; using AAS the reproducibility at the lowest values reported is no better than 50%.

We must therefore conclude that each of the techniques discussed has advantages and disadvantages and that no one offers the perfect solution to the problem of measurement of chromium in biological materials. On balance AAS is the only practical technique since the apparatus is (comparatively) inexpensive and (comparatively) easy to use, it can be automated, and it is feasible to process 10 or more samples per hour.

Reference values for chromium in human material

At present it is not possible to define the "normal" chromium levels for a number of reasons, including physiological variation which will be discussed later, but more importantly the lack of reference materials with defined chromium content. Such materials must be validated by at least two techniques based on different principles one of which involves no manipulation other than measurement - ideally INAA - and should cover the anticipated range of values. They must correspond in matrix and chromium content to the human material to be analysed and obviously must be physically identical to the sample, in effect "standard" human serum and urine. When such material becomes available it will then be possible to correlate the reference values given by different analysts. Several investigators have cited values for the NBS standard yeast powder, which has a defined chromium content, as a means of verification of their results. This is not a realistic situation since the chromium content of this material is at least three orders of magnitude greater than that of serum or urine. It is thus apparent that contamination which would give a serum or urine level ten times too high would not be detected with the yeast standard. A further complication is the different chemical composition and physical state of the yeast compared to human material. The same objections can be posed to the use of the bovine liver powder. A point to be borne in mind is that the processing of the standard to ensure stability may change the state of the chromium to a lesser or greater proportion of non-protein bound volatile material. This could have implications when the sample is charred as in AAS or oxidised by heat.

Bearing in mind these caveats we may consider the values cited in the literature and the advantages and disadvantages of the analyses performed on the different milieux.

Serum chromium : There is no general agreement as to the serum chromium level in healthy adults. This cannot simply be explained by different methodologies since values obtained for a standard material (NBS bovine liver) showed a 50 fold variation in content when analysed by 30 different workers all using NAA (7). One assumes that the false high values were due to contamination, but whether the false low values were due to loss of chromium during processing or simply inaccuracy is not clear.

Assuming that the differences reported are due to contamination in the higher values (and not due to loss of chromium in the lowest values !) then the serum chromium would appear to be less than 1 ppb and probably less than 0.5 ppb. The variation between different analysts has been recen-

tly reviewed and suggestions made to ameliorate the situation (8). The
major variation with serum or plasma chromium is due to contamination
during sample collection. The caveat that the analysis is only as good as
the sample applies particularly to this analysis. The blood must be with-
drawn without contact with chromium by using a plastic catheter or plati-
nium needle and collected into an acid washed container. Since chromium
free heparin is available commercially either serum or plasma may be used.
The advantage of serum as a measure of chromium status is that it has a
more or less standard matrix and, presumably, is less subject to short-
term changes than urine chromium. The disadvantage are the lower level
than urine, the difficulty of collection and the lack of knowledge as to
how the serum level relates to tissue levels and to metabolic activity.

Urine chromium : there is better agreement as to the normal value
for urine chromium. It has been shown that there is little variation in
the chromium/creatinine ratio during normal activity and so single samples
are probably a reasonable index up the 24 hrs excretion providing that the
urinary creatinine out put constant with time (9). Urine from adult subjects can be collected directly into an acid-washed container, for children this may pose a severe problem because of the necessity for the use
of devices to collect urine. The advantage of urinalysis is the higher
level of chromium, the ease of collection and that the urine chromium level appears to be better related to physiological changes. The outstanding disadvantage is the highly variable matrix which makes analysis more
difficult, particularly using AAS.

Hair chromium : As with other trace elements, chromium is present at
higher concentration in hair than in other body tissues. Compared to serum chromium, there has been remarkably good agreement between investigators as to the hair chromium level in healthy adults, and the values have
changed little during the last 10 years or so (10).

This analysis is particularly suitable for NAA, for other techniques
it has been widely used because of the ease of sampling, storage and the
high level which makes analysis by ICP-AES particularly attractive, and
certainly far more hair samples have been analysed for chromium than serum or urine. The hair is evidently open to environmental contamination
from airborne particles and hair will take up metal ions from aqueous
solution but this is not so great a problem as with zinc or selenium.
There is disagreement as to the optimum manner of sample preparation, and
it behooves the analyst to verify the validity of his own procedure. Hair
analysis has come under increasing criticism, not necessarily for the
scientific validity of the analyses, but because the analyses are performed commercially and under these conditions the analyst has little knowledge of the history of the sample nor the contact with physician necessary to interpret the results. Hair analysis thus offers the advantage of
simplicity and ease of sample collection and storage. The outstanding
disadvantage is that changes in hair chromium level cannot be directly
related to tissue levels or physiological changes - but this applies
equally well to serum and urine analysis.

In conclusion - although hair, urine and serum have each their advocates ; at present, in individual humans not exposed to chromium, none of
the analyses can be related to the tissue chromium level nor to the physiological activity of the metal.

The biochemistry of chromium

There are four stable valence states of chromium ; zero, divalent,
trivalent and hexavalent. In biological milieux the element will be ion-
nised, the hexavalent ion is an oxidising agent and will be rapidly re-

duced to the trivalent ion, particularly by small molecules with free SH groups (11) the reverse reaction is not possible in vivo.

The hexavalent ion passes across cell membranes and is reduced within the cell to the trivalent ion which, in the case of the erythrocyte, remains within the cell bound to hemoglobin. Whether or not the hexavalent ion has a significanttly long existence in vivo to participate in animal metabolism is a moot point ; it is not possible to perform chemical speciation at the sub ppb level to confirm this.

It has been suggested that the divalent form may be physiologically significant, this appears unlikely since this is unstable in solution in the presence of oxygen and so there is little chance that the metal would be ingested as divalent ion, and the reducing potential of the Cr^{2+}/Cr^{3+} couple strongly favours the oxidation of the divalent ion. The weight of the evidence thus strongly favours the existence of chromium in vivo as the trivalent ion. Evidently this metal cannot function as electron donor or acceptor in oxido-reduction reactions. The daily utilisation of chromium is roughly comparable to that of vitamin B12 and it is tempting to assume that the metal participates, as does the vitamin, in one or two specific enzyme reactions but such chromiumdependent enzymes have not yet been identified. While some reports have appeared purpoting to demonstrate a biochemical activity of chromium they implicated either unrealistically high chromium level or non-physiological conditions and so, while interesting and possibly indicative, are not, of themselves, conclusive.

Conclusion : that chromium exists in vivo as the trivalent ion which does not participate in oxido-reduction reactions, chromium will bind to macromolecules but biochemical reactions implicating the metal have not been identified.

The physiological activity of chromium

Until recent years there has been little interest in Cr as a component of living tissue. While it had been recognised for years that industrial exposure to hexavalent Cr was associated with tissue damage and malignancy, there was little concern for the trivalent form of the element which was considered to be intert. This was supported by demonstration of very low toxicity in acute studies on animals.

The discovery in 1957 of the glucose tolerance factor (GTF), a substance present in brewers yeast, and subsequent demonstration that chromium is an integral part of the GTF has provoked great interest in the role of this metal in mammalian physiology (12). The early work on isolated tissues from experimental animals showed that inorganic chromium had little biological activity, and that the metal had to present as GTF. Furthermore, in isolated tissue cell preparations, GTF showed biological activity only in the presence of added insulin, for example oxidation of glucose by adipocytes. Experimental animals fed diets deficient in chromium resulted in impaired glucose tolerance (lower glucose dissapearance rate) and evantually a diabetes-like syndrome. In both experimental animals and man less than 1% of a physiological dose of inorganic chromium is absorbed, as against up to 20% of yeast chromium. GTF was identified as being a complex of a chromium atom bound to two molecules of nicotinic acid, a molecule each of glycine, glutamic acid and cysteine (14). It has been proposed that the latter three amino acids are present as glutathione. A major problem with GTF is its instability, the material from yeast decomposes during purification, and while synthetic material with the composition of GTF and GTF activity can be prepared, attempts to crystallise the material to perform crystallographic study resulted in decomposition. More recently, the concept of a unique GTF molecule has been challenged, and it appears probably that other low molecular water soluble chromium complexes will have GTF activity (14). A number of publications have appeared showing that GTF itself is not necessarily an essential dietary component and inorganic chromium administered to animals previously made deficient will restore normal glucose tolerance, it is evident that animals can synthetize GTF, in fact certain strains of genetically obese mice have been shown to respond to GTF but not to inorganic chromium, indicating a genetic defect in GTF synthesis (15).

Studies on the uptake of radioactive chromium in animals, and more recently in man (16, 17) have demonstrated the existence of at least 3 body pools of short (< 24 hrs), medium (∿ weeks) and long (∿ months) (figure 2). By analogy to other trace elements chromium will be absorbed as a low molecular weight chelate. Within the blood the metal is held to be transported bound to transferrin (18) ; if this is so then it should be transported specifically into erythrocyte precursors (as is iron) and the level within the erythrocytes should be greater than in the plasma, which is not so in animals. In view of the physiological implications the binding to transferrin should be confirmed. In animals, 3 days after intraperitoneal administration of radioactive chromium, non-protein bound metal was still present in plasma (19). This may represent the fraction excreted in the urine.

There are two particular aspects of the physiology of chromium which merit our attention, the form in which chromium, released from the tissues, is transported in the circulation. If this is protein bound, then it is more likely to be taken up by the liver and excreted in the feces, if non-protein bound then it will be excreted in the urine. Also, the actual level of circulating chelators may influence the excretion as has been shown to occur with high circulating levels of amino acids or oligo peptides (17, 18). It would also be of great interest to determine the origin of the increased urinary and plasma chromium occuring in response to increased circulating insulin.

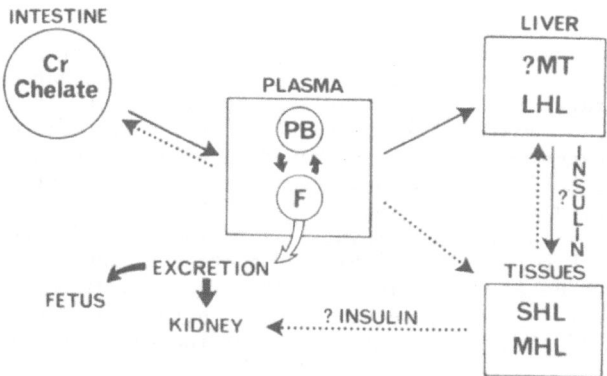

Figure 2 : Metabolism of chromium
The solid arrows represent probable pathways, the open arrow the major excretory pathway and dotted arrows possible pathways :
LHL - long half life ; MHL - medium half life ; SHL - short half life ; MT - metallothionein

The clinical and experimental evidence strongly supports a physiological function for chromium as a cofactor for insulin activity. It has been suggested that chromium forms a bridge between sulfur atoms in the insulin molecule and its' membrane receptors ; and if one assumes that all circulating chromium can so function, then there is an approximately equimolecular ratio between circulating insulin and chromium. However, it must be emphasized that a great deal of experimentation on the combination of insulin and membrane receptors, both in cells and isolated membranes, has been carried out without chromium (22). Therefore, while chromium may increase the insulin-receptor affinity, it is not necessary for the binding.

Conclusion : that chromium exists in at least 3 body pools, their relationship has not yet been established. Chromium most probably functions as a cofactor for insulin, although it is not necessary for insulin-receptor binding.

Clinical studies : the most convincing studies demonstrating clinical deficiency of chromium were carried out on malnourished infants, where a marked improvement in glucose tolerance following microgram quantities of chromium was recorded (23). Subsequent studies have confirmed the effect on glucose tolerance and also a significantly greater gain in weight of the infants. It is unfortunate that no measurements of serum or urine chromium were undertaken but the conditions probably precluded adequate sample collection and also this was before the current analytical techniques had been developed. Because of the postulated role of chromium several investigators undertook clinical assessment of the effect of chromium in subjects with impaired glucose tolerance. The results of these studies can best be described as equivocal, even those who reported benefit this only occured in some subjects, and in a non-predictable manner. It must be stressed that all clinical studies of chromium or GTF in regard to diabetes or to impaired glucose tolerance have shown this unpredictable response and so the interest in chromium as a therapeutic agent to improve glucose tolerance is minimal at present. Other have claimed better response by using yeast preparations with high GTF activity, leading to the hypothesis that the poor results obtained with inorganic chromium were due to inability to transform the metal into GTF. The problem here is that yeast is a complex mixture and under the circumstance the activity has not been confirmed as due to GTF. The observation that serum cholesterol was often lowered following supplementation, together with the observation that animals made deficient in chromium may develop hypercholesterolemia and atherosclerotic changes prompted a continuing interest into chromium deficiency as a risk factor in humans. Carefully controlled studies have demonstrated a cholesterol-lowering effect in subjects with normal or moderately raised serum cholesterol following administration of yeast GTF preparations and also chromium (25, 26). The decrease in serum cholesterol was comparable to that produced by low cholesterol diets.

Conclusion : that supplementation with inorganic chromium or yeast preparations rich in chromium has resulted in improvement in impaired glucose tolerance and/or reduction in serum cholesterol levels in some individuals in a non-predictable manner.

Dietary chromium

In recent years the availability of more reliable measurement of chromium in biological material has brought about a number of changes in our ideas concerning the element. The chromium content of representative diets has been reported, with widely varying results, although the variation is much less than that reported for serum (27, 28). Other studies of the bioavailability of chromium in foodstuffs show the total content of the metal may be less significant than the chemical form (29).

Chromium is unique among the essential trace metals in that 80% or more of a physiological dose is excreted in the urine. As with serum chromium, the values reported for urinary chromium have shown a sharp decrease in recent years and appear to be less than 1 ppb ; this gives a daily excretion of about a microgram or less. Based on estimation of daily turnover and absorption a daily intake of 50-200 micrograms has been recommended in the USA. Our actual knowledge of the chromium content of foods and how this may be altered by preparation, preservation and cooking do not permit the actual prescription of diets which would furnish this RDA.

Recent measurements of intake and excretion of chromium in apparently healthy adults have shown a mean intake considerably below the minimum recommended intake, and a turnover of less than 1 µg (30, 31).

From the work of Tipton and Schroeder it appears that the soft tissue content of chromium is a few milligrams which is equivalent to several years daily turnover of the element (32). Under these circumstances and in view of the extremely small daily requirements it is difficult to see how deficiency can be so common as has been suggested. By analogy to the rapid development of zinc deficiency in fetuses of pregnant rats which occurs within a few days after a zinc deficient diet is begun and long before the tissue content is depleted, it is possible that only a small fraction of chromium is biologically available.

Conclusion : the daily requirements of chromium to balance output appear to be less than 1 µg. In the very few studies conducted on healthy subjects self-selected diets appeared adequate to prevent deficiency.

THE PHYSIOLOGICAL SIGNICANCE OF MEASUREMENTS OF CHROMIUM IN BODY TISSUES

Accepting a normal serum chromium level of 0.05-0.5 ppb and urinary chromium level of less than 1 ppb than there have been fewer than one thousand such analyses reported which makes it difficult to draw any hard and fast conclusions.

It must be assumed that for chromium, as for other trace elements, there exists a "concentration window" a level at which there is full physiological activity. It must also be assumed that levels below this window will be associated with symptoms of deficiency and levels superior will be associated with symptoms of toxicity. Such concentration windows have been established for iron and selenium in experimental animals but not yet for chromium.

At present, the values reported for serum and urine chromium, even the most recent very low levels show the same wide range as for serum iron but there is no information as to the variation within the individual except following supplementation with chromium. Some individuals with impaired tolerance to glucose have shown improvement following chromium but this was not related to serum or urinary chromium levels nor could it be predicted from the glucose tolerance. It is clear, therefore, that there is no physiological measurement which indicates chromium deficiency. There is evidence that an association exists between serum chromium and serum insulin levels, this needs further investigation (33).

Factors which have been shown to influence the levels of chromium in body tissues include intake, exercice, reduced glucose tolerance, pregnancy and malnutrition.

Chromium intake : In terms of normal dietary intake the evidence is conflicting. Two carefully conducted studies were carried out in which diets and urinary output were measured, with approximately equal results. In one these was an excellent correlation between intake and excretion, in the other no significant correlation. The actual dietary content of chromium would appear to be 50 μg or less per day for subjects taking the usual mixed diet. How this would be influenced by major deviations such as vegan diets and diets with a higher content of unprocessed foods has not yet been described. Supplementation by oral inorganic chromium results both in increased serum and urinary chromium levels (34).

Exercice : In experimental animals there was a decrease in tissue chromium concentration (particularly the heart) with age, this did not occur in animals subjected to forced exercice (treadmill) (35). This could not be attributed to greater intake. In human studies, there was a marked increase in urinary chromium excretion following exercice (Running) (36).

Reduced glucose tolerance : serum and urinary chromium have been demonstrated to be related inversely to glucose level and also inversely to insulin level (37). The older reports showing changes in serum and urine chromium are open to criticism on the grounds of unacceptably high values.

Hair chromium was low in malnourished adults (38) and is often low during pregnancy, whether this represents a physiological adaptation or an actual deficiency due to loss of chromium to the fetus is, at present, an open question (39, 40, 41).

Assuming a total soft tissue content of 2 mg and a circulating chromium level of 1 ppb then this is much less than 1% of the total and so obviously major changes in serum chromium do not necessarily reflect tissue status. The same applies to the urinary chromium level.

Conclusion : Some changes in chromium values in response to metabolic stimulus (exercice, pregnancy) were sufficiently constant as to merit further investigation. The relationship chromium/glucose tolerance/insulin needs much further analysis, as does chromium/cholesterol before any conclusions may be drawn.

Chromium toxicity

It is not possible to perform chemical speciation at the level of tissue chromium, but, on the grounds of probability it will be assumed that tissue chromium is in the trivalent state. Historically, chromium toxicity is associated with the hexavalent ion, both from industrial exposure and accidental or intentional ingestion. The effects were noted according to the organ exposed - for example lung and skin. One can postulate two levels of toxicity, that associated with the reduction of chromate and for which a number of factors may be involved such as oxidation of NADH, glutathione and free radical formation. The second toxic effect, that due to the trivalent chromium ion is pertinent to the theme of this conference. Chromates are relatively liposoluble and pass within the cells where reduction takes place. Acute toxicity to chromates has been associated specifically with renal damage (42). In a recent study of welders exposed to metal fumes it was shown that there is rapid urinary excretion following exposure to hexavalent chromium and slower excretion with the trivalent metal (43). Urinary excretion increased with exposure to a limit, indicating that the excretion mechanism is saturable. Urinary excretion was also related to enzyme and protein levels as indices of tissue damage but not in a direct linear fashion.

Trivalent chromium will not participate in oxido-reduction processes nor in free radical formation and the tissue levels are not such as to implicate direct interaction or competition with other metals. However, combination with DNA resulting in mutagenesis is a strong possibility and it may well be that this is the toxic mechanism one must postulate for chromium containing metal implants.

Metallic implants in both animals and humans are associated with a tissue reaction in the immediate vicinity, this tissue was found to be rich in chromium (and other metals), and there was evidence of accumulation in organs (44). The chemical species were not identified but, on the basis of chromium chemistry it is most probable that the metal would be released as the trivalent ion chelated with metabolic acids or amino acids.

Several publications have cited increased chromium levels in urine or serum following implantation of a metallic prosthesis (45, 46, 47).

However, the values reported were quite different and so this illustrates a fundamental problem, when a difference is reported between a reference population and test population, and the values reported for the reference population are much higher than those found by other investigators can one truly accept the difference as significant ?

Evidently there are two levels of interest, one being the effect of released metal ions in the area surrounding the implant where they would be at high concentration and where one could anticipate tissue damage and also the effect of a general accumulation of the metal. Once again, it is hazardous to extrapolate from the very few subjects where tissue analysis has been performed to the tens of thousands of people with metallic implants.

Conclusion : that toxicity to chromium may occur (a) from the reduction of hexavalent to trivalent forms (b) as the results of leaching of metal ions from implants (c) specific target organs such as the kidney. Reliable information on tissue levels is necessary in order to assess the effect of metal implants on the adjacent tissue and on total body burden.

This breif survey had, as objective, to underline the difficulties encountered in the study of this element ; for a more detailed view the reader is referred to the following reviews :

For an excellent and up-to-date review see :

Nieboer E et al (1984), Effects of chromium on health, Ontario Ministry of Labour, 400 University Ave, 8th Floor, Toronto Ontario M7A 1T7, Canada.

Also :

Mertz W (1969), Chromium occurence and function in biological systems. Phys. Rev. 49 : 163-239 (earlier studies)

Chromium in Nutrition and Metabolism (1979). D. Shapcott and J. Hubert eds. Elsevier, New York.

Chromium in Nutrition and Disease (1980). G. Saner ed. AR Liss, New York.

Shapcott D. (1982) "Essential trace mineral deficiencies" in Nutrition and Killer Disease, J. Rose ed. Noyes Publications New Jersey (hair analysis).

Anderson RA(1985). Chromium supplementation : effects on glucose tolerance and lipid metabolism in "Trace Elements in Health and Disease". Almiquist and Wiksell, Stockholm Sweden.

References

1. Veillon C, Wolf WR and Guthrie BE (1979) Determination of chromium in biological materials by stable isotope dilution. Anal Chem 51 : 1022.
2. Simonoff M et al. (1984) Low plasma chromium in patients with coronary and heart diseases. Biol Trace Element Res 6 : 431
3. Guthrie BE, Wolf WR and Veillon C (1978) Background correction and related problems in the determination of chromium in urine by graphite furnace atomic absorption spectrometry. Anal Chem 50 : 1900
4. Brodie KG and Routh MW (1984) Trace analysis of lead in blood, aluminium and maganese in serum and chromium in urine by graphite furnace atomic absorption spectrometry. Clin Biochem 17 : 19
5. Shapcott D et al. (1977) The measurement of volatile chromium in biological materials. Clin Biochem 10 : 178
6. Lewis SA, O'Haver TC, Harnly JM (1985) Determination of metals at the microgram per liter level in blood serum by simultaneous multielement atomic absorption spectrometry with graphite furnace atomisation. Anal Chem 57 : 2
7. Parr RM (1977) Problems of chromium analysis in biological materials : An international perspective with special reference to results for analytical quality control samples. J Radioanal Chem 39 : 421
8. Versieck J (1984) Trace element analysis - A plea for accuracy. Trace Elements in Medicine 1 : 2
9. Gurson C and Saner G (1978) Urinary chromium excretion diurinal changes and relationship to creatinine excretion in healthy and sick individuals of different ages. Am J Clin Nutr 31 : 1162
10. Shapcott D (1982) Essential trace mineral deficiencies in "Nutrition and Killer Diseases" J Rose ed. Noyes Publications Park Ridge NJ
11. Connett PH and Wetterhahn KE (1983) Metabolism of the carcinogen chromate by cellular constituents in "Inorganic Elements in Biochemistry Vol 54 p 93 Springer-Verlag Berlin
12. Schroeder HA (1968) The role of chromium in mammalian nutrition. Am J Clin Nutr 21 : 230
13. Mertz W et al. (1974) Present knowledge of the role of chromium. Fed Proc 33 : 2275
14. Gonzalez-Vergara E, Hegenauer J and Saltman P (1982) Biological complexes of chromium : A second look at the glucose tolerance factor Fed Proc 41 : 286
15. Tuman RW, Bilbo JT and Doisy RJ (1978) Comparison and effects of natural and synthetic glucose tolerance factor in normal and genetically diabetic mice. Diabetes 27 : 49
16. Jain R et al. (1981) Tissue chromium exchange in the rat. Am J Clin Nutr 34 : 2199
17. Lim TH, Sargent T and Kusubov N (1983) Kinetics of trace element chromium (III) in the human body. Am J Physiol 244 : R445
18. Sayato et al. (1980) Metabolic fate of chromium compounds. J Pharm Dyn 3 : 17
19. Shapcott D et al. (1977) The measurement of volatile chromium in biological materials. Clin Biochem 10 : 178
20. Carr G and Wilkinson AW (1979) The urinary excretion of iron and chromium by children with burns and scalds. Clin Chimica Acta 96 : 73
21. Fell GS, Hall SD and Shenkin A (1979) Chromium requirements during intravenous nutrition in "Chromium in Nutrition and Metabolism". D Shapcott and J Hubert, eds. Elsevier, New York
22. Czech M and Massague J (1982) Submit structure and dynamics of the insulin receptor. Fed Proc 41 : 2719

23. Hopkins LL et al. (1968) Improvement of impaired carbohydrate utilisation by chromium (III) in malnourished infants. Am J Clin Nutr 21 : 203

24. Schroeder H et al. (1970) Chromium deficiency as a factor in atherosclerosis. J Chron Dis 23 : 123

25. Riales R (1979) The influence of brewers yeast on lipoprotein cholesterol concentrations in "Chromium in Nutrition and Metabolism". D Shapcott and J Hubert, eds. Elsevier, New York

26. Riales R and Albrink MJ (1981) Effect of chromium chloride supplementation on glucose tolerance and serum lipids including high density liloprotein of adult men. Am J Clin Nutr 34 : 2670

27. Guthrie BE (1975) Chromium, manganese, copper, zinc and cadmium content of New Zealand foods. N.Z. Med J 82 : 418

28. Kirkpatrick DC and Coffin DE (1971) The trace metal content of representative Canadian diets in 1970 and 1971. Can Inst Food Sci Technol J 7 : 56

29. Toepfer EW et al. (1973) Chromium in foods in relation to biological activity. J Agric Food Chem 21 : 69

30. Bunker VW (1984) The uptake and excretion of chromium by the elderly. Am J Clin Nutr 3_ : 797

31. Kozlovsky AS, Hallfrisch J and Anderson RA (1984) Chromium intake, excretion and absorption by adult subjects consuming self-selected diets. Fed Proc 43 : 471

32. Tipton IH and Cook MJ (1963) Trace elements in human tissue. Part III, Adult subjects from the United States. Health Phys 9 : 103

33. Liu VJK and Abernathy RP (1982) Chromium and insulin in young subjects with normal glucose tolerance. Am J Clin Nutr 35 : 661

34. Anderson RA, Bryden NA and Polansky MM (1985) Serum chromium of human subjects : effects of chromium supplementation and glucose. Am J Clin Nutr 41 : 571

35. Vallerand AL et al. (1984) Influence of exercise traning on tissue chromium concentrations in the rat. Am J Clin Nutr 39 : 402

36. Anderson RA et al. (1982) Effect of exercise (running) on serum glucose, insulin, glucagon, and chromium excretion. Diabetes 31 : 212

37. Anderson RA (1985) Chromium supplementation : Effects on glucose tolerance and lipid metabolism in "Trace Elements in Health and Disease, Skandia International Symposium" Almquist and Wiksell. Stockholm, Sweden

38. Vobecky J et al. (1980) Hair and urine chromium content in 30 hospitalised female psychogeriatric patients and mentally healthy controls. Nutr Rep Internat 22 : 49

39. Shapcott D et al. (1980) Hair chromium at delivery in relation to age and number of pregnancies. Clin Biochem 29 : 129

40. Hambidge KM and Rodgerson DO (1969) A comparison of hair chromium levels of milliparous and parous women. Am J Obstet Gynecol 103 : 320

41. Saner G (1981) The effect of parity on material hair chromium concentration and changes during pregnancy. Am J Clin Nutr 34 : 853

42. Sharma BK, Singhal PC and Chugh KS (1978) Intravascular hemolysis and acute renal failure following potassium dichromate poisoning. Postgraduate Med J 54 : 414

43. Mutti A et al. (1979) The role of chromium accumulation in the relationship between airborne and urinary chromium in weldess. Int Arch Occup Environ Health 43 : 123

44. Ferguson AB et al. (1962) Characteristics of trace ions released from metal implants in the rabbit. J Bone Joint Surg 44A : 323

45. Coleman RF, Herrington J and Scales JT (1973) Concentration of wear products in hair, blood and urine after total hip arthroplasty. Lancet II : 1873

46. Pazzaglia UE (1983) Metal determination inorganic fluids of patients with stainless steel hip arthropplasty. Acta Orthop Scand 54 : 574

47. Hildebrand HF et al. (1985) Orthopedic implants and corrosion products. Ultrastructure and analytical studies of 65 patients. This workshop.

ANALYSIS OF COBALT, CHROMIUM AND NICKEL IN

BIOLOGICAL MATERIALS - STATE OF THE ART

Markus Stoeppler

Institute of Applied Physical Chemistry (ICH-4)
Nuclear Research Center (KFA) Juelich
P.O. Box 1913, D-5170 Juelich, F.R.G.

INTRODUCTION

Since approximately 1975 a steady progress in trace
analytical chemistry is to be noticed. This is due to a
remarkable methodological improvement which has led to the
identification and elimination of many sources of error
from sampling and sample treatment preceeding the analytical determination step. Results of these improvements are
significantly lower normal (i.e. reference) levels for numerous elements as supposed before (Versieck and Cornelis,
1980; Versieck, 1985). This paper thus summarizes the most
important aspects and up to now known facts for reliable
analysis of cobalt, chromium and nickel regarded as toxi-
cologically significant metals in the use of orthopedic
implants (Friberg et al, 1985; Sunderman, 1986).

"NORMAL" OR REFERENCE LEVELS FOR COBALT, CHROMIUM AND NICKEL

From recent investigations it is obvious that the
contents of cobalt, chromium and nickel are much lower
than given in former years. To illustrate this, in table
1 is shown the decrease of values reported for chromium
in urine from Versieck (1985). This is an excellent example
for the rapid improvement of analytical methodology and
expertise.

Table 1.   Values for the Urinary Excretion
of chromium (μg/24 h) reported
by various workers during the last
decade (quoted after Versieck, 1985)

7.2	(Wolf et al, 1974)
3.1	(Gürson et al, 1978)
3.9	(Punsar et al, 1977)
0.8	(Guthrie et al, 1979)
0.32	(Veillon et al, 1979)
0.16	(Anderson et al, 1982)

Thus, at present it is generally accepted that reference levels for cobalt, chromium and nickel in body fluids are at the low μg/l or even ng/l level, if proper sampling with utmost care for minimizing contamination (use of PTFE cannules for whole blood and of scrupulously cleaned polyethylene bottles for urine, Aitio and Järvisalo, 1984) is done. This is depicted in table 2.

Table 2. Average endogenous Co, Cr and Ni contents in body fluids of occupationally nonexposed persons from recent investigations

Values in μg/l (whole blood, serum and plasma)
μg/g creatinine (urine)

Element	Urine	Serum/Plasma	Whole blood
Co	< 0.4	< 0.1	< 0.1
Cr	< 1.0	< 0.2	< 0.5 ?
Ni	≤ 2.0	< 0.4	< 0.5

From these data it is obvious that reliable analysis of these elements at normal (reference) levels is quite difficult and that even studies with slightly exposed persons may pose problems if not all necessary precautions are taken in every step from sampling to analysis.

Up to now no determinations in human tissues have been performed, applying adsorption pulse voltammetry (ADPV). Thus, the determination of cobalt and nickel in muscle tissue, liver and kidney from beef as has been recently performed in our Institute (Narres et al, 1984) may nevertheless give some ideas about levels of these elements in soft tissues if applied with utmost care from sampling (dissection) to analysis. A summary of these values is given in table 3. These values are at the same order of magnitude found in Sunderman's laboratory for various human materials (Sunderman, 1980; Sunderman et al, 1985).

Table 3. Cobalt and Nickel in animal tissues (beef) determined with ADPV after complete wet digestion[+] (ranges in μg/kg)

Matrix	Cobalt	Nickel
Muscle Tissue	1.5 - 18	1.5 - 15
Liver	40 - 270	0.4 - 16
Kidney	10 - 150	2.6 - 13

[+] The applied method attained under the given working conditions determination limits around or below 1 μg/kg for both elements

The need for scrupulously performed sampling and sample treatment has been addressed in many recent papers and amply discussed. Therefore here only some of the most detailed papers on this subject may be quoted (e.g. Sunderman, 1980; Stoeppler, 1980; Versieck and Cornelis, 1980; Stoeppler, 1983; Versieck, 1984 and 1985; Katz, 1985; Shapcott, 1980). Since, there is, however, some recent progress to be noticed if analytical methodology is concerned, this will be summarized for cobalt, chromium and nickel in order to give an updated overview on the present potential for the analysis of these elements in biological samples.

As an introduction to this chapter, typical detection limits for cobalt, chromium and nickel, applying different analytical methods are given from present knowledge and use of commercially available instrumentation. In some cases, particular techniques - e.g. multiple injection in GFAAS - provide values that are somewhat lower than indicated in the table. These are marked with <. As usual, for routine analysis a limitation in GFAAS is the injected volume. Thus, the table is based on a volume of 50 μl of non interfering analyte solution. This also applies to Total-reflection X-Ray Fluorescence (TXRF) a new, very sensitive version of X-Ray Fluorescence that is now gaining importance for trace and even ultratrace analysis in various fields (Michaelis et al, 1985). For voltammetry (Adsorption Differential Pulse Voltammetry, ADPV), flame AAS and ICP-AEAS (Inductively Coupled Plasma Atomic Emission Spectrometry) typical sample solution amounts are < 1 to a few ml.

Table 4. Detection limits for Co, Cr and Ni for some trace analytical methods, remarks see text, values in μg/l (μg/kg for NAA)

Method	Cobalt	Chromium	Nickel
ICP-AES	< 3	< 3	3
AAS-flame	15	5	5
GFAAS	0.15	< 0.3	< 0.3
TXRF	0.1	0.4	0.1
ADPV	≤ 0.002[1]	< 0.1[2]	≤ 0.002
NAA	0.03[2]	0.03[3]	< 5[4]

Remarks:

[1] For the determination of Co and Ni the adsorption of metal glyoximates at the hanging mercury drop was used (Nürnberg, 1983, Ostapczuk et al, 1984)

[2] Adsorption of a distinct chromium complex at the hanging mercury drop. This is however at present only feasible in aqueous or very dilute samples after oxidation of Cr(III) to Cr(VI) (Golimowski et al, 1985

[3] Radiochemical NAA (Versieck et al, 1978)

[4] Instrumental NAA

Subsequently, for the elements treated in this paper, some information about methodological improvements is given.

Cobalt

For cobalt, it appears that only Radiochemical Neutron Activation Analysis (RCNAA) has been able to provide reliable data for lower contents as far as former studies are concerned (e.g. Versieck et al, 1978, Co in serum). With the introduction of Adsorption Voltammetry, initially only using dimethylglyoxime as the chelating agent, a powerful method entered the scene. An additional gain in reliability and detection power could be obtained with the commercial introduction of the square wave mode (Osteryoung and Osteryoung, 1985). Applying this approach, it was easily possible, after proper sampling and a complete wet $HNO_3/HClO_4$ digestion, to confirm the rather low values found earlier with RCNAA in human serum (Ostapczuk et al, 1984). Another improvement of ultratrace voltammetric cobalt determination was the replacement of dimethylglyoxime by furil dioxime which significantly enhanced the sensitivity for cobalt (Nieboer, 1984). This is of particular importance if cobalt besides a great excess of nickel has to be analyzed, since this reagent enhances the detection power for cobalt more than twofold as compared to nickel.

At slightly elevated or generally higher levels, however, also modern graphite furnace techniques with pyrolytically coated graphite tubes and Zeeman background correction, for plasma and whole blood after deproteinization with warm nitric acid and centrifugation (Christensen et al, 1983) can be used. For the determination of slightly elevated cobalt levels in urine a method has been recently proposed by Angerer and Schaller (1983) which in principle can be used also for nickel, see table 5.

Table 5. DFG (German Science Foundation) GFAAS Procedures for Cobalt and Nickel in Urine

---

Principle: Direct Extration of Co and Ni from slightly acidified Urine using Hexamethylene Ammonium/Hexamethylene Dithiocarbamidate (HMA/HMDC).

Solvent: Diisopropylketone and Xylene

Enrichment Depending on Metal Content up to 1 : 8

Injected Volume up to 50 µl
Atomization at 2600 °C for Cobalt and 2700 °C for Nickel, Miniflow 30 ml. Peak Height Evaluation against Calibrating Solutions)

Cobalt: Extraction at pH 4.4
        Detection Limit 0.1 µg/l
        Imprecision around 10 %

Nickel: Extraction at pH < 4
        Detection Limit 0.2 µg/l
        Imprecision around 10 %

Another promising method, however, at lower levels always requiring decomposition is TXRF (Michaelis, 1985). Thus, from the methodological point of view it can be stated that a good selection of methods is presently available for the sensitive determination of all levels of cobalt in biological materials (Elinder et al, 1986). In this context it is also of importance to mention that chelate Gas-Chromatography can be used for the multielement analysis of higher metal contents (i.e. cobalt and nickel among others). A recent paper dealing with this technique gives absolute detection limits at the ng level and a determination limit of 50 ng (absolute) for cobalt in real samples (Meyer and Neeb, 1985).

Nickel

In general, the methods already described for cobalt, particularly ADPV and GFAAS, can be used for nickel as well except NAA which is less sensitive (Sunderman, 1980; Stoeppler, 1980, 1984 a, 1984 b). However, the normal (reference) levels are not in all cases as low as for cobalt so that particularly GFAAS procedures often can be applied directly or after a simple sample treatment for serum and whole blood with nitric acid and heat (Sunderman et al, 1984). This is particularly true for the application of Zeeman corrected GFAAS (Stoeppler, 1984 b) which, if combined with oxygen ashing and multiple sample injection, is able to analyze nickel contents as low as 0.15 μg/l in serum and approx. 0.3 μg/l in whole blood samples (Mohl et al, 1986). Recently it has been shown that a rapid direct Zeeman GFAAS method could be successfully applied for nickel in body fluids exposed and control groups (Sunderman et al, 1986). Besides the voltammetric procedure already mentioned for cobalt and nickel determination in tissues down to approximately 1 μg/kg (Narres et al, 1984) Sunderman et al (1985) have developed a method for the determination of nickel in tissue homogenates based on stringent precautions to minimize nickel contamination during tissue sampling and processing. This method is summarized in table 6.

Table 6. New Procedure for Nickel Determination with GFAAS in Tissue Samples

- Tissue dissection with metal free obsidian knives

- Homogenization in polyethylene bags by use of a stomacher blender

- Wet digestion of 1 ml tissue homogenate with a $HNO_3/HClO_4/H_2SO_4$ mixture

- Nickel measurement in digests with P.E. Zeeman 5000 and pyrotubes, atomization at 2600 $^\circ$C (argon miniflow, 30 ml/min), evaluation of integrated absorbance against aqueous calibration solutions

- Detection limit: 10 μg/kg (d.w.) coefficient of variation 7 - 15 %

Chromium

As for cobalt radiochemical neutron activation analysis has turned out for years to be the most sensitive and reliable approach (Versieck et al, 1978) and is still a valuable reference method (Cornelis, 1985).

Improvements in AAS techniques (Veillon et al, 1980, 1982; Slavin, 1981; Ping et al, 1983, Slavin and Carnrick, 1985) made this technique quite reliable for many routine applications.

Table 7 summarizes some chromium properties and the potential of various AAS techniques for its determination.

Table 7.   Chromium Analysis in Biological
           Materials I

- Chromium mainly occurs in the trivalent state
- Due to its ubiquitous appearance contamination precautions are very important
- Routine method: AAS
  For lower levels: Zeeman GFAAS
- For higher levels (implantation studies): N_2O flame
- Detection limit approx. 0.15 $\mu g/L$; pyrotubes or L'vov Platform with matrix modification $(Mg(NO_3)_2)$ (STPF-concept) charring temp. 1600 $^{\circ}C$, atomization temp. 2500$^{\circ}C$ and 20 μl injections of 1+1 diluted urine
- using five sequential injections the detection limit can be lowered to approx. 0.02 $\mu g/L$

The application of another independent analysis technique, stable isotope dilution (Veillon et al, 1979 and 1984), see table 8, provided the basis for the at present reached quite acceptable accuracy and precision of total chromium determinations in biological materials, at least if skilled laboratories are considered (Ottaway and Fell, 1986).

Table 8.   Isotope dilution mass spectrometric
           reference method for chromium deter-
           mination in urine or serum

Spiking of sample with ^{50}Cr solution, ashing under contamination controlled conditions (LTA)

Dissolution in buffer (NH_4 acetate)

Chelation with TTA (Trifluoroacetylacetone) and solvent extraction using n-hexane

Measurement of the ratio $^{50}Cr/^{32}Cr$ by gas chromatography/mass spectrometry using dual ion monitoring

With this approach average chromium values around 0.3 $\mu g/l$ have been found in urine

---

Since for chromium determination and discrimination of species is very important, Lewalter (1985) has proposed a method that gives some information about biochemical actions of Cr(VI) as a rapid screening method in occupational exposure. The scheme of this method is given in table 9.

Table 9. Proposed method for chromium determination in stabilized whole blood, plasma/serum and erythrocytes

- Dilution of the matrix with Triton-X-100, injection into a graphite furnace (20 ,ul) evaluation by standard addition/peak height
- Charring: 1300 °C, atomization: 2700 °C (gas stop)
- Detection limits:
  whole blood: 0.9 ,ug Cr/l
  serum/plasma and erythrocytes: 0.5 ,ug Cr/l
- Imprecision around 15 %
- The measurement of Cr in erythrocytes allows a differentiation if there is not enough reduction capacity of plasma in cases of heavy exposure to (Cr (VI) since only Cr(VI) can penetrate the membranes of erythrocytes

SPECIATION OF METALS IN BIOLOGICAL FLUIDS

The identification of metals as organometallic compounds, bound to proteines or the detection and quantification of its oxidation state is a task certainly deserving more attention in the future. From a recent compilation (Cornelis, 1986) an overview - example chromium - is given in table 10.

Table 10. Example for chromium speciation in biological fluids

I.   Separation of proteines e.g. by HPLC or other chromatographic techniques and determination of chromium in the respective fractions

II. Determination of the oxidation state in which Cr is transformed or bond to macromolecular compounds. This mainly is difficult to achieve due to the numerous interacting compounds. If, however, free ions can be found, one can treat the analyte solution with resins that bind either Cr(VI) anions or Cr(III) cations

Using the direct Zeeman effect, solid and also numerous liquid materials, can be analyzed without any pretreatment. This method is now increasingly used for a selection of elements and will be probably applied in the future also for cobalt, chromium and nickel in numerous biological materials and for homogeneity studies, which are important in the development and characterization of reference materials (Kurfürst, 1984; Kurfürst et al, 1984; Langmyhr and Wibetoe, 1985).

Table 11. Properties of solid sampling with Zeeman-Platform GFAAS

- Sample amounts from 0.1 - 10 mg
- Local concentration differences
- Rapid screening
- Homogeneity studies
- Reference method

Detection range (direct Z-GFAAS)

Element	(ng)	Resonance line (nm)
Cobalt:	0.1 - 40	240.7
Nickel:	0.2 - 15	232.0
Chromium:	0.05 - 5	359.3
	1 - 40	429.0

QUALITY CONTROL AND REFERENCE MATERIALS

Above some information have been already given about reference methods that might be radiochemical NAA for voltammetry or atomic absorption spectrometry. In cases where e.g. AAS is the main method, voltammetry could be used as reference method and vice versa.

As far as presently available reference materials are concerned, these are listed in Table 12; a compilation including most of commercially available RMs has been published recently (Muramatsu and Parr, 1985). A second generation of biological reference material (serum) recently has been prepared by Versieck et al (1986). From preliminary results this material will appearently provide the lowest possible levels for Co, Cr and Ni which is extremely important for further studies on normal (reference) levels.

Table 12. Content of Co, Cr, and Ni in various biological reference materials (RM's) and certified reference Materials (CRM's)
Values in /ug/kg, Reference Values ()

Material	Type	Co	Ni	Cr
NIES Human Hair	CRM	(0.1)	1.8 ±0.1	1.4 ±0.2
NIES Mussel (Myt. Ed)	CRM	(0.37)	0.93±0.06	0.63±0.07
IAEA Animal Muscle	RM	-	(1)	
Behring, Urine 1	RM	-	0.0035	0.0043
Urine 2	RM	-	0.030	0.026
Urine 3	RM	-	0.070	0.050
Blood	RM	-	(0.008)	-
NBS Bovine Liver 1577 a	CRM	0.21	-	-
NBS Rice Flour 1568		0.02	-	-
Nycomed: Serum 105	RM	-	(0.003)	-
Urine 208	RM	(10)	(0.040)	(0.024)

FUTURE PROSPECTS

For future studies in various biological and environmental materials an increased application of multielement methods appears to be absolutely necessary. Besides the continued use of already proven techniques, additional multi-element methods are now under detailed study that could be applied very soon for rapid and reliable fingerprinting in many research and surveillance tasks. Besides the already mentioned methods INAA, RCNAA and TXRF, promising approaches might be the extension of isotope dilution with thermal ionization to more than ten elements as an accurate and very precise reference method (Hilpert and Waidmann, 1986), possibly simultaneous multielement atomic absorption spectrometry with graphite furnace atomization (Harnly et al, 1984, Lewis et al, 1985) and/or furnace atomic nonthermal excitation spectrometry (FANES) as a very promising and most probably very sensitive multielement technique (Falk et al, 1984).

Finally ICP-MS coupling, now under thorough study in several expert laboratories around the world, despite its extreme high costs of investment might constitute an ultimate approach in studies of particular importance if other methods cannot provide the required utmost detection power reliability and/or element coverage (Gray, 1985).

REFERENCES

Aitio, A., and Järivsalo, J., 1984, Collection, processing
and storage of specimens for biological monitoring
of occupational exposure to toxic chemicals, Pure Appl.
Chem., 56:549

Angerer, J. and Schaller, K.H., 1983, "Cobalt", in: Analy-
tische Methoden, Band 2, Analysen in biologischem
Material, D₁-D₃ and 1-9, 7. Lieferung, Deutsche Forschungsgemeinschaft, Bonn, F.R.G.

Christensen, J.M., Mikkelsen, S. and Skov, A., 1983,
"A direct determination of cobalt in blood and urine
by Zeeman Atomic Absorption Spectrophotometry", in:
Brown, S.S. and Savory, J. eds. Chemical Toxicology
and clinical chemistry of metals, Academic Press,
London, New York

Cornelis, R., 1985, Trace element studies in the biosphere
with neutron activation analysis, J. Trace and Microprobe techniques, 2-4:237

Cornelis, R, Chromium, in: M. Stoeppler (ed) Trace metal
analysis in biological specimens, PSG Publishing Company
of Massachusetts, Littleton, MA, USA

Elinder, C.G., Ostapczuk, P., Stoeppler M., 1986, "Cobalt"
in: Stoeppler, M., ed. Trace metal analysis in biological Specimens, PSG Publishing Company of Massachusetts, Littleton, MA, USA

Falk, H., Hoffmann, E., and Luedke, C.H., 1984, A comparison
of furnace atomic nonthermal excitation spectrometry
FANES) with other atomic spectroscopic techniques,
Spectrochim. Acta, 39B:383

Fernandez, F.J., Bohler, W., Beaty, M.M., and Barnett, W.,
1981, Correction for high background levels using
Zeeman effect, atomic spectrosc. 2:73

Friberg, L., Nordberg, G.F., and Vouk, V.B., eds., 1985,
"Handbook on the Toxicology of Metals", 2nd edition,
Elsevier/North Holland Biomedical Press, Amsterdam-
New York-Oxford

Golimowski, J., Sigg, L., Valenta, P., and Nürnberg, H.W.,
1985, "Chromium levels in Europan inland waters", in:
Proc. International Conference, Heavy metals in the
environment, Athens, ed. T.D. Lekkas, CEP Consultants,
Edinburgh

Gray, A.L., 1985, The potential of ICP source mass spectrometry in: B. Sansoni, ed. Instrumentelle Multielementanalyse, p. 277, VCH-Verlagsgesellschaft, Weinheim

Harnly, J.M., Miller-Ihli, N.J., and O'Haver, T.C., 1984,
Simultaneous multielement atomic absorption spectrometry
with graphite furnace atomization, Spectrochim. Acta
39B:305

Hilpert, K., and Waidmann, E., 1986, Multi-element determination in environmental samples by mass spectrometric
isotope dilution analysis with thermal ionization,
Part I: Pine needles, Fresenius Z. Anal. Chem. (in press)

Katz, S.A., 1985, Collection and preparation of biological
tissues and fluids, Int. Biotechnology Lab. 3: June, 10

Kurfürst, U., 1984, Untersuchungen über die Schwermetallana-
lyse in Feststoffen mit der direkten Zeeman-Atomabsorp-
tions-Spektroskopie, Diss., Univ. Bremen, Bundesrepublik
Deutschland

Kurfürst, U., Grobecker, K.H., and Stoeppler M., 1984, Homogeneity studies in biological reference and control materials with solid sampling and direct Zeeman AAS in: P. Brätter, P. Schramel, eds., Trace element-analytical chemistry in medicine and biology, Vol. 3, p. 591, Walter de Gruyter & Co., Berlin-New York

Langmyhr, F.J., and Wibetoe, G., 1985, Direct analysis of solids by atomic absorption spectrophotometry, Proc. analyt. atom. Spectrosc. 8:193

Lewalter, J., 1985, personal communication

Lewis, S.A., O'Haver, T.C., and Harnly, J.M., 1985, Determination of metals at the microgram-per-liter level in blood serum by simultaneous multielement atomic absorption spectrometry with graphite furnace atomization, Anal. Chem. 57:2

Michaelis, W., Prange, A., Knoth, J., 1985, Applications of TXRF in multielement analysis, in: B. Sansoni ed., Instrumentelle Multielementanalyse, p. 109, VCH-Verlagsgesellschaft, Weinheim

Mohl, C., Bagschik, U. and Stoeppler, M., 1986, to be published

Muramatsu, Y., and Parr, R.M., 1985, Survey of currently available reference materials for use in connection with the determination of trace elements in biological and environmental materials, IAEA/RL/128

Narres, H.-D., Valenta, P., and Nürnberg, H.W., 1984, Voltammetric determination of heavy metals in meat and organs of slaughter cattle, Fresenius Z. Anal. Chem., 179:440

Nieboer, E., 1984, personal communication

Nürnberg, H.W., 1983, Potential and application of voltammetry in the analysis of toxic trace metals in body fluids, in: S. Facchetti (ed.) Analytical techniques for heavy metals in biological fluids, p. 209, Elsevier, Amsterdam

Ostapczuk, P., Valenta, P., Stoeppler, M., Nürnberg, H.W., 1984, Voltammetric determination of nickel and cobalt in body fluids and other biological materials, in: S.S. Brown, J. Savory eds., Chemical Toxicology and Clinical Chemistry of Metals, p. 61, Academic Press, London

Osteryoung, J.G., Osteryoung, R.A., 1985, Square wave voltammetry, Anal. Chem. 57:101A

Ottaway, J.M. and Fell, G.S., 1986, Critical evaluation of analytical methods for the determination of trace elements in various matrices, Part IV, Determination of chromium in biological materials, Pure Appl. Chem., in press

Ping, L., Matsumoto, K., and Fuwa, K., 1983, Determination of urinary chromium levels for healthy men and diabetic patients by electrothermal atomic absorption spectrometry, Anal. Chim. Acta, 147:205

Shapcott, D., 1986, Measurement of the uptake of chromium in humans in: Proc. Advanced research workshop "Biological incidences of Co-Cr-Ni-alloys used in orthopaedic surgery and stomatology", Plenum Publishing Company, London, U.K.

Slavin, W., 1981, Determination of chromium in the environment and in the work place, atomic spectroscopy, 2:8

Slavin, W., and Carnrick, G.R., 1985, A survey of applications of the stabilized temperature platform furnace Zeeman correction, atomic spectroscopy, 6:157

Stoeppler, M., 1980, Analysis of nickel in biological materials and natural waters", in: J.O. Nriagu (ed.) Nickel in the environment, P. 661, John Wiley, New York

Stoeppler M., 1983, Analytical aspects of sample collection, sample storage and sample treatment, in: P. Brätter, P. Schramel, eds., Trace element-analytical chemistry in medicine and biology, Vol. 2, p. 909, Walter de Gruyter, Berlin-New York

Stoeppler, M., 1984, Analytical chemistry of nickel", in: F.W. Sunderman et al, (eds.) Nickel in the human environment, p. 459, IARC Sci. Publ. No. 53

Stoeppler, M., 1984 a, "Analytical chemistry of nickel," in: Sunderman et al, (eds.) Nickel in the human environment, IARC Sci. Publ. No. 53, p. 459, Int. Agency for Research on Cancer, Lyon, France

Stoeppler, M., 1984 b, "Recent improvements for nickel analysis in biological materials" in: Brätter P., and Schramel, P., Trace element-analytical chemistry in medicine and biology, Vol. 3, p. 539, Walter de Gruyter, Berlin-New York

Stoeppler, M., 1986, Aanalytical methods and quality control for trace metal determinations - A critical review of the state of the art, in: Proceedings of the Rochester Conference "The scientific basis and practical applications of biological monitoring of toxic metals", June 1986, in press

Sunderman, F.W., Jr., 1980, Analytical biochemistry of nickel, Pure Appl. Chem., 52:527

Sunderman, F.W., Jr., Crisostomo, M.C., Reid, M.C., Hopfer, S.M., and Nomoto, S., 1984, Rapid analysis of nickel in serum and whole blood by electrothermal atomic absorption spectrophotometry, Ann. Clin. Lab. Sci., 14:232

Sunderman, F.W., Jr., Marzouk, A., Crisostomo, M.C., and Weathersby, D.R., 1985, Electrothermal atomic absorption spectrophotometry of nickel in tissue homogenates, Ann. Clin. Lab. Sci., 15:299

Sunderman, F.W., 1986, Toxicological aspects of the metals Ni-Cr-Co in: Proc. Advanced research workshop", Biological incidences of Co-Cr-Ni alloys used in orthopaedic surgery and stomatology", Plenum Publishing Company, London, U.K.

Sunderman, F.W., Jr., Hopfer, S.M., Crisostomo, M.C. and Stoeppler, M., 1986, Rapid analysis of nickel in urine by electrothermal atomic absorption spectrophotometry, Ann. Clin. Lab. Sci., 16:219

Veillon, C., Wolf, W.R., and Guthrie, B.E., 1979, Determination of chromium in biological materials by stable isotope dilution, Anal. Chem. 51:1022

Veillon, C., Guthrie, B.E., and Wolf, W.R., 1980, Retention of chromium by graphite furnace tubes, Anal. Chem. 52:457

Veillon, C., Patterson, K.Y., and Bryden, N.A., 1982, Direct determination of chromium in human urine by electrothermal atomic absorption spectrometry, Anal. Chim. Acta 136:233

Versieck, J., 1984, Trace element analysis - A plea for accuracy, Trace elements in medicine, 1:2

Versieck, J., 1985, Trace elements in human body fluids and tissues, CRC Crit. Rev. Clin. Lab. Sci, 22:97

Versieck, J., Hoste, J., Barbier, F., Seyaert, H., De Rudder, J., Michels, H., 1978, Determination of chromium and cobalt in human serum by neutron activation analysis, Clin. Chem. 24:303

Versieck, J., and Cornelis, R., 1980, Normal levels of trace elements in human blood plasma or serum, Anal. Chim. Acta 116:217

Versieck, J., Hoste, J., Vanballenberghe, L., De Kesel, A., and Van Tenterghem, D., 1986, Collection and preparation of a second generation biological reference material for trace element analysis, in: Proc. 7th Int. Conf. on modern trends in activation analysis, Copenhagen, June 23-27

PHENOMENOLOGY OF THE TRACE ELEMENT BURDENING OF THE HUMAN ORGANISM BY THE
IN-BODY CORROSION OF Co-Cr-Ni-ALLOYS AS REVEALED BY NEUTRON ACTIVATION
ANALYSIS

R. Michel[1], F. Löer[2], M. Nolte[3], M. Reich[3], and J. Zilkens[4]

1 Zentraleinrichtung für Strahlenschutz der Universität Hannover, Hannover, 2 Abteilung für Orthopädie an der RWTH
Aachen, Aachen, 3 Abteilung für Nuklearchemie am Institut für
Biochemie der Universität zu Köln, Köln, 4 Rheinisch-Orthopä-
dische Landesklinik Viersen-Süchteln, Viersen, F.R.G.

INTRODUCTION

Metal implants in form of Co-Cr-Ni-alloys have found wide spread
applications in orthopaedic surgery and in odontology. Among the alloys
used one can distinguish two main classes according to their Co and Fe
contents, respectively. One comprises the so-called stainless steels on an
iron basis with differing amounts of Cr, Ni and Mo, mostly according to
AISI 316L. This type of material only contains minor amounts (ca. 0.1 %) of
cobalt as impurity and not as a specified alloying constituent (e.g. Michel
and Zilkens, 1978). The other class contains alloys for which Co is one of
the basic constituents (up to 65 %). They will be called Co-Cr-alloys
throughout this paper. For a detailed survey on the various individual
alloy types and their specifications see the work of Ungethüm and Winkler
(1984) and of Willert and Semlitsch (1980).

The application of all metal implants is complicated by the corrosion
of the alloys by the highly aggressive body fluids. This corrosion can
cause local tissue reactions called metallosis (Contzen and Broghammer,
1964; Zilkens, 1981) as well as allergic reactions (MacKenzie et al., 1967;
Tinckler, 1972; Symeonides et al., 1973; Samitz and Katz, 1975; Vernon-
Roberts and Freeman, 1977), which even might enforce the removal of the
implant. Toxic effects have been described by Mital and Cohen (1968),
Pappas and Cohen (1968), Jones et al. (1975), Taylor and Marks (1978), Abel
and Ohnsorg (1979), Harms and Mäusle (1980), Rae (1975 and 1981). Cancerogenic actions of the corrosion products have been discussed by Mc Dougall
(1956), Oppenheimer et al. (1956), Delgardo (1958), Dube and Fisher (1972),
Heath and Daniel (1962), Sundermann (1977) and Tayton (1980).

A further discussion about the possible risk introduced by the application of medical implants comes from the fact that practically all the
constituents of todays metal implants áre essential elements, which in
normal tissues and body fluids are present only in trace or ultratrace
concentrations. Due to their masses the metal implants represent pools of
trace elements in the patients body which by corrosive dissolution are
mobilized. They can affect the trace element balance of the surrounding
tissues as well as that of the entire organism.

Trace element analysis of tissues and body fluids is a necessary pre-requisite for the investigation of the interactions of the human organism with metal implants. Quantitative determinations of the concentration levels of the implant constituents in the patients body have to provide data bases which document the changes of trace element concentrations due to the implant corrosion. At the same time the chemical analyses have to describe the trace element status of the patients as complete as possible in order to look for interactions of corrosion products with other essential elements. Finally, by connecting the analytical results with the individual case histories a basis should be provided for an assessment of the risk introduced by the application of metal implants.

METHODS AND MATERIALS

To meet these scientific goals a number of analytical requirements have to be fullfilled. First, it is necessary to use a multi-element technique or a combination of single-element techniques which allow to determine in one sample all the required elements. This is desirable in order to determine interelement relationships and, moreover, to be able to detect unexpected effects. Though accurate results should be a matter of course for chemical analysis experience demonstrates strikingly that many analytical investigations are seriously hampered by methodological problems. In particular, for trace and ultratrace analyses for many elements in human tissues and body fluids literature reveals severe discrepancies (Iyengar et al., 1978) which mostly have to be attributed to systematic errors rather than to a "normal" spread of human trace element levels, see Versieck (1985) for a detailed discussion. Therefore, from the accuracy point of view only those investigations are acceptable in which it is documented that distinct measures of quality control have been continously performed and which allow the reader to judge about the the quality of the analytical procedure on the basis of results for suitable standard reference materials.

Since most of the alloy constituents, in particular Cr, Co and Ni, are present in ultratrace concentrations in tissues and body fluids, only analytical methods with extreme sensitivity are applicable. The choice of the techniques is further complicated by the necessity to look for a method which at the same time has extremely low blanc values. Too high analytical blancs can completely hide the effects searched for.

A final requirement for the analytical procedures is that they should also allow to determine the speciation of the trace elements investigated. Though it is common knowledge that the chemical state will be decisive for the action of an element in biological media, the demand to determine the speciation of trace elements is still widely futuristic and only minor but extremely important attempts have been made up to now. Most of the todays methods are not able to provide information on the speciation of trace elements.

In order to completely describe the interaction of metal implants with the patients organism quite different types of materials have to be analysed. The prime source of tissue samples for analysis is surgery on occasion of change or removal of implants, either as planned or wanted action or as a consequence of implant failure as e.g. loosening of totalendoprostheses (TEPs). Here, mostly the availability of tissues is restricted to connective tissues either from the nearest proximity (contact tissue) or from some distance where the surgery necessarily has to affect the tissues, e.g. fascia lata in case of hip joint surgery. Distant tissues and organs only are available from autopsies. Unfortunately, it is quite difficult to find deceased implant bearers which as an additional condition should not have suffered from a long term disease followed by intense medical therapy. One must be aware that the latter may alter severely the trace element levels of tissues and organs; see Michel (1984) for a survey on such effects.

Easily available materials which can be obtained by not or just slightly invasive methods are blood and blood constituents, urine, hair and nails. The sampling of all these materials exhibits severe contamination problems. In case of blood and urine they can be overcome by sophisticated measures. Hair and nails, on the other hand, are affected by a variety of extraneous influences which make it hard to unrevel the effects caused by the implants from those of external trace metal exposures.

Because trace analysis of biological materials still is a quickly developing field of science, there is a severe danger of mismatching results from different sources. This handicap, particularly applies to the so-called "normal" trace element levels. As has been discussed in detail by Versieck (1985) just for very few materials and for only a small number of elements it is possible to give reliable estimates of normal concentrations. On the other hand, the comparison of analytical results obtained for implant bearers with those of "normal" individuals is a conditio sine qua non. At the present state of knowledge one must demand that in order to draw decisive conclusions it is necessary to analyze all the materials investigated for implant bearers also for non-implant-bearing persons. If there are no analyses of non-exposed persons originating from the same laboratory obtained by identical methods there is practically no chance to evaluate the quality of the findings.

Up to now no analytical technique has been found which fullfills all requirements and the knowledge about the chemical interaction between organism and implant is still not satisfactory.

Instrumental neutron activation analysis is one of the best-suited analytical methods for the required analyses. As a multielement technique it allows for the simultaneous determination of nearly all alloy constituents and of a number of further essential trace elements thus giving a comprehensive survey on the trace element status of the patient. It's inherent low blanc values, it's excellent accuracy are further advantages. However, there are some limitations with regard to sensitivity which only can be overcome by applying time consuming and difficult radiochemical techniques.

The purpose of this paper is to survey the role of NAA for the investigation of the trace element burdening of the human organism by metal implants. A general review on trace element anlysis and its significance for biocompatibility testing will soon be published elsewhere (Michel, 1986). With respect to other analytical methods used for describing the impact of the corrosion of implants on the patients body we refer to the work of other authors in this proceedings (e.g. Black, Hildebrand, Ostapczuk and Stoeppler).

During the last years we performed in our laboratory comprehensive studies of the changes of normal trace element levels in man and animal due to the corrosion of Cr-Ni(AISI 316L)- and Co-Cr- alloys used for ostheosynthesis, alloarthroplasties and total joint replacement in orthopaedic surgery. This paper allows to give a survey on the data obtained so far and to combine them with a presentation of the todays state of knowledge in this field of trace element research. However, due to the limited space of this publication we will only deal with the results obtained for human tissues and body fluids and will restrict the discussion to the three elements chromium, cobalt and nickel. For a discussion of animal experiments and of other trace elements see Michel (1986). Details of our analytical procedure and a report on the detection limits achieved, on quality control and blanc values is given elsewhere (Hofmann et al., 1982; Michel et al., 1984 and 1986).

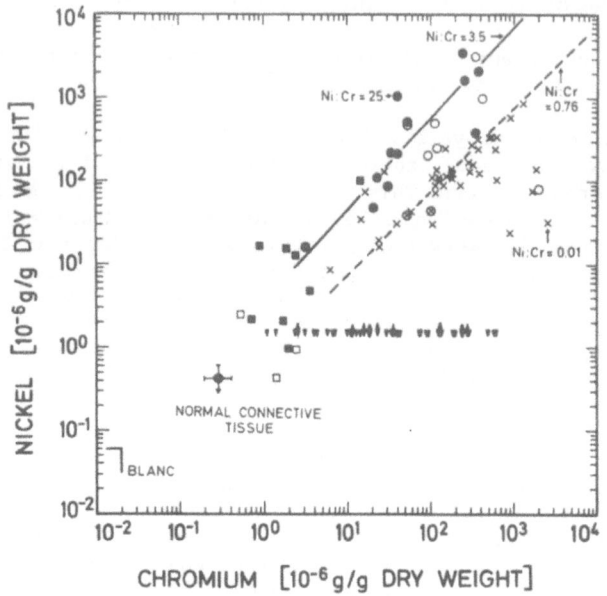

Fig. 1. Survey on the results obtained for tissues from the vicinity of Cr-Ni-implants (AISI 316L). Data for connective tissue from the nearest proximity of the implants (contact tissue) are from Lux and Zeisler (1972) (X and ♦), from Blettenberg (1975) (○) and from Michel and Zilkens (1978) (●). Data for fascia lata are from Lux and Zeisler (1972) (◑ and ▼), from Blettenberg (1975) (□) and from Michel and Zilkens (1978) (■). For some samples of contact tissue (♦) and fascia lata (▼) Lux and Zeisler (1972) were not able to detect Ni. Those data points have been plotted at their Ni detection limits.

ANALYSIS OF TISSUES FROM THE VICINITY OF THE IMPLANTS

Historically, most of the existing experimental data are related to tissues from the surroundings of the implants. For such tissues analyses reveal extreme burdening by corrosion products, the actual tissue concentrations, however, being influenced by both the type of corrosion and the mobility of the particular elements in the body. Hence the data will be looked at separately for the different alloys used.

For stainless steel implants, according to AISI 316L, results for human connective tissues from the nearest proximity of the implants and from fascial tissues were reported by Lux and Zeisler (1972), Schuster et al. (1973), Blettenberg (1975) and Michel and Zilkens (1978). Fig. 1 gives a survey on all the data for Cr and Ni given by these authors. For both

elements partially extreme concentrations of corrosion products are observed, exceeding the normal concentrations by more than three orders of
magnitude. However, the degree of burdening by extraneous trace elements
varies strongly and no correlation of the duration of implantation with the
tissue concentrations has been found. The burdening with corrosion products
is not limited to the connective tissue from the nearest vicinity of the
implants (contact tissue), but is to a smaller degree observed also for
samples of fascia lata. This latter material is 5 - 8 cm away from the
implants, the results thus pointing to the existence of long distance
effects and transport processes.

One remarkable point is, that the ratios of Cr and Ni in the tissues
show strong variability, ranging from Ni:Cr=25 to Ni:Cr=0.01. High Ni:Cr
ratios have been explained by a selective dissolution of implant constituents in the course of a passivation of the implant surfaces (Hofmann et
al., 1981). Cr:Ni ratios near to that of the steel (Cr:Ni=0.76) point to
the existence of more drastic corrosion including wear and massive dissolution without passivation. This is mainly observed for tissues from the
surroundings of plates and nails used for osteosyntheses after traumatic
events as they were mostly investigated by Lux and Zeisler (1972) and
Schuster et al. (1973). Similiar findings from animal experiments were also
reported by Michel et al. (1980) and Zlkens et al. (1981). Extremely low
Ni:Cr ratios can be caused by two effects. First, Cr-rich christallites
were observed by Lux and Zeisler (1972), most probably representing wear
from Cr-rich surface layers produced either by stress to a previously
passivated surface or by Cr-rich inclusions of the steel. Secondly, a
higher solubility and mobility has been claimed for Ni (and Co) than for Cr
considering the chemical attributes of these elements (Zitter, 1976). The
diversity of the actual findings in the tissues from the surroundings of
the implants demonstrates that the burdening of the tissues by corrosion
products is a multifactorial occurence revealing strong differences in the
damage of individual implants, the action of a variety of corrosion processes and the importance of the biochemical attributes of the implant constituents.

Besides for Cr and Ni all the investigations cited showed similiarly
severe enrichments of corrosion products for the elements Mo and Fe. Even
Co, which in stainless steel is present only as an impurity, showed elevated concentrations in tissues from the proximity of the implants (Michel
and Zilkens, 1978). Moreover, it turned out that the burdening of tissues
by corrosion products even can influence the concentrations of other (essential) elements. This has been shown for Zn and K by Lux and Zeisler
(1972) and by Zilkens et al. (1981).

For Co-Cr-alloys INAA was used to investigate human tissues from the
proximity of the implants by Evans et al. (1974), Schniewind et al. (1975),
Miehlke et al.(1981), Ohnsorge et al. (1978), and by our group (Hofmann et
al., 1982; Michel et al.,198 ; Löer et al., 1984). Schniewind et al. (1975)
analyzed the contact tissue of Blount's Vitallium clips. They found the
constituents of the implants to be enriched by several orders of magnitude
in the adjacent tissues relative to the tissues of non-burdened patients.
Also Evans et al. (1974) showed that in the vicinity of five metal-metal
prostheses of the hip joint high concentrations of Co and Cr were to be
observed. For metal-polyethylene prostheses they described significantely
lower but still enriched concentrations for 3 patients. Miehlke et al.
(1981) investigated serum, synovia and synovial fluid after application of
artificial knee-joints using INAA. For serum they observed no enrichments
(compare discussion below), but in the samples of synovia and synovial
fluid definite enrichments of Co and Cr were found.

Actually, the wide spread application of Co-Cr-alloys as totalendo-prostheses (TEP) for joint replacement and the frequent necessity to re-place non-functionating or loosened TEP's has made available a large amount of samples for investigation. In our laboratory up to now samples of arti-cular capsule have been analyzed from 46 patients and those of fascia lata from 22 patients, thus allowing to give a comprehensive survey on the tissue burdening with corrosion products caused by Co-Cr-alloys. Fig. 2 shows the results obtained for Co and Cr in samples of articular capsule and fascia lata. All the prostheses investigated were metal-polyethylene combinations with shaft and head made of Co-Cr-alloy and an acetabulum made of polyethylene. The observed Co and Cr concentrations observed in the tissues are seriously affected by corrosion products, the actual values exceeding the normal ranges by up to 3 orders of magnitude. A detailed statistical analysis (Michel et al., 1984) showed that for all patients a burdening with corrosion products can be seen, the actual distributions of Co and Cr being completely different for "normal" and "implant" group. As in the case of stainless steel no correlation of the actual burdening with the duration of the implantation was seen and again the element ratios do not reveal those of the implant materials. One observes a distinct chromium surplus in the surrounding tissues relative to the other alloy constitu-ents. This can only be explained by differences in the transport velocity of the elements in the body, since from tracer studies of the corrosion of Co-Cr-alloys in Ringer's solution a higher release of Co and Ni than of Cr by a selective solution process has to be adopted (Hofmann et al., 1982; Michel et al.,1984).

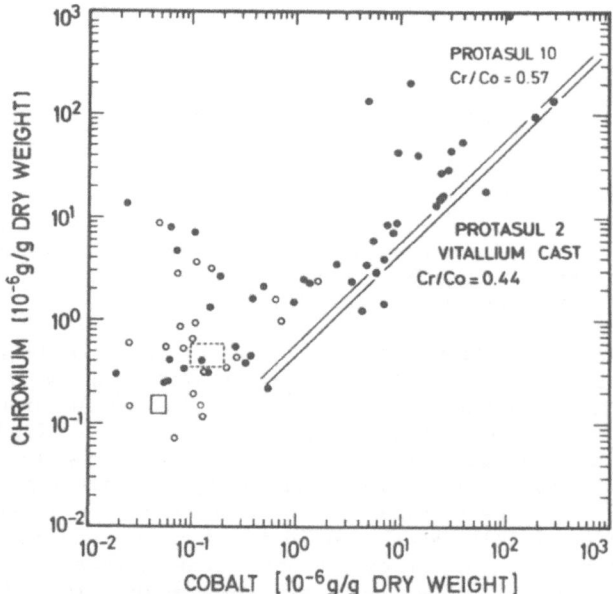

Fig. 2. Cr and Co in tissues from the vicinity of TEP's of the hip joint made of Co-Cr-alloys. The data are for articular capsule (●) and for fascia lata (O). The ranges (mean +- s.d.) of concentrations in normal articularcapsule and fascia lata are given as rectangles with full lines and dotted ones, respectively.

Due to the (usually) small amount of Ni in Co-Cr-alloys the Ni concentrations of the tissues (Fig. 3) do not show such extreme enrichments as observed for stainless steel implants. However, Ni-contents exceeding the normal values by two orders of magnitude are still found. A judgement about the element ratios Co:Ni (Fig. 3) and Cr:Ni (Fig. 4) shurely is complicated by the fact that the TEP's partially are made of two types of Co-Cr-alloys which are used for the construction of head and shaft, respectively. They have approximate compositions of (ca. 33% Co, 19-21% Cr, 33-37% Ni, 9.5-10.5% Mo, 1% Ti) and (ca. 60% Co, 27-30% Cr, 5-6% Mo, less than 2% Ni). But even if one considers these differences the actual tissue concentrations can only be understood if one combines the action of complex corrosion processes with the individual mobilities of the constituting elements in the body. A detailed discussion of this point is given elsewhere (Michel et al., 1984).

Also for Co-Cr-alloys not only the elements specific for the implants are affected. For Zn in fascia lata concentrations higher by a factor of two were observed in the normal than in the implant group. This demonstrates that there are consequences of the corrosion processes affecting the entire trace element metabolism rather than the vicinity of the implants only. Also the trace element balance of other essential trace elements may be changed. So for Cs in articular capsule a decrease in the implant group relative to the normal one is observed.

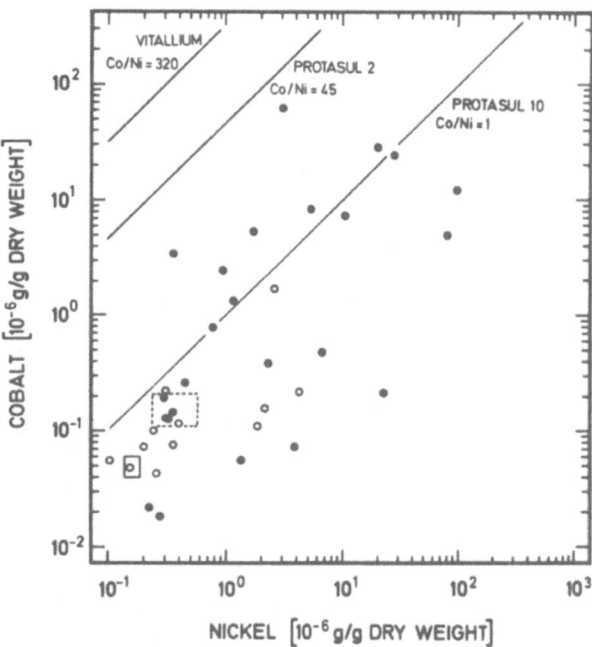

Fig. 3. Co and Ni in tissues from the vicinity of TEP's of the hip joint made of Co-Cr-alloys. The data are for articular capsule (●) and for fascia lata (O). The ranges (mean +- s.d.) of concentrations for normal articular capsule and fascia lata are given as rectangles with full lines and dotted ones, respectively.

Even further exogenous trace elements are brought into the patients body in the course of total joint replacement as a consequence of the use of bone cements for the fixation of the implants. Extreme burdening of tissues by Ba, Zr and Hf was observed as a consequence of the application of TEP's of the hip joint. Articular capsule as well as fascia lata are heavily polluted by these elements originating from additives to the bone cements for X-ray contrast. A detailed investigation of this effect showed that the bone cements are by no means inert in the body, but are subject to material damage and transport phenomena (Löer et al., 1983). In the context of implant corrosion it is, however, to be noted that there is a slight correlation (R=0.37) between the Co- and Zr-contents of the tissues, pointing to a possible correlation between the loosening of the implants, destruction of the cements and corrosion processes (Fig. 5).

Considering the necessitiy to adopt transport and metabolic processes to explain the concentrations of the corrosion products in the tissues from the vicinity of the implants, the necessity to investigate distant tissues, organs and body fluids becomes obvious. But before going to this topic some data on the tissue burdening after application of ceramic-metal prostheses shall be made. These types of TEP's having a shaft made of Co-Cr-alloy, while head and acetabulum are made of Al_2O_3-ceramics. These prostheses are supposed to result in a lower burdening of the tissues by corrosion products. The investigation of samples from articular capsule and of fascia lata from 9 patients after application of such ceramic TEP's indeed revealed lower metal concentrations (Fig. 6). The highest Cr and Co concentrations exceed the ranges of normal values by factors of hundred, only.

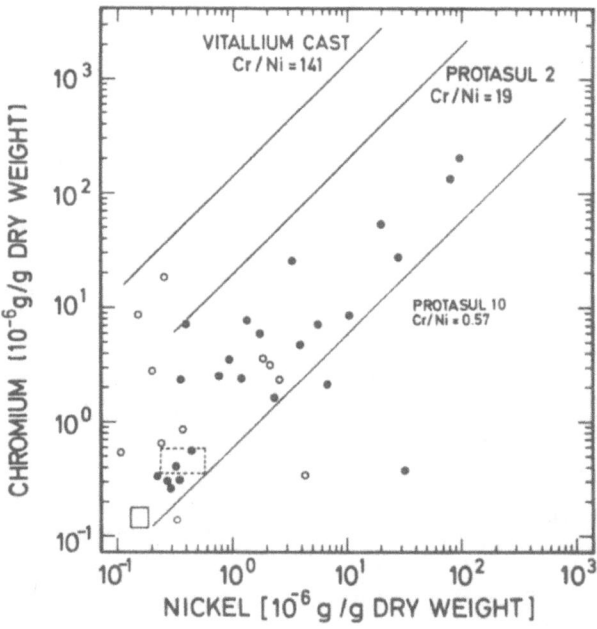

Fig. 4. Cr and Ni in tissues from the vicinity of TEP's of the hip joint made of Co-Cr-alloys. The data are for articular capsule (●) and for fascia lata (O). The ranges (mean +- s.d.) of concentrations for normal articular capsule and fascia lata are given as rectangles with full lines and dottedones, respectively.

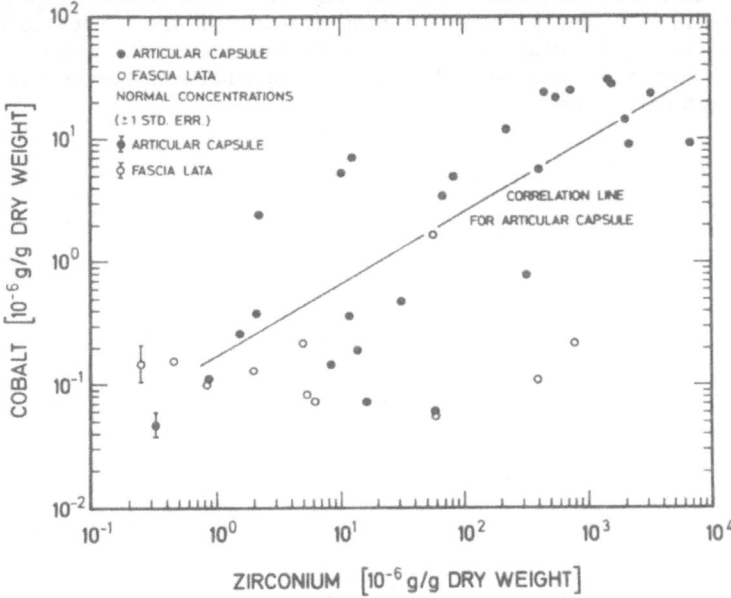

Fig. 5. Correlation of Co and Zr in human articular capsule (●) and in fascia lata (○) after application of TEP's of the hip joint made of Co-Cr-alloys.

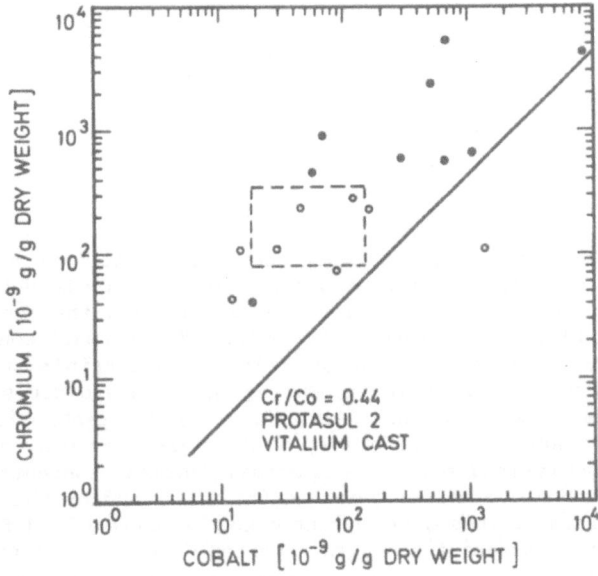

Fig. 6. Cr and Co in tissues from the vicinity of metal-ceramics TEP's of the hip joint. The data are for articular capsule (●) and for fascia lata (○). The ranges (mean +- s.d.) of concentrations in normal articular capsule are given as a full line rectangle.

However, a detailed analysis (Löer et al., 1985) showed that this advantage is countervailed by the presence of high amounts of Al in the tissues (most probably as wear products). Up to now no discussion of the medical consequences has been made whether the Al burdening counterbalances the decrease of Cr, Co and Ni in the tissues.

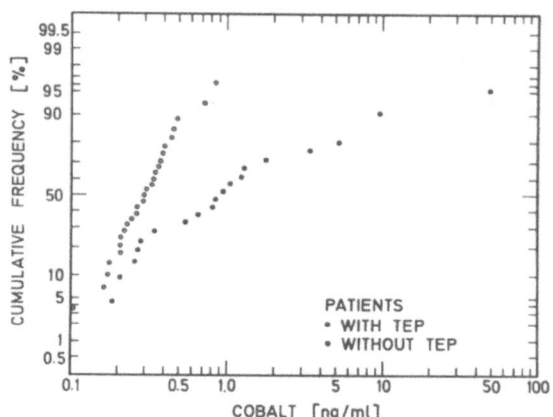

Fig. 7. Distribution of Co in human serum from patients with (●) and without (○) implants made of Co-Cr-alloys.

ANALYSIS OF HUMAN SERUM

The analysis of blood and blood constituents from patients with metal implants is of particular interest since it should allow to detect the impact of corrosion products on the total organism. However, the analysis of body fluids is a still widely unresolved problem. The "normal" concentrations as given in literature (Iyengar et al., 1978) show a terrible scatter for most elements contained in the implants. During an intense discussion lasting for several years (Cornelis and Versieck, 1980; Versieck, 1985) it became evident that the apparent variability of most elements has to be attributed to systematic analytical errors, the actual "normal" concentrations of elements as e.g. Cr, Co and Ni being at the lower end of the reported ranges in literature. This development makes it extremely difficult to evaluate and to judge about the data reported up to now. In particular all those conclusions have to be doubted which were drawn on the basis of comparisons with "ranges of normal literature data" taken e.g. from the compilation of Iyengar et al. (1978). It is a necessary prerequisite for investigations involving serum or whole blood that the results can only be compared with normal values analyzed in the same laboratory under identical conditions.

Coleman et al. (1973) were the first to report about the investigation of hair, blood and urine of patients with TEP's made of Co-Cr-alloys by INAA. They found enriched Co and Cr contents in blood as well as in the urine of these patients, in particular of those with metal-metal prostheses. They analyzed also blood of non-implant-bearers and found a mean cobalt concentration of 0.22 (range 0.07 to 0.49) ng/g dry weight, which is relatively near to values adopted today (Versieck, 1985). However, they gave normal Cr concentrations exceeding 1 ng/g dry weight which in the light of recent evaluations (Cornelis and Versieck, 1980) point to severe contamination problems.

Abeln (1977), Ohnsorge et al. (1978) and Abeln et al. (1980) described INAA of human serum after application of TEP's made of Co-Cr-alloys and studied trace element burdening as a function of time. These authors observed for metal-metal prostheses enriched Co-concentrations in serum. For metal-polyethylene prostheses no significant results were obtained. The problematics of their work definitely lies in the determination of normal concentrations. Abeln (1977) compared the findings of the implant patients exclusively with ranges of literature normal values. In the later publication of this work (Ohnsorge et al., 1978; Abeln et al., 1980) they compared their data with analyses of normal serum performed in the same laboratory. They reported normal values for Co between 1 and 2 ng/ml for Co and between 5 and 40 ng/ml for Cr (Ohnsorge et al., 1978) much too high and revealing severe methodological problems. However, the results found for implant bearers with McKee-Farrar metal-metal prostheses showed enrichments of up to two orders of magnitude for Co, being still highly significant. For Cr the results were not significant, as they were not for metal-polyethylene prostheses of the Charnley-Müller type.

In 1981 Miehlke et al. analyzed serum and synovial fluid from patients with metal-metal and polyethylene-metal knee-prostheses. They reported normal concentration ranges of 0.16 to 79 ng/g dry weight for Co and 0.47 to 84 ng/g dry weight for Cr. By comparison with these today obsolete normal concentrations they concluded that no changes of Co and Cr serum levels do occur in their patients.

In view of the unsatisfactory and contradictory results reported in literature in our laboratory a study was initiated to investigate human blood and serum from patients with metal-polyethylene TEP's made of Co-Cr-alloys. In this study the time dependence of trace element levels after implantation as well as the comparison of collectives of patients with and without implants is investigated. The details of the analytical procedure will be described elsewhere (Michel et al., 1986). Using a thermal neutron flux of 6×10^{13} cm^{-2} sec^{-1} for an irradiation time of 36 h the detection limits (LD according to Currie (1968)) of our INAA procedure for CR were 2.3 ng/ml and 20. ng/ml for Ni. Both elements could not be detected in human serum neither for patients with nor without implants. These elements can only be detected via radiochemical neutron activation analysis, thus pointing to the limits of INAA. For Co, however, we were able to analyze serum of both groups of patients. First results will be given here. Fig. 7 gives a comparison of the serum Co concentrations found in patients with and without implants. The implant group consisted of 19 patients (5 male, 14 female) aged between 49 and 91 years. The duration of implantation ranged between 7 and 216 month. 7 of the patients had prostheses at both hip joints. The normal group consisted of 27 patients (11 male, 16 female) aged between 21 and 79 years. While the normal group showed a linear normal distribution of Co with a mean of 0.33 +- 0.16 ng/ml, range 0.16 to 0.85 ng/ml, the implant group revealed a logarithmic normal distribution with a mean of 1.13 x: 4.3 ng/ml, range 0.18 to 70 ng/ml. Thus also for metal-polyethylene prostheses of the hip joint significant enrichments of Co due

to the action of corrosion products have been demonstrated. Up to now, no
correlation of the Co concentrations with the duration of the implantation
has been found. In general, the up to now results are similiar to those
found for the tissues from the vicinity of the implants (Michel et al,
1984) in so far that there is a wide variation of the burdening of serum as
well as of connective tissues ranging from normal concentrations to enrichments by up to several orders of magnitude.

The same observation is also made when investigating the time dependence of the Co burdening in serum. For this study the serum trace element
concentrations of 10 patients (4 male, 6 female), aged 63 to 79 were followed up to 90 d. Serum samples were taken at the evening of the day before
surgery, at the morning before surgery and at days 4, 7, 11, 14, 18 and 90
post operation. Also in this study we observed cases where the serum Co
level raised p.o. significantely, e.g. Fig. 8, and cases where it remained
practically normal, e.g. Fig. 9. From the medical point of view up to now
no differences in the results of these implantations have been observed.
Surely, all the cases have to be followed up. Our results show – together
with the findings for Cr and Ni of Hildebrand, Ostapczuk and Stoeppler in
this proceedings – that there are definite long-distance effects of the
implant corrosion which influence the trace element balance of the entire
organism.

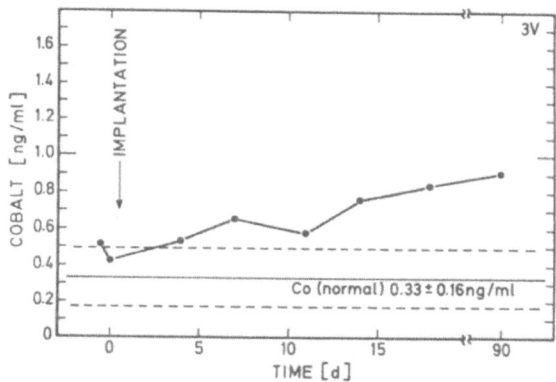

Fig. 8. Cobalt concentration of human serum as a function of time before and
after implantation of a TEP of the hip joint made of Co-Cr-alloy.

ANALYSIS OF ORGANS

In order to estimate long-term effects of implant corrosion in the
body the investigation of organs as possible storage tissues is essential.
However, as pointed out above, the analysis of organs from implant bearers
is seriously hampered by the extremely low availability of suitable autopsy
material. Without going into details of investigations not involving neu-

tron activation techniques, see Michel (1986) for a detailed discussion, only two groups reported about the analysis of organs of implant bearers by NAA. Jones et al. (1975) were the first to describe enriched Co values for various organs after application of Co-Cr-alloys. A second case was reported by Minski and Dobbs (1980) and by Dobbs and Minski (1980). They used INAA to investigate hair, tissues, organs and body fluids from an 81 years old female which had a metal-metal prostheses for 14 years and a polyethylene-metal one for 5.5 years. Unfortunately, these authors did not give any data for normal trace element concentrations, nor did they report on the analysis of standard reference materials. They compared their data with those of "Reference Man" (ICRP, 1975) and stated Co and Cr enrichments in lung, kidney, liver and spleen up to fifty times "Reference Man" values.

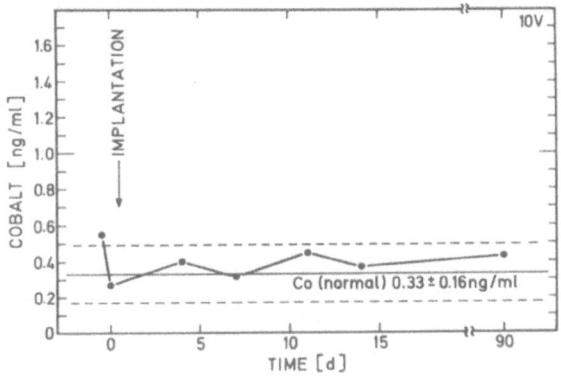

Fig. 9. Cobalt concentration of human serum as a function of time before and after implantation of a TEP of the hip joint made of Co-Cr-alloy.

After several years of search for adequate autopsy material we succeeded in our laboratory to receive samples from two patients (1 male, 1 female), aged 81 and 63 years, which had TEPs of the hip joint for 4 and 5 years, respectively. In order to allow for a sound interpretation of the analytical results a study of normal values was initiated involving 5 persons. The details of sampling and the analytical procedure will be described elsewhere (Michel et al., 1986). Up to now only the data for the first deceased implant bearer are finished and will be shortly discussed here. The complete results will be reported in due course (loc. cit.).

Tables 1 and 2 give the results for Co and Cr obtained for liver, spleen, aorta and heart. Though these data up to now can only be compared with literature values the Cr concentrations of liver and aorta and the Co concentrations found in liver and spleen are significantely higher than all values reported in literature. It has to be mentioned, however, that the data compiled by Iyengar et al. (1978) are not evaluated. From the present

Table 1. Cobalt in human autopsy material of a deceased implant bearer compared with "normal" values from literature. The data from literature are taken from: Iyengar et al. (1978)(Ref. 1), Zeisler et al. (1983)(Ref. 2) and ICRP (1975) (Ref. 3).

	COBALT [ng/g DRY WEIGHT]		
	WITH IMPLANT	NORMAL	
LIVER	15 200 ± 100	150 ± 36	Ref. 2
	19 100 ± 100	215.	Ref. 3
SPLEEN	16 000 ± 4600	23 – 1080	Ref. 1
		157.	Ref. 3
HEART	171. ± 2.	47 – 770	Ref. 1
	42.8 ± 0.7	111.	Ref. 3
	224. ± 2.		
AORTA	92. ± 1.	66 – 373	Ref. 1
	97. ± 1.	< 60. – 300.	Ref. 3

Table 2. Chromium in human autopsy material of a deceased implant bearer compared with "normal" values from literature. The data from literature are taken from: Iyengar et al. (1978)(Ref. 1), Zeisler et al. (1983)(Ref. 2)and ICRP (1975) (Ref. 3).

	CHROMIUM [ng/g DRY WEIGHT]		
	WITH IMPLANT	NORMAL	
LIVER	1130. ± 70.	396. ± 396.	Ref. 2
		< 4.5 – 158	Ref. 3
SPLEEN	180. ± 36.	31 – 1251	Ref. 1
		5.7 – 202.	Ref. 3
HEART	89. ± 3.3	40. – 476.	Ref. 1
		< 3.3 – 288	Ref. 3
AORTA	3860. ± 100.	90. – 1400.	Ref. 1
	40 900. ± 3500.	< 4.3 – 600.	Ref. 3

opinion the actual ranges of normal concentrations are likely to be at the lower end of the reported ranges. This assumption is also supported by the (evaluated) data for "Reference Man".

In spite of the fact, that a final conclusion can only be drawn after completing the entire study, including the analyses of normal organs, it can already be stated that there do exist enrichments of corrosion products in human organs. In view of the present results, as well as of the earlier published ones, it seems highly desirable to initiate further systematic studies on the influence of implant corrosion on the total organism. Considering the extensive use of metallic devices in the human body the unique chance to study the exposure of man to metals via a very unusual pathway should not be given away.

ACKNOWLEDGEMENT

This work was supported by the Deutsche Forschungsgemeinschaft, Bonn-Bad Godesberg. Our thanks are due to the Kernforschungsanlage Jülich GmbH for making available the neutron irradiations and to the staff of the reactors MERLIN and DIDO for their kind cooperation.

REFERENCES

Abel, J., and Ohnesorg, F.K., 1979, Toxikologie der Spurenelemente, in: Spurenelemente, E. Gladtke, G. Heimann, and J. Eckert, eds., Thieme, Stuttgart: 185.

Abeln, M., 1977, Spurenelemente im Serum nach Implantation von Kobalt-Chrom-Totalendoprothesen der Hüfte - Untersuchungen mit Hilfe der Neutronenaktivierungsanalyse, Thesis, Universität zu Köln.

Abeln, M., Ohnsorge, J., and Kasparek, K., 1980, Biochemische Untersuchungen der legierungsspezifischen Spurenelemente von Kobalt-Chrom-Hüftgelenktotalendoprothesen und AO-Winkelplatten im implantatnahen und und implantatfernen Gewebe sowie im Serum mit Hilfe der Neutronenaktivierungsanalyse, in: "Grenzschichtprobleme der Verankerung von Implantaten unter besonderer Berücksichtigung von Endoprothesen", M. Jäger, M.H. Hackenbroch, and H.J. Refior, eds., Thieme, Stuttgart.

Blettenberg, G., 1975, Metalloseuntersuchungen nach AO-Plattenosteosynthesen mit Hilfe der Instrumentellen Neutronenaktivierungsanalyse, Thesis, Universität zu Köln.

Coleman, R.F., Herrington, J., and Scales, J.T., 1973, Concentration of wear products in hair, blood, and urine after total hip replacement, Bri. Med. J., 1:527.

Contzen, H., and Broghammer, H., 1964, Korrosion und Metallose, Bruns' Beitr. klin. Chir., 208:75.

Cornelis, R., and Versieck, J., 1980, Critical evaluation of the literature values of eighteen trace elements in human serum or plasma, in: Trace element analytical chemistry in medicine and biology, P. Brätter, and P. Schramel, eds., Walter de Gruyter, Berlin: 587.

Currie, L.A., 1968, Limits for qualitative detection and quatitative determination, Anal. Chem., 40:586.

Delgado, E.R., 1958, Sarcoma following a surgical treated fractured tibia, Cli. Orthop. Rel. Res., 12:315.

Dobbs, H.S., and Minski, M.J., 1980, Metal ion release after total hip replacement, Biomaterials, 1:193.

Dube, V.E., and Fisher, D.E., 1972, Haemangioepithelioma of the leg following metallic fixation of the tibia, Cancer, 30:1260.

Evans, E.M., Freeman, M.A.R., Miller, A.J., and Vernon-Roberts, B., 1974, Metal sensitivity as a cause of bone necrosis and loosening of the prostheses in total joint replacement, J. Bone Jt. Surg., 56B:626.

Harms, J., and Mäusle, E., 1980, Biokompatibilität von Implantaten in der Orthopädie, Hefte zur Unfallheilkunde, 144, Springer, Berlin.

Heath, J.C., and Daniel, M.R., 1962, The production of malignant tumors by cobalt in the rat: intrathoracic tumors, Br. J. Cancer, 16:473.

Hofmann, J., Michel, R., Holm, R., and Zilkens, J., 1981, Corrosion behaviour of stainless steel implants in biological media, Surface and Interface Anal., 3:110.

Hofmann, J., Wiehl, N., Michel, R., Löer, F., Zilkens, J., 1982, Neutron activation studies on the in-body corrosion of hip joint prostheses of Co-Cr-alloys, J. Radioanal. Chem., 70:85.

ICRP, 1975, ICRP-23: Report of the task group reference man, Pergamon Press, London.

Iyengar, G.V., Kollmer, W.E., and Bowen, H.J.M., 1978, The elemental composition of human tissues and body fluids, Verlag Chemie, Weinheim.

Jones, D.A., Lukas, H.K., O'Driscoll, M., Price, C.H., and Wibberley, B., 1975, Cobalt toxicity after McKee hip arthroplasty, J. Bone Jt. Surg., 57B:289.

Löer, F., Zilkens, J., Michel, R., Freisem-Broda, G., Bigalke, K.H., 1983, Gewebebelastungen mit Körperfremden Spurenelementen durch Röntgenkontrastmittel der Knochenzemente, Z. Orthop., 121:255.

Löer, F., Zilkens, J., Schmidt, E., Nolte, M., Reich, M., and Michel, R., 1986, Release of trace elements from ceramic-metal endoprostheses of the hip joint, in: Proc. 5th Europ. Conf. on Biomaterials, Paris, September 4-6, 1985, in press.

Lux, F., and Zeisler, R., 1972, Instrumentelle Multi-Element-Aktivierungsanalyse von biologischem Gewebe und ihre Anwendung zur Metallose-Untersuchung, Z. Anal. Chem., 261:314.

Mc Dougall, A., 1956, Malignant tumor at site of bone plating, J. Bone Jt. Surg., 38B:709.

MacKenzie, A.W., Aitken, C.V.E., and Risdill-Smith, R., 1967, Urticaria after insertion of Smith-Petersen Vitallium nail, Br. Med. J., 4:36.

Michel, R., 1984, Changes of trace element concentrations of human tissues and body fluids due to therapeutic and diagnostic treatment, Fresenius Zeitschr. Anal. Chem., 317:451.

Michel, R., 1986, Trace metal analysis in biocompatibility testing, CRC Critical Rev. on Biocompatibility Testing, in preparation.

Michel, R., and Zilkens, J., 1978, Untersuchungen zum Verhalten von Metallspuren im umgebenden Gewebe von AO-Winkelplatten mit Hilfe derNeutronenaktivierungsanalyse, Z. Orthop., 116:666.

Michel, R., Hofmann, J., Holm, R., and Zilkens, J., 1980, Zum Übertritt von Korrosionsprodukten aus Stahlimplantaten in das Kontaktgewebe, Z. Orthop., 118:793.

Michel, R., Freisem-Broda, G., Hombach, V., and Tauchert, M., 1983, Systematic errors and quality assurance in the trace element analysis of (normal) human serum, in: Trace element analytical chemistry in medicine and biology, P. Brätter and P. Schramel, eds., Walter de Gruyter, Berlin: 833.

Michel, R., Hofmann, J., Löer, F., and Zilkens, J., 1984, Trace element burdening of human tissues due to the corrosion of hip-joint prostheses made of cobalt-chromium alloys, Arch. Orthop. Trauma. Surg., 103:85.

Michel, R., Nolte, M., Reich, M., Löer, F., Zilkens, J., 1986, Neutron activation analysis of human tissues, organs and body fluids to describe the interaction of orthopaedic implants made of Co-Cr-alloys with the patients organism, Proc. 7th Int. Conf. Modern Trends in Activation Analysis, June 23-27, 1986, Copenhagen, in press.

Miehlke, R., Henke, G., and Ehrenbrink, H., 1981, Kobalt- und Chrom- Konzentrationen in der Synovialflüssigkeit und in Blutserum nach Implantation von Kniegelenkendoprothesen mit Hilfe der Neutronenaktivierungsanalyse, Z. Orthop., 119:767.

Minski, M.J., and Dobbs, H.S., 1980, Determination of trace element concentrations in particular tissues contaminated by stainless steel implants, in: Trace element analytical chemistry in medicine and biology, P. Brätter and P. Schramel, eds.,Walter de Gruyter, Berlin: 339.

Mital, M., and Cohen, J., 1968, Toxicity of metal particles in tissue culture, J. Bone Jt. Surg., 50A:547.

Ohnsorge, J., Abeln, M., Zilkens, J., 1978, Spurenelementkonzentrationen ver-
    schiedener Gewebe, nachgewiesen mit der Neutronenaktivierungsanalyse,
    Z. Orthop., 116:607.
Oppenheimer, B.S., Oppenheimer, E.T., Danishefsky, I., and Stout, A.P.,1956,
    Carcinogenic effect of metals in rodents, Cancer Res., 16:439.
Pappas, A.M., and Cohen, J., 1968, Toxicity of metal particles in tissue
    culture, part I, J. Bone Jt. Surg., 50A:535.
Rae, T., 1975, A study on the effects of particulate metals of orthopaedic
    interest on murine macrophages in vitro, J. Bone Jt. Surg., 57B:
    444.
Rae, T., 1981, The toxicity of metals used in orthopaedic prostheses, J.
    Bone Jt. Surg., 63B:435.
Samitz, M.H., and Katz, S.A., 1975, Nickel dermatitis hazards from prosthe-
    ses, J. Dermatol., 92:287.
Schniewind, E.O., Kasparek, K., and Ohnsorge, J., 1975, Untersuchungen des
    Kontaktgewebes von Blount'schen Vitalliumklammern mit Hilfe der Ins-
    trumentellen Neutronenaktivierungsanalyse, Z. Orthop., 113:209.
Schuster, J., Lux, F., and Zeisler, R., 1973, Untersuchungen der Metallose
    durch Neutronenaktivierungsanalyse, Mschr. Unfallheilk., 76:537.
Sunderman, F.W., 1977, Metal carcinogenesis, in: Advances in modern toxi-
    cology of trace elements, R.A. Goyer, and M.A. Mehlman, eds., Hemi-
    sphere Publishing Corp., Wahington: 257.
Symeonides, P.P., Paschacoglu, C., and Papageorgiou, S., 1973, An allergic
    reaction after internal fixation of a fracture using a Vitallium
    plate, J. Allergy Clin. Immunolog., 51:251.
Taylor, A., and Marks, V., 1978, Cobalt: a review, J. Hum. Nutr., 32:165.
Tayton, K.J.J., 1980, Ewing's sarcoma at the site of a metal plate, Cancer,
    45:413.
Tinckler, L.F., 1972, Nickel sensitivity to surgical skin clips, Brit. J.
    Surg., 59:745.
Ungethüm, M., and Winkler-Gniewek, W., 1984, Metallische Werkstoffe in der
    Orthopädie und Unfallchirurgie, Thieme, Stuttgart.
Vernon-Roberts, B., and Freeman, M.A.R., 1977, The tissue response to total
    joint replacement prostheses, in: The scientific basis of joint
    replacement, S.V. Swanson, and M.A.R. Freeman, eds., Pitman Med.
    Publ., Turnbridge Wells: 86.
Versieck, J., Trace elements in human body fluids and tissues, CRC Criti-
    cal Review Clin. Lab. Sci., 22:97.
Willert, H.G., and Semlitsch, 1980, M., Biomaterialien und orthopädische Im-
    plantate, in, Ortopädie in Praxis und Klinik, 2nd ed., Vol. II,
    Allgemeine Orthopädie, A.N. Witt, H. Rettig, K.F. Schlegel, M.
    Hackenbroch, and W. Hupfauer, eds., Thieme, Stuttgart:22.1.
Zeisler, R., Harrison, S.H., and Wiese, S.A., 1983, The Pilot National
    Environmental Specimen Bank - Analysis of human liver specimen, NBS
    Spec. Pub. (U.S.) 656, U.S. Dept. of Commerce, Washington.
Zilkens, J., 1981, Metallose, eine erweiterte Definition, Thieme, Stuttgart.
Zilkens, J., Löer, F., Michel, R., and Hofmann, J., 1981, Tierexperimentelle
    Untersuchungen zur Frage der Gewebebelastungen mit Korrosionsproduk-
    ten bei Verwendung von V4A-Implantaten, Z. Orthop., 119:760.
Zitter, H., 1976, Schädigung des Gewebes durch metallische Implantate, Z.
    Unfallheilk., 79:91.

CHAIRMEN SUMMARY: FUNDAMENTAL ASPECTS

F.W. Sunderman Jr and M. Stoeppler

Based upon epidemiological studies of occupational exposures certain
nickel and chromium compounds and carcinogenesis bioassais of nickel,
chromium and cobalt compounds in animals, as well as the outcome of
in vitro tests for mutagenesis and genotoxicity, there is some reason to
suspect that release of nickel, cobalt and chromium corrosion and wear
particles from implanted prostheses could possibly constitute a carcinogenic hazard. Carcinogenic bioassais of cobalt and nickel metal powders
in rodents have been positive, although chromium metal has given negative
results under comparable experimental conditions. Only a few cases of
sarcomas have been reported in patients at the sites of orthopaedic
prostheses, so that the clinical evidence of carcinogenic hazards from
Ni-Co-Cr prostheses is not convincing.

However, the reporting of such sarcomas may be incomplete. Everyone
agreed that an international registry of prostheses associated malignancies
is desirable to monitor the possible occurence of neoplastic complications
of implanted devices.

Recent findings that Ni and Co compounds can induce lipid peroxidation in animals opens a new avenue to investigate the pathogenesis of local
reactions to metal prostheses. Therefore measurements of lipid peroxidation
products (e.g. malonaldehyde and related thiobarbituric acid chromogens,
as well as lipid conjugated dienes and hydroperoxidase) in tissues adjacent
to metal prostheses would be appropriate as possible indice of local lipid
peroxidation.

The litterature data on the toxicity of chromium show that there is
no real evidence of risk for human from chromium in the zero and trivalent
oxidation state, except of chromium allergy by occupational exposure and
in patients with failures of long-term metal-to-metal prostheses. Results
in experimental animals, however, indicate a potency of unwanted health
effects, i.e. subacute lethality and contact hypersensitivity induced by
injected Cr(III)-compounds, and weak carcinogenic effects elicted by
implantation of both Cr^0 and Cr(III).

In contrast, from other investigations with Cr(VI), it appears that
Cr(VI) containing compounds may be responsable for some unwanted health
effect especially by their genotoxicity. Further investigations, however,
are needed on toxicity as well as biovailability of these compounds in
biological milieu.

There was general agreement that neither method applied to metal
analyses can be regarded as error-free. The most important factor is staff
experience and ability to handle the posed problems. In view of these
statements present findings for so-called normal levels in unexposed
persons can be regarded as provisional only. In many cases further confir-
mation by other methods and other laboratories is required since normal
levels are extremely important for the assessment of threshold levels and
risk estimations etc...

From this doubtless further method evaluation from the basis of
reliability (accuracy, precision, comparability, standardization ---->
e.g. national bodies and I.U.P.A.C) is required.

In this context the urgent need for development and current application of appropriate - i.e. adopted - control and reference material
(example : Versiech's serumpool from normal subjects) has been mentioned
and the intensification of present efforts strongly supported. Only
with the aid of those "zero-level" materials trace metal laboratories
might be able to improve trace analytical approaches from the view of the
most important goals as is already shown above (accuracy etc...).

Possibly anvironmental specimen banks becoming now operational in a
few countries (U.S.A, F.R.G, Japon, Sweden, Canada) will play a role in
providing some of the required matrices.

For the near future one can be moderately optimistic due to some methodological progress (multi-dement approaches and direct analyses e.g. by solid sampling etc...).

Last not least it should be mentioned how important interdisciplinary efforts like the present meeting are !

TISSUE RESPONSE

In vivo CORROSION OF A COBALT-BASE ALLOY AND ITS

BIOLOGICAL CONSEQUENCES

Jonathan Black

University of Pennsylvania

Philadelphia, Pennsylvania

INTRODUCTION

Implants used in modern orthopaedic surgery are fabricated from three alloy systems with different majority or base components. These alloy systems are iron based high chromium alloys, collectively termed stainless steels, cobalt based alloys and titanium based alloys. Stainless steels and titanium alloys are used for fabrication of internal fixation devices. Cobalt based alloys are not widely used in this application due to their high material and fabrication costs and their relative lack of ductility. Such devices are considered to be temporary implants; long term biological effects should not be a factor in their routine use.

The stainless steels were used in partial and total joint replacement devices by early investigators such as Charnley[1] and McKee[2]. Today, this alloy system has been largely supplanted by cobalt based alloys, commonly (but in error) called "vitallium*". This alloy system was apparently introduced into orthopaedics in the 1930's by Venable and Stuck[3] in fixation devices and soon afterwards was adopted by Smith-Petersen for his mold arthroplasty[4]. By 1950, McKee was using the alloy system in total joint replacements.[2] Today this alloy system has come to dominate joint arthroplasty applications. In the past decade, titanium and titanium based alloys have begun to be used, initially because of their believed superior fatigue endurance limits and arguments suggesting a closer "match" to the elastic properties of supporting tissues.

Today, more than 90% of devices being placed in patients have a component fabricated from a cobalt based alloy. This alloy is expected to dominate total joint replacement fabrication for the next ten years and it is probable that for at least twenty years, the majority of joint replacement implants in place in patients will include this material. It is convenient to speak more simply about total hip replacement devices (THR's), since together they probably equal or exceed in number the sum of all other metal containing partial and total joint replacement devices.

*Trademark, Howmedica, Inc., Rutherford, NJ

COMPOSITION OF COBALT BASED ALLOYS

A variety of alloys of varying compositions are in use. However, the vast majority are included within the limits of two compositional standards of the American Society for Testing and Materials (ASTM)[5,6] (Table 1).

These two alloys are similar, with a cobalt base, chromium to confer corrosion resistance and a refractory metal, either tungsten or molybdenum, to increase strength, as well as a variety of trace element components to control these and other properties. The cast alloy, F 75 and a similar alloy, F 799, with an essentially identical composition, which is used for forged and powder metallurgical fabrication of the so called "fine grain" alloys, are the most commonly used today.

There are other cobalt based alloys used in implant applications but they have sufficiently similar compositions that they would be expected to release the same materials, albeit in different relative amounts, corresponding to alloy component activities.

BIOLOGICAL RESPONSE

Cobalt based alloys have a very favorable combination of mechanical and chemical properties which has led to their present dominance of the commercial fabrication of THR components. Nothing written here should be taken as deprecation of the major contributions that these alloys have made to the revolutionary progress in total joint replacement in the last three decades.

In fact, there is a consensus that they "...(have) been shown to produce a well characterized level of local biological response following long term clinical use..."[5]. In my view, this fairly describes the situation with regard to the local tissue (implant site) response in a majority of patients.

Until recently, this consensus has been an adequate and appropriate basis for the use of these alloys in THR's. However, three factors suggest that we must now carefully re-evaluate it:

1. Typical patient age at the time of conventional THR surgery is probably declining. Charnley reported [8] that the "the highest age incidence was in the seventh decade" among his 582 patients operated on between November 1962 and December 1965. Recent reports of conventional THR surgery include 108 procedures performed before age 45[9] and 58 procedures before age 50[10]. Another recent report[11], which concludes that "...conventional hip arthroplasty is a highly successful treatment in active patients between forty and sixty years old who have osteoarthritis of the hip.", cites other studies reporting 168 procedures before age 30. Earlier operative age leads to longer exposure to implanted devices, adding approximately eight-nine years of life expectancy for each decade decrease in age at surgery below 60[12].

2. There is broad interest in porous or fiber coated, uncemented implants. Galante[13] has summarized this field, including the pioneering clinical studies of Judet and his associates[14] and of Lord[15]. These devices, both experimental and clinical, have in common high specific surface areas and the absence of a cement mantle. Both of these features probably result in higher release rates of corrosion products than from conventional implants.

3. Laboratory and clinical investigations are gradually producing an understanding that focusing on the acute and short term local implant site biological responses may overlook significant sources of risk in the use of implants. It is possible that cobalt based alloys, used as chronic implants, may mediate systemic and remote site metabolic, bacteriologic, immunologic and carcinogenic processes[16] and that these processes may have long latency times.

Thus, the question of the biological consequences of total joint replacement can be seen to be a large and diverse one. Appropriate considerations include the short, intermediate and long term, local, systemic and remote effects, the relative contributions of polymeric, ceramic and metallic materials, of corrosion, dissolution, wear and other degradation mechanisms, and so on.

While this field of enquiry is broad, no clear trends have emerged. We are not able to evaluate the relative risks inherent in the use of each material type except in the presence of the most obvious acute adverse effects. Each material or alloy represents a particular set of possible risks. I do not intend to make value judgements concerning materials selection but will restrict my discussion to two topics:

*The release of degradation products by cobalt based alloys, taking F 75 type alloys as an exemplar.

*The biological consequences of this release.

DEGRADATION OF F 75 TYPE ALLOYS IN VIVO

In vivo corrosion rate

The modern F 75 casting alloy represents a generalization and gradual refinement of the alloy "Vitallium" introduced by Austenal Laboratories in 1929 for dental applications. Its "tasteless" characteristic, reflecting its' low corrosion rate, was partly responsible for its' adoption for dental applications and its' resultant introduction into orthopaedic device fabrication. However, alloys of this type have definite, non-zero corrosion rates. Steinemann[17] has calculated that the in vivo corrosion current, reflecting the dissolution and reformation rate of the passive surface layer, is approximately $10^{-2} - 10^{-3}$ uA/cm^2. This corresponds to a typical surface mass loss (release rate) of 0.15 ug/cm^2 per day. Before considering processes which can increase this release rate, let us enquire as to whether this is a large or small amount.

A typical metallic femoral THR component has a surface area of 200 cm^2 [18] and, thus a passive release rate of 30 ug per day. In Table 2, we can see how this compares, in patients, to body content and daily dietary intake of the major elements in the F 75 alloy.

It is clear that these are not large quantities, in the short term. However, it is known that in the long term, these elements, or more properly, complexes containing them, can be transported and stored in remote organs. Ferguson et al. [22] implanted cylinders of this alloy in rabbits and analysed various tissues 6 and 16 weeks later. The implants exhibited implant surface area to animal body weight (SA/BW) ratios near 3X.

The SA/BW ratio has been introduced[23] to permit scaling and comparison of various experiments. It reflects the ratio of the surface area of the typical femoral component of a THR (200 cm^2) to the weight of a standard patient (70Kg) and has a value of 2.86 cm^2/Kg when exposure is 1X or equivalent to normal THR implantation. Higher exposures, resulting in higher doses per unit time, are expressed as multiples of this number. While this ratio permits some form of inter-experiment comparison, it overstates the expected physiological effect of a particular exposure to typical test animals, since they have significantly higher basal metabolic rates per unit body mass than do patients.

Systemic distribution of corrosion products

Although Ferguson et al.[22] noted that changes in ion concentration associated with implantation of cobalt based alloy implants were minimal, they reported that there were significant increases in the chromium and cobalt content of the muscle surrounding the implant, indicative of corrosion, and in the cobalt and nickel content of the lung, spleen and kidneys of some animals, reflecting transport, storage and excretion.

That study was done more than twenty years ago, under somewhat primitive and uncontrolled conditions. In our laboratory, I have directed a series of experiments which have cast additional light on the release and transport of corrosion products from this alloy. In these studies we have made use of 50-100 micron diameter microspheres of F 75 alloy, with the exception of two studies performed with total joint replacement devices.

Initial studies[24] in vitro have demonstrated that that presence of serum proteins can raise release rates of cobalt and nickel from F 75 alloy five to ten fold, through a variety of specific and nonspecific binding processes. In addition, we found evidence that nickel forms interface regulated organometallic complexes. Studies on stainless steel have led to similar conclusions concerning chromium.

A first implant study[24], placing F 75 alloy intramuscularly in rats at 10X, 100X and 1000X, demonstrated significant serum nickel but not cobalt concentration elevations after ten days. Chromium content was not determined in this study.

A comparison of SA/BW effects on serum corrosion product concentrations in rabbits[25,26] with those seen in patients[27], for periods up to six months, suggests that SA/BW ratios of up to 100X may be required in the rat and rabbit to simulate patient postconventional THR exposure (1X), due to the higher metabolic effort and superior liver function in these small animals.

A second implant study[28], placing F 75 alloy intramuscularly in rats at 30X and 300X, revealed previously unsuspected greater than 10 fold post-operative elevations in serum chromium and cobalt concentrations. The cobalt concentration decreases rapidly but chromium remains elevated beyond thirty days post-operatively, in agreement with concentration elevations beyond six months found in patients after conventional stainless steel THR implantation[29].

A third implant study[30], utilizing the same model in rats, but examining chromium release at 10X, 100X and 300X as long as 120 days post-implantation, again demonstrated this post-operative serum chromium concentration elevation and suggested a pattern of chromium uptake and delayed release. In this study, we also found an SA/BW related inhibition of weight gain and a related pattern of morbidity and mortality secondary to expression of adventitial lung infection.

A further study in this series examined the possibility of suppressing ion release by use of pyrolitic carbon surface coating[31]. Implantation in the same rat model at 100X for 120 days, with various carbon coating thicknesses up to 1 micron, actually produced accelerated release, due to coating processing and structural defects. In one group of animals, implant site (gluteus maximus) ion concentrations rose sufficiently to produce undifferentiated fibrosarcomas in three animals, probably secondary to cobalt release.

While these studies are of interest in that they document release and distribution of and biological response to corrosion products from F 75 alloy, the use of microsphere implants at elevated SA/BW ratios may properly be criticized as being unrepresentative of human clinical experience. During the time when these studies were being conducted, we had the opportunity to examine a more clinically related situation.

We were able to obtain serum, clot and tissue specimens up to one year post-insertion from cats which received a human F 75/poly (etheylene) (UHMWPE) finger prosthesis as a knee joint replacement[32]. This device, fabricated from the same materials and to the same standards as human clinical devices, represented a geometric exposure of 0.4X. We were able to analyse for cobalt and nickel. These animals showed significant serum concentration elevations of both elements, increasing with implant time, as well as demonstrating tissue accumulations, reflective of body burden elevation. Fractionation and analysis of serum proteins demonstrated non-specific binding to albumin as well as specific binding to high-molecular weight proteins. A pattern of high molecular weight protein synthesis, reflective of a biological response, was also seen.

Conclusions concerning in vivo corrosion of F 75 alloy

From these studies, we can reach the following conclusions:

1. F 75 corrodes sufficiently under in vivo conditions to elevate serum metal concentrations in both animals and patients.

2. Since metal storage in tissues is dependent either upon instantaneous or time averaged serum concentration, there must be resultant increases in body burdens of metals.

3. Adventitious observations in animals suggest that corrosion products at these serum and tissue concentrations are biologically active.

4. Inferential evidence, from both in vitro and in vivo studies, suggests the presence of unexpected, more active forms of metals, such as hexavalent chromium (Cr[VI]).

Factors which may increase metallic ion release in patients:

Before turning to the biological implications of metallic ion release, let me point out several factors related to clinical use of F 75 which suggest that (human) in vivo release rates may be higher than predicted by the previous calculations. These factors include physical disruption of the passive layer during articulation, dissolution of released metallic wear debris, accelerated release due to the presence of serum proteins and increased effective surface areas in uncemented, porous coated devices. The magnitude and individual contribution of each of these effects is somewhat indeterminant, but together they may increase release rates 10 to 20 fold[33].

BIOLOGICAL CONSEQUENCES

Infection

Infection associated with clinical implants is generally viewed as grave[34]. Treatment of well established infection frequently requires the removal of the implant, thus adding the rigors of re-implantation to the problems of infection management and treatment. In bony milieus, such as the hip joint, re-implantation after treatment of infection, while becoming more frequent, is generally regarded as being less successful than primary THR insertion[35].

Infections in implant sites appear to be due either to contamination during surgery or to hematogenous "seeding" secondary to a systemic bacteremia. Animal studies suggest that the presence of implant materials generally reduces critical infectious titers in either immediate[36] or secondary[37] infection. The persistence of local infection may be related to geometric factors[38], protection of bacteria by the glycocalyx[39] or to reduction in macrophage number and/or activity[40,41]. Finally, it is possible that iron and chromium releasing implants may suppress "nutritional" immunity processes which normally act to deny iron to iron dependent bacteria[42,43].

If the relationship between infection and metallic implants depends upon the products released from the implant and not on the physical presence of the implant at the site of infection, then we might be encouraged to consider the possibility of infection elsewhere; that is, systemically or at remote sites. This is suggested by an animal study previously cited[30] where expression of an adventitial lung infection was monotonically related to the SA/BW ratio for F 75 spheres planted remotely in soft tissue location.

It is well known that patients with implants, like all other patients, develop both local and systemic infections. This is frequently assumed, particularly in the post-operative period for older more debilitated patients, to be related to the challenge of surgery. However, is there the possibility that such implants act to generally suppress the immune system and thus render infections easier to initiate and perhaps more difficult to treat? This seems like a subject amenable to epidemiological study; as of now there appears to be little interest in such studies.

General Immune Response

Resistance to infection is one specialized attribute of the action of the immune defense system of the body. The more general problem of response of the immune system to metallic implants has been considered with regard to both local and remote effects.

One of the local problems associated with THR implantation is loosening, secondary to changes in local tissue quantity and/or quality. It has been related to a direct clastic response to wear debris[44,45] as well as to bone necrosis secondary to local manifestations of a type IV hypersensitivity to corrosion products[46], most likely to their chromium and nickel content. However, this point remains controversial[47] and it is probable that several factors of a more mechanical nature (micromotion, strain incompatibility, etc.) play major roles in loosening of metallic prosthesis[48].

However, it is well established that metallic implants can invoke symptoms in the presence of previously established B-cell and T-cell mediated hypersensitivities, in both animal studies and in human clinical populations[49]. While there is some debate concerning the nature of implant site effects, remote site responses, including urticaria, have been reported to be precipitated by chromium bearing implants and to resolve upon implant removal. Furthermore, studies of patients from whom metallic implants have been removed[50], in comparison to epidemiological expectations[51] from studies of non-implanted individuals[52], suggests that implants themselves can cause immune system sensitization; that is, that reported symptoms are not solely the result of challenging already established hypersensitivities. More recent studies of patients with modern cemented polymer-metal THRs produce less certain results[53], perhaps due to the lower metal release rates of such devices.

Oncogenesis

Of all biological processes, oncogenesis, the production or promotion of tumors, is of the most concern to biomaterials investigators, since the general public shows a strong aversion to the subject and a desire to avoid exposure to potential oncogenic agents. This desire has been reflected repeatedly in legislative and executive actions of local and national authorities in many countries. For this reason, I will discuss this biological response in somewhat more detail.

Genetic, physical and dietary factors and exposure to ionizing radiation and organic and inorganic agents all apparently play a part in producing neoplastic transformation. Metals and metal bearing compounds have long been recognized as chemical oncogenic agents.

Our knowledge of the oncogenic potential of metals and their compounds comes from only two sources: animal injection and exposure experiments and human epidemiological studies. In deciding that an agent is oncogenic in animals, strict criteria must be applied, such as those proposed by Furst[54]. Few agents exhibit behavior that fully satisfies these strict criteria. Thus, established chemical oncogenesis in animals is rare. However both chromium and its compounds[55,56] and nickel and its compounds[57] are potent carcinogens in animals. These findings are supported by our observations[31] and those of others[58,59] of tumorgenesis by cobalt based alloy implants in animals.

In man, due to the epidemiological nature of the studies, certain knowledge of chemical oncogenesis is even more rare. However, the US Public Health Service[60] includes chromium and certain chromium compounds among only 22 agents and chemical processes known to be carcinogenic in man and nickel and certain nickel compounds among only 95 substances which may be reasonably anticipated to be carcinogenic in man. Although still questionable, due to the relative rarity of their industrial use, cobalt and cobalt compounds are also suspected[61] human carcinogens.

A few comments are in order about chromium. Chromium may be released from implants in either a trivalent (Cr[III]) or hexavalent (Cr[VI]) oxidation state[62]. The distinction is important, since Cr[III] is relatively inactive biologically, since it is effectively unable to penetrate cell membranes. Cr[VI] on the other hand, enters cells easily and is reduced to Cr[III]. Possibly as a result, Cr[VI] is a potent chemical oncogen. Despite the demonstration of the release of Cr[VI] from metallic chromium under physiological (in vitro) conditions[63], until recently, based upon thermodynamic equilibrium considerations, it was assumed that Cr[III] was the predominant form released in vivo[64].

We have been unsuccessful in an initial attempt to detect Cr[VI] in patients after F 75/UHMWPE THR implantation[62]. However, in vitro studies on stainless steel suggest that the presence of serum proteins during corrosion results in formation of the more oxidized Cr[VI][65]. Theoretical considerations[66] reflect this possibility in vivo; this has been supported by the observation[67] that compounds containing trivalent chromium which fail to cause mutagenesis (indicative of possible oncogenic activity) in vitro in the absence of oxidizing agents, are capable of producing tumors in vivo, probably secondary to Cr[III] to Cr[VI] conversion[68]. The valence of chromium released by F 75 in vivo is not currently known but the possibility that significant amounts of Cr[VI] are present cannot be ruled out. The observation that urine/serum chromium concentration ratios in patients with chromium bearing implants rise with implantation time[69] is consistent with the production of Cr[VI].

Implant oncogenesis

Tumors might be expected at the site of implants, since degradation product concentrations will be high there. Experimental studies have demonstrated that F 75 wear debris are oncogenic in rodents[58]. Solid implants of F 75 did not elicit implant site tumors in rats over a two year exposure[70] (possibly related to a very low dosage (SA/BW ≈ 0.15X)) in that particular study; however, solid implants of related but higher nickel content alloys are capable of producing local fibrosarcomas[59]. Implant site tumors are well recognized in clinical veterinary practice, associated with the use of stainless steel internal fixation devices[71-76]. The tumors seen include both osteosarcomas and fibrosarcomas. The chemical oncogenic agent is probably either nickel or Cr[VI], or some organometallic complex containing these moieties, although other factors, such as chronic infection seem to be involved[75].

Until 1984, there were only six reports of implant site tumors in the human clinical orthopaedic literature[77-82]. As in animals, 4 of these are associated with mild and stainless steel fracture fixation devices while two are associated with F 75 type alloy[79,82].

During 1984, three separate reports of tumors associated with total hip replacements appeared[83-85] (Table 3). In addition, there are anecdotal reports[86] of three other cases similar to that reported by Swann[84].

These cases must be evaluated against the background of the millions of THRs which have been implanted in patients. However, I would like to make two comments:

1. The finding of a chronic histiocytic response to THRs is not uncommon. It has been suggested that this may, in fact, be a frequent or general response to total joint replacement[87]. This should be a matter of concern in light of the association of malignant fibrous histiocytoma with bony infarcts[88] and the possibility that such tumors may have their origins in malignant transformation of previously benign fibroses[83,89,90]. In the case of implant sites with their fibrous capsules and interfacial membranes, this risk may be increased by the propensity for histiocytes to phagocytize and thus concentrate corrosion precipitates and wear debris[91]. Additionally, we should be concerned by a report that the presence of a foreign body, in an animal model, may serve as a tumor promoter, even if it is not intrinsically oncogenic[92].

2. The failure to find significant numbers of tumors at the site of total joint replacement may be related to three issues:

i. Primary tumors of bone and cartilage are not common. All bone and cartilage tumors, both primary and secondary, account for only 0.22-0.25% of all reported tumors, judging from statistics for the US[93], England[94] and Sweden[95]. This corresponds to a probable incidence rate of just over one per million adult population per year. Despite the current high rates of total joint replacement, it is unlikely that there are more than several million total joint recipients world wide. In light of this low intrinsic sensitivity to neoplastic transformation and the (relatively) small population at risk, implant site tumors appear fairly unlikely to have been observed.

ii. Tumors may be occuring in more sensitive, remote organs such as lungs, kidneys, liver and spleen. There can be no question that oncogenic elements released by F 75 accumulate in such remote organ sites in man. Dobbs et al.[96] performed a whole body assay on the body of a patient who had received two THRs, 5 1/2 and 14 years respectively prior to death, and reported concentrations of nickel up to 10 times and cobalt and chromium up to 50 times accepted standard values in the lungs, kidneys, liver and spleen. Although concentrations in serum were not determined, urine concentrations of these metals were found to be 2 to 6 fold elevated compared to cited normal values. This picture is consistent with modest serum concentration elevations and continuing remote site storage. Thus it is probably more instructive to look for tumors in organs which are known to accumulate metals and which demonstrate higher intrinsic tumor incidence rates than bones, such as the lungs, liver and pancreas.

These results suggest that we ought to survey total joint replacement patients for tumor incidences in soft tissues and organs remote to the implant site. It cannot be denied that such tumors occur, since 1/4 of all adults will eventually die from primary or secondary disease related to oncogenesis[93-95]. However, it is not known whether such tumors occur with higher frequencies after total joint replacement, as no epidemiological studies have been performed.

iii. The long latency times expected for chemical carcinogenesis, for instance 21 years (range: 10-30) for chromium and 23-27
years (range: 5-40) for nickel (estimated from industrial exposure[97])
suggest that true local and remote site tumors should only begin to be
seen 15 - 20 years after implant insertion. Note that, although latency
declines with increasing dose, the brief postoperative periods cited in
Table 3 for implant site tumors after THR cast some doubt on the connection between those reports and the implants involved. Therefore, remote
site implant related oncogenesis may only become a clinical factor as
significant patient populations survive 15 - 20 years after total joint
replacement.

IMPLICATIONS

I propose that we should agree upon the following general
propositions:

1. There is adequate evidence to conclude that there is a
finite, non-zero rate of complications, including oncogenesis, in
humans, associated with the implantation of F 75 alloy.

2. This risk increases with increasing dosage and increasing
period of exposure.

3. It is prudent to act on these conclusions while awaiting a
definite determination of risk.

Since there are risks of adverse biological consequences
resulting from the use of F 75 type alloys, prudence dictates the
careful assessment of alternate approaches before high specific surface
area implants are implanted, without cement, especially in younger
patients.

These are difficult questions that arouse strong opinions and
reactions. What is needed is careful, even handed consideration,
supported by accelerated efforts in animal research and human epidemiology. Such an approach will resolve our present uncertainties about
the magnitude of adverse biological effects associated with the clinical use of F 75 and related alloys and define rules for their continued
use in surgical implants.

TABLE 1
Composition of Principle Cobalt Base Alloys[5,6]

Cast Co-Cr-Co Alloy   Wrought Co-Cr-W-Ni
ASTM designation: F 75   ASTM designation: F 90
ISO designation: 5832/4   ISO designation: 5832/5

    Major alloy additions:
        Cr: 27-30 (wt. %)   Cr: 19-21 (wt. %)
        Ni: 1   (max.)   Ni: 9-11
        Mo: 5-7   W: 14-16
    Trace components:
        Fe: 0.75 (max.)   Fe: 3   (max.)
        Si: 1   (max.)   Si: 0.4   (max.)
        Mn: 1   (max.)   Mn: 1-2
         C: 0.35 (max.)   C: 0.05-0.15
                                            P: 0.04 (max.)
                                            S: 0.03 (max.)
    Alloy base:
        Co: bal. (≈ 65 wt.%)   Co: bal. (≈ 50 wt.%)

TABLE 2
Elemental Content and Absorption

	Chromium	Cobalt	Nickel
Serum concern.[18,19] (ug/l):	0.2	0.3	5
Total body burden[20] (ug):	6000	1200	1000
Dietary intake[20] (ug/day):			
in food:	100	50	300-600
absorbed(%):	?	?	9-18[3]
Implant release (ug/day):	9	18	0.3
(ug/year):	3285	6570	110

TABLE 3
Tumors Associated with Total Hip Replacement Prostheses
in Humans

Authors	Device	Tumor	Time of diagnosis (Post-Implantation)
Bago-Granell et al.[83]	Charnley-Muller (F 75?/UHMWPE)	M.F.H.	2 years
Swann[84]	McKee-Farrar (F 75?/F 75?)	M.F.H	3 1/3 years
Penman et al.[85]	Ring (F 75/F 75)	O.S.	5 1/4 years

M.D.H.=malignant fibrous histiocytoma   O.S.=osteogenic sarcoma

REFERENCES

1. Charnley, J.: Low Friction Arthroplasty of the Hip, (Springer-Verlag, Berlin, 1979).

2. McKee, G.K.: Total hip replacement - Past, present and future. Biomaterials 3: 130, 1982.

3. Venable, C.S. ans Stuck, W.G.: Electrolysis controlling factor in the use of metals in treating fractures. J. Am. Med. Assoc. 111: 1349, 1938.

4. Smith-Petersen, M.N.: Evolution of mould arthroplasty of the hip joint. Clin. Orthop. 134: 5, 1978.

5. ( ): Standard specification for cast cobalt-chromium-molybdenum alloy for surgical implant applications. Designation: F 75-82. 1984 Annual Book of ASTM Standards, Vol. 13.01: Medical Devices, (American Society for Testing and Materials, Philadelphia, 1984), p. 13.

6. ( ): Standard specification for wrought cobalt-chromium-tungsten-nickel alloy for surgical implant applications. Designation: F 90-82. 1984 Annual Book of ASTM Standards, Vol. 13.01: Medical Devices, (American Society for Testing and Materials, Philadelphia, 1984), p. 17.

7. ( ): Standard specification for thermomechanically processed cobalt-chromium-molybdenum alloy for surgical implant applications. Designation: F 799-82. 1984 Annual Book of ASTM Standards, Vol. 13.01: Medical Devices, (American Society for Testing and Materials, Philadelphia, 1984), p. 325.

8. Charnley, J. and Cupic, A.: Results of low friction arthroplasty of the hip. Nine and ten year followup. Clin. Orthop. 95: 9, 1978.

9. Dorr, L.D., Takei, G.K. and Conaty, J.P.: Total hip arthroplasties in patients less than forty-five years old. J. Bone Joint Surg. 65A: 474, 1983.

10. Collis, D.K.: Cemented total hip replacement in patients who are less than fifty years old. J. Bone Joint Surg. 66A: 353, 1984.

11. Ranawat, C.S., Atkinson, R.E., Salvati, E., and Wilson, P.D. Jr.: Conventional total hip arthroplasty for degenerative joint disease in patients between the ages of forty and sixty years. J. Bone Joint Surg. 66A: 745, 1984.

12. ( ): Vital Statistics of the United States - 1980, Life Tables, Vol. II, Sec. 6, DHHS(PHS)84-1104 (U.S. Government Printing Office, Washington, 1984), p. 13.

13. Galante, J.O.: Overview of current attempts to eliminate methylmethacrylate. The Hip - 1983 Proceedings of the 1983 Open Meeting of the Hip Society C.V. Mosby, St. Louis 1984, p. 181.

14. Judet, R., Siguier, M. and Brumpt, B.: A noncemented total hip prosthesis. Clin. Orthop. 137: 76, 1978.

15. Lord, G., and Bancel, P.: The madreporique cementless total hip arthroplasty. New experimental data and a seven-year clinical follow-up study. Clin. Orthop. 176: 67, 1983.

16. Black, J.: Systemic effects of biomaterials. Biomaterials 5: 11, 1984.

17. Steinemann, S.G.: Corrosion of surgical implants - in vivo and in vitro test. Evaluation of Biomaterials, ed. by G.D. Winter, J.L. Leray, K. de Groot, (John Wiley & Sons Ltd, Chichester, 1980), p. 1.

18. Taylor, D.M.: Trace metal patterns and disease. J. Bone Joint Surg. 55B: 422, 1973.

19. Versieck, J. and Cornelis, R.: Normal levels of trace elements in human blood plasma or serum. Anal. Chim. Acta 116: 217, 1980.

20. Cornelis, R.: Chromium revisited. in: Proceedings: 3rd. International Workshop: Trace Element Analytical Chemistry in Medicine and Biology, ed. by P. Bratten and P. Schramel, (DeGruyter, Berlin, in press, 1986).

21. Luckey, T.D. and Venugopal, B.: Metal Toxicity in Mammals, Vol. 1: Physiological and Chemical Basis for Metal Toxicity, Vol. 1, (Plenum Press, New York, 1977).

22. Ferguson, A.B., Akahoshi, Y., Laing, P.G. and Hodge, E.S.: Characteristics of trace ions released from embedded metal implants in the rabit. J. Bone Joint Surg. 44A: 323, 1962.

23. Woodman, J.L.: Organometallic Corrosion Products: An In Vivo and In Vitro Comparison, Ph.D. Dissertation, University of Pennsylvania, Philadelphia, 1980.

24. Woodman, J.L., Black, J. and Jiminez, S.A.: Isolation of serum protein organometallic corrosion products from 316LSS and HS-21 in vitro and in vivo. J. Biomed. Mater. Res., 18: 99, 1983.

25. Smith, G.K.: Systemic Transport and Distribution of Iron and Chromium from 316L Stainless Steel Implants, Ph.D. Dissertation, University of Pennsylvania, Philadelphia, 1982.

26. Smith, G.K., Black, J.: Estimation of in vivo 316L Stainless steel corrosion rate from blood transport and accumulation data, ASTM STP 859: Corrosion and Degradation of Implant Materials, ed. by A.K. Fraker and C.D. Griffin, (American Society in Testing and Materials, Philadelphia, 1985), p.223.

27. Black, J., Maitin, E.C., Gelman, H. and Morris, D.: Serum concentrations of chromium, cobalt and nickel after total hip replacement: A six month study. Biomaterials 4: 160, 1983.

28. Keogel, A. and Black, J.: Release of corrosion products by F 75 cobalt base alloy in the rat. I: Acute serum elevations. J. Biomed. Mater. Res. 18: 513, 1984.

29. Pazzaglia, U.E., Minoia, C., Ceciliani, L. and Riccardi, C.: Metal determination in organic fluids of patients with stainless steel arthroplasty. Acta Orthop. Scand. 54: 574, 1983.

30. Wapner, K.L., Black, J., Morris, D.: Release of corrosion products by F 75 cobalt base alloy in the rat. II: Morbidity apparently related to chromium release in vivo. J. Biomed. Mater. Res. (in press, 1986).

31. Oppenheimer, P.H., Morris, D.M., Konowal, A.K., Clark, C.C., and Black, J.: Effect of carbon coatings on in vivo release of Cr, Co & Ni from F 75 alloy. Trans. Soc. Biomater. 7: 130, 1984.

32. Woodman, J.L., Black, J. and Nunamaker, D.N.: Release of cobalt and nickel from a new total finger joint prosthesis made of vitallium. J. Biomed. Mater. Res. 17: 655, 1983.

33. Black, J.: Metallic ion release and its relationship to oncogenesis. The Hip - 1985 (Proceedings of the 1985 Open meeting of the Hip Society) (C.V.Mosby Co., St. Louis, in press, 1986).

34. Sugarman, B. and Young, E.J. eds.: Infections Associated with Prosthetic Devices (CRC Press, Boca Raton, 1984).

35. Carlsson, A.S., Josefsson, G. and Lindberg, L.: Function of fifty-seven septic, revised and healed total hip arthroplasties. Acta Orthop. Scand. 51: 937, 1980.

36. Merritt, K., Shafer, J.W. and Brown, S.A.: Implant site infection with porous and dense materials. J. Biomed. Mater. Res. 13: 101, 1979.

37. Blomgren, G.: Hematogenous infection of total joint replacement. An experimental study in the rabbit. Acta Orthop. Scand. (Suppl.) 187: 1, 1981.

38. Black, J.: Biological Performance of Materials: Fundamentals of Biocompatibility. (Marcel Dekker, New York, 1981), p. 96.

39. Gristina, A.G. and Costerton, J.W.: Bacterial adherence to biomaterials and tissue. J. Bone Joint Surg. 67A: 264, 1985.

40. Rae, T: A study of the effects of particulate metals of orthopaedic interest on murine macrophages in vitro. J. Bone Joint Surg. 57B: 444, 1975.

41. Rae, T.: The toxicity of metals used in orthopaedics prostheses. J. Bone Joint SUrg. 63B: 435, 1981.

42. Weinberg, E.D.: Iron and susceptibility to infectious disease. Science 184: 952, 1974.

43. Weinberg, E.D.: Iron and infection. Microbiol. Rev. 42: 45, 1978.

44. Willert, H.G. and Semlitsch, M: Reaction of the articular capsule to wear products of artificial joint prostheses. J. Biomed. Mater. Res. 11: 157, 1977.

45. Revell, P.A., Freeman, M.A.R. and Weightman, B.W.: The response of articular tissues to polyethylene wear debris from prosthetic joints. Arch. Orthop. Traum. Surg. 91: 167, 1978.

46. Evans, E.M., Freeman, M.A.R., Miller, A.J. and Vernon-Roberts, B: Metal sensitivity as a cause of bone necrosis and loosening of the

prosthesis in total joint replacement. J. Bone Joint Surg. 56B: 626, 1974.

47. Brown, G.C., Lockshin, M.D., Salvati, E.A. and Bullough, P.G.: Sensitivity to metal as a possible cause of sterile loosening after cobalt chromium total hip replacement arthroplasty. J. Bone Joint Surg. 59A: 164, 1977.

48. Lee, A.J.C. and Ling, R.S.M.: Loosening, in: Complications of Total Hip Replacement Surgery, ed. R.S.M. Ling (Churchill-Livingstone, Edinburgh, 1984), p. 110.

49. Merritt, K. and Brown, S.A.: Hypersensitivity to metallic biomaterials. in: Systemic Aspects of Biocompatibility, Vol. II, ed. D.F. Williams (CRC Press, Boca Raton, 1981), p. 33.

50. Merritt, K. and Brown, S.A.: Metal sensitivity reactions to orthopaedic implants. Int. J. Dermatol. 20: 89, 1981.

51. Fregert, S. and Rorsman, H.: Allergy to chromium, nickel and cobalt. Acta Derm.-Venereol. 46: 144, 1966.

52. Deutman, R., Mulder, T.J., Brian, R. and Kemp, H.B.S.: Incidence of metal sensitivity before and after total hip arthroplasty. J. Bone Joint Surg. 59A: 862, 1977.

53. Carlsson, A.S., Magnusson, B. and Moller, H.: Metal sensitivity in patients with metal-to-plastic total hip arthroplasties. Acta Orthop. Scand. 51: 57, 1980.

54. Furst, A.: An overview of metal carcinogenesis. Adv. Exp. Med. Biol. 91: 1, 1978.

55. Léonard, A. and Lauwerys, R.R.: Carcinogenicity and mutagenicity of chromium. Mutat. Res. 76: 227, 1980.

56. International Agency for Research on Cancer (IARC), Chromium and Chromium Compounds, in: (IARC Monographs on the Evaluation of the Carcinogenic Risk of Chemicals to Humans, Vol. 23, IARC, Switzerland, 1980), p. 205.

57. Léonard, A., Gerber, G.B. and Jaquet, P.: Carcinogenicity, mutagenicity and teratogenicity of nickel. Mutat. Res. 87: 1, 1981.

58. Heath, J.C., Freeman, M.A.R. and Swanson, S.A.V.: Carcinogenic properties of wear particles from prostheses made in cobalt-chromium alloy. Lancet 1: 564, 1971.

59. Memoli, V.A., Woodman, J.L., Urban, R.M. and Galante, J.O.: Malignant neoplasms associated with orthopedic implant materials. Trans. Orthop. Res. Soc. 7: 164, 1982.

60. ( ): Third Annual Report on Carcinogenesis, December 1982. U.S. Department of Health and Human Services, Public Health Service. (U.S. Government Printing Office, Washington, 1982).

61. Sax, N.I.: Cancer Causing Chemicals, (Van Nostrand Reinhold, New York, 1981).

62. Bartolozzi, A. and Black, J.: Chromium concentrations in serum, blood clot and urine from patients following total hip arthroplasty. Biomaterials 6: 2, 1985.

63. Grogan, C.H.: Experimental studies in metal carcinogenesis. VIII. On the etiological factor in chromate cancer. Cancer 10: 625, 1957.

64. Mertz, W.: Chromium: An ultra-trace element. Chem. Scripta. 21: 145, 1983.

65. Merritt, K., Wortman, R.S., Millard, M. and Brown, S.A.: XPS analysis of 316 LVM corroded in serum and saline. Biomat., Med. Dev., Art. Org. 11: 115, 1983.

66. Rogers, G.T.: In vivo production of hexavalent chromium. Biomaterials 5: 244, 1984.

67. Petrilli, F.L. and De Flora, S: Oxidation of inactive trivalent chromium to the mutagenic hexavalent form. Mutat. Res. 58: 167, 1978.

68. Maltoni, C.: predictive value of carcinogenesis bioassays. Ann. N.Y. Acad. Sci. 271: 431, 1976.

69. Hildebrand, H.F., Ostapczuk, P., Mercier, J.F., Stoeppler, M., Roumazeille, B. and Decoulx, J.: Orthopaedic implants and corrosion products. Ultrastructural and analytical studies of 65 patients. in: Biocompatibility of Co-Cr-Ni Alloys, eds. H. F. Hildebrand, M. Champy (Plenum, New York, 1986).

70. Gaechter, A., Alroy, J., Andersson, G.B.J., Galante, J., Rostoker, W., and Schajowicz, F.: Metal carcinogenesis. A study of the carcinogenic activity of solid metals alloys in rats. J. Bone Joint Surg. 59A: 622, 1977.

71. Banks, W.E., Morris, E., Herron, M.R. and Green, R.W.: Osteogenic sarcoma associated with internal fracture fixation in two dogs. J. Am. Vet. Med. Assoc. 167: 166, 1975.

72. Harrison, J.W., McLain, D.L., Horn, R.B., Wilson, G.P. III, Chalman, J.A. and McGowan, K.N.: Osteosarcoma associated with metallic implants. Report of two cases in dogs. Clin. Orthop. 116: 253, 1976.

73. Sinibaldi, K., Rosen, H., Liu, Si-K. and DeAngelis, M.: Tumors associated with metallic implants in animals. Clin. Orthop 118: 257, 1976.

74. Herring, M.E., Smith, G.K. and Nunamaker, D.M.: Eight cases of implant-associated osteosarcoma in the canine. J. Am. Vet. Med. Assoc. (In press, 1986).

75. Stevenson, S., Hohn, R.B., Poller, O.E., Fetter, A.W., Olmstead, M.L. and Wind, A.P.: Fracture-associated sarcoma in the dog. J. Am. Vet. Med. Assoc. 180: 1189, 1982.

76. Knecht, C.D. and Priester, W.A.: Osteosarcoma in dogs: A study of previous trauma, fracture and fracture fixations. J. Am. Anim. Hosp. Assoc. 14: 82, 1978.

77. Mcdougall, A.: Malignant tumor at the site of bone plating. J. Bone Joint Surg. 38B: 709, 1956.

78. Delgado, E.R.: Sarcoma following a surgically treated fractured tibia. A case report. Clin. Orthop. 12: 315, 1958.

79. Dube, V.E. and Fisher, D.E.: Hemangioendothelioma of the leg following metallic fixation of the tibia. Cancer 30: 1260, 1972.

80. Tayton, K.J.J.: Ewing's sarcoma at the site of a metal plate. Cancer 45: 413, 1980.

81. McDonald, I.: Malignant lymphoma associated with internal fixation of a fractured tibia. Cancer 48: 1009, 1981.

82. Dodion, P., Putz, P., Amiri-Lamraski, M.H., Efira, A., de Martelaere, E. and Heimann, R.: Immunoblastic lymphoma at the site of an infected bone plate. Histopathology 6: 807, 1983.

83. Bagó-Granell, J., Aguirre-Canyadell, M., Nardi, J. and Tallada, N.: Malignant fibrous histiocytoma of bone at the site of a total hip arthroplasty. A case report. J. Bone Joint Surg. 66B: 38, 1984.

84. Swann, M.: Malignant soft-tissue tumour at the site of a total hip replacement. J. Bone Joint Surg. 66B: 629, 1984.

85. Penaman, H.G. and Ring, P.A.: Osteosarcoma in association with total hip replacement. J. Bone Joint Surg. 66B 632, 1984.

86. R.S.M. Ling: Personal communication, 1984.

87. L. Riley: Orthopaedic Research Seminar, University of Pennsylvania, 1983.

88. Heselson, N.G., Price, S.K., Mills, E.E.D., Conway, S.S.M and Marks, R.K.: Two malignant fibrous histiocytomas in bone infarcts. J. Bone Joint Surg. 65A: 116, 1983.

89. Bogumill, G.P. and Schwamm, H.A.: Orthopaedic Pathology, (W.B. Saunders, Philadelphia, 1984).

90. Capanna, R., Bertoni F., Bacchini, P., Bacci, G., Guera, A and Campanacci, M: Malignant fibrous histiocytoma. Cancer 54: 177, 1984.

91. Mirra, J.M., Marder, R.A. and Amstutz, H.C.: The pathology of failed total joint arthroplasty. Clin. Orthop. 170: 175, 1982.

92. Vasiliev, J. M. and Moizhess, T.G.: Tumorgenicity of sarcoma cells is enhanced by the local environment of implanted foreign body. Int. J. Cancer. 30: 525, 1982.

93. (   ): Cancer Facts & Figures 1984, (American Cancer Society, New York, 1984).

94. (   ): Cancer Statistics - Regular Series, MB 1, No. 12, Office of Population & Census Surveys, (Her Majesty's Printing Office, London, 1980).

95. (   ): <u>Cancer Incidence in Sweden</u>, National Board of Health and Welfare, The Cancer Registry, (Tryckindustri, Solna, 1983).

96. Dobbs, H.S. and Minski, M.J.: Metal ion release after total hip replacement. Biomaterials 1: 193, 1980.

97. Schottenfeld, D. and Haas, J.F.: Carcinogens in the workplace. CA 29: 144, 1979.

RELATION BETWEEN METAL CORROSION AND ELECTRICAL POLARIZATION

Surendra Singh* and Harcharan Singh Ranu**

*Department of Biomedical Engineering, Rensselaer
Polytechnic Institute, Troy, New York 12180, U.S.A.
**Department of Biomedical Engineering, Louisiana Tech
University, Ruston, Louisiana 71272, U.S.A.

INTRODUCTION

The increasing use of metals as implant materials in orthopaedic
surgery has stimulated a considerable amount of interest among researchers
and orthopaedic surgeons enabling the study of corrosion, wear, hypersensitivity, toxcity and carcinogenicity (Furst and Haro, 1960; Mears, 1979;
Woodman, 1980). Partly, the corrosion Phenomena is due to the metal
implant ionization in living tissue (Ferguson et al, 1960; Lautenschlager
et al, 1974). Corrosion is a complex electrochemical deterioration of
metals (Clark and Hickman, 1953). Futhermore, corrosion is referred to as
a combination of oxidation and reduction of reactions at the implant-
tissue interface. The basic reactions in corrosion are the removal of
positive metal ions from their positions in the metal crystal lattice.
This is called an anodic reaction. Similarly, the freed electrons, as a
result of removed cations, also react to oxygen and hydrogren molecules in
tissues to form hydroxyle. This is defined as cathodic reactions at the
metal-tissue interface (Mears, 1979).

More recent investigations have shown that electrochemical reaction
at implant-tissue interface is the source of metal-ion release into
biological-tissue fluid (Woodman et al, 1984; Ferguson et al, 1960 and
Mears, 1979). The cathodic reaction represents reaction of dissolved oxygen in solution and electric current flows equivalent to the amount of ion
release at the anode. Thus electrical quantities, such as electrical
potential, charge and current between corroded (anode) and non-corroded
zones (cathode) can be measured to estimate the magnitude of metal implant
corrosion at the implant-tissue interface. The resulting polarization
current or charge or potential and their decay rate may be used as an
estimate of corrosion currents on metal carrying both the anodic and
cathodic reactions to deduce the corrosion-rate (Mears, 1979).

Several authors have studied electrical polarization to determine
metal implant-corrosion at implant-tissue interface (Bowden et al, 1954;
Ferguson et al, 1960). Bowden et al (1954) studied the effect of metallic
transfer on the implant corrosion for stainless steel and vitallium
screws. Clark and Hickman, (1953) measured anodic back EMF for several
metals and reported a correlation between the anodic back EMF (ABE) and
the loss of weight in implants due to corrosion in vitro. Ferguson et al
(1960) implanted titanium and zirconium in protein-containing oxygenated

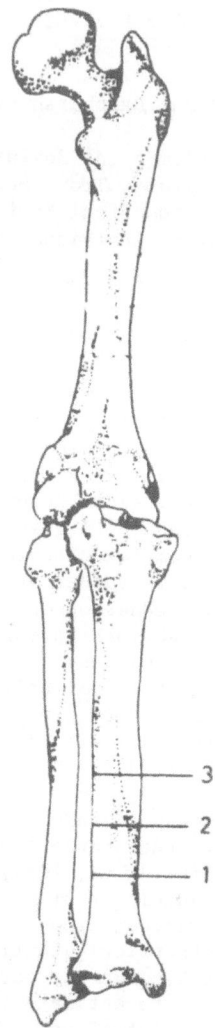

Figure 1. Electrode implantation in adult rabbit bone.

living tissue and reported a larger concentration of titanium around titanium implants than the zirconium. Brettle (1971) measured the typical
anodic polarization curve for stainless steel, titanium and titanium
alloys and cobalt-based alloys. The corrosion of implants and tissue
metallois for stainless steel, cobalt-chromium alloys titanium and its
alloys has also been measured. Brunski (1982) studied the corrosion
resistance of pure titanium in dental implant materials. In more recent
work, Lautenschlager et al (1974) examined the effect of external current
density on the corrosion of implantable stainless steel. Their
(Lautenschlager et al, 1974) results show that interface-corrosion defects
did not appear in passive samples until approxiamtely 300mV was applied
following 800mV stimulation. Similarly, Woodman et al (1984) examined the
corrosion behaviour of cobalt-chromium alloys, nickel and titanium using
electron microprobe analysis. Thus, these results appear to be and
suggest the need to use electrical quantity to understand the basic mechanism of the implant-corrosion in tissues. With this object in mind, an
attempt was made to measure electric potential between the cathodic
(uncorroded area) and the anodic (corroded area adjacent to tissue) area
using in vivo rabbit tibia applying stainless steel, platinum and silver
metal wires.

MATERIALS AND METHODS

The experiments were performed on adult rabbits weighing about 2.0
kg. The method of implanting electrodes was similar as mentioned earlier
by (Singh, 1982). Metal electrodes (1 mm diameter) were drilled through the
bone with their tips lying in the bone marrow (Figure 1). Lowest
electrode '1' was insulated all along its length and other electrodes
(2,3) were implanted without insulation (Figure 1). High-speed stainless
steel, silver and platinum were used to prepare the electrodes. Electrodes of each metal were implanted in six rabbits. The surgery and implantation were performed under standard aseptic conditions. Animals were
given doses of strepto-penicillin to maintain an aseptic environment. No
measurement were performed for about one week after the day of surgery.

Bioelectric potential was recorded using a high input impendance
($> 10^8 \Omega$) voltmeter, taking the lowest electrode '1' as the reference
electrode, bioelectrical potentials were measured between electrodes (1,2)
and (1,3). This was repeated for each animal of three groups. The
average values of bioelectric potentials for different metals are given in
Table 1.

Table 1.

Metal	Biopotentials (mV)	N
Platinum	+800	6
Silver	+280	6
Stainless Steel	-300	6

RESULTS AND DISCUSSION

From Table 1, it can be seen that the least biopotential, and so,
metal corrosion was found for stainless steel electrodes. Figure 2 shows
that electrochemical activities at the bone marrow-metal interface can
also be changed by varying the external circuit. From figure 2, it is
clear that electrical polarization current was maximum for an external
resistance of 1 MΩ. Figure 3 shows that the electrical polarization at
bone-electrode interface also changed with time. Since there is probably

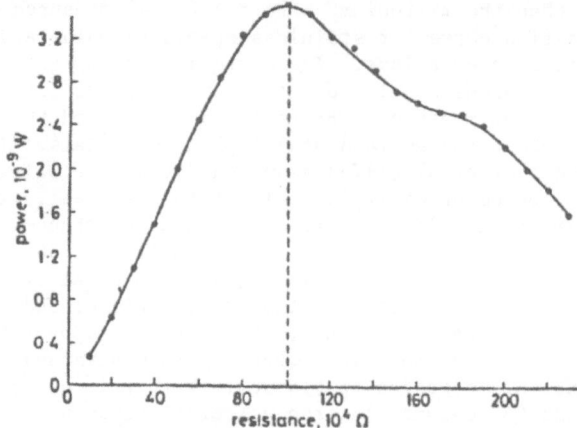

Figure 2. Available power against resistance for electrode positions (1,3).

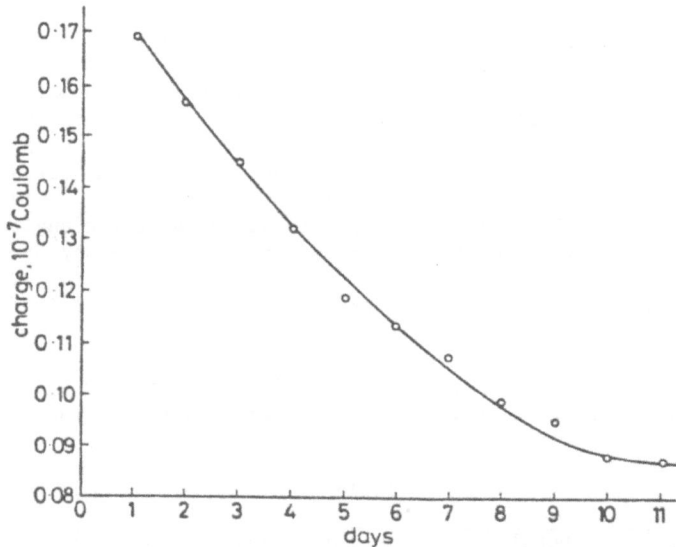

Figure 3. Decay of charge with time for electrode positions (1,3) when the stimulation was stopped after seven days.

a direct relationship between electrochemical reactions at the bone-metal interface and metal-corrosion. This suggests that chemical reaction (both oxidation and reduction processes) at the interface and the resultant electrical activities may be indicative of the fact that the rate of metal-corrosion decreased with time (figure 3). Currently we are extending our study for other materials by measuring polarization potential, current, and their decay with time.

REFERENCES

Bowden, F.P., Williamson, J.B.P. and Laing, P.S., 1954, Metallic transfer in screwing and its significance in bone surgery. Nature (London), 173:520.

Brettle, J., 1971, Stress corrosion cracking of Surgical implant materials. Atomic Weapons Research Establishment (Aldermaston) Report GR/44/83/73.

Brunski, J.B., 1982, In vivo forces on endosseous dental implants, II-transducer and telemetry. J. Dent. Res. 61:282.

Clark, E.G.C. and Hickman, J., 1953, An investigation into the correlation between the electrical potentials of metals and their behaviour in biological fluids. J. Bone and Joint Surg., 35-B:467.

Ferguson, A.B. Jr, Laing, P.G., Hode, E.S., 1960, The ionization of metal implants on living tissues. J Bone and Joint Surg, 42A:77.

Furst,A and Haro, H., 1969, A survey of metal carcenogenesis, proc. Exp Tumor Res. 12:102.

Lautenschlager, E.P., Sarkeas, N.K., Achasya, A, Galante, J.O, Rostoker, W., 1974, Anodic polarization of porous metal fibers, J. Biomed Mater Res, 8:189.

Mears, D.C., 1979, The dissolution of implantable materials. In: Materials and Orthopaedic Surgery, Chapter 4, pp 107, (Baltimore, The Williams & Wilkins Company, Publishers).

Singh, S., 1982, Ultrasound generation and propagation in hard tissues. Ph.D. Thesis. Jawaharlal Nehru University, New Delhi, India.

Woodman, J.L., 1980, Organometallic corrosion product: an in vivo and in vitro comparison Ph.D. (dissertation), University of Pennsylvania, Philadelphia.

Woodman, J.L., Jacob, J.J., Galante, J.O. and Urban, R.M., 1984, Metal ion release from Titanium based prosthetic segmental replacements of long bones in baboons: A long term study. J. Orthop Res 1:421.

IN VITRO AND IN VIVO CELLULAR RESPONSE TO SEMI-PRECIOUS AND AMORPHOUS ALLOYS

Pierre Guiraldenq[1], Henri Mazille[2], Jacques Brugirard[3],
Martine Bouvier[3], and Henri Magloire[3]

[1]Ecole Centrale de Lyon, BP 163, 69131 Ecully, France
[2]INSA-Lyon, 69621 Villeurbanne, France
[3]Fac Dentaire, 69372-Lyon, France

INTRODUCTION

In the last few years, a large variety of new alloys has been proposed for orthodontic applications and investigations concerning their biocompatibility have to be carried out.

In this way, two types of materials have been used in the present work:

- Two new semi precious alloys have been tested, due to the high price of noble metals and presumed, but discussed, toxicity of chromium nickel alloys.

- A new amorphous ribbon, choosen because of its typical structure with an homogeneous composition, without metallurgical defects, such as inclusions, grain boundaries or precipitates. Laser equipments may allow us in the future to elaborate by superficial melting and fast quenching amorphous or, at least, microcrystalline layers.

In order to complete preliminary results obtained with these materials, two methods have been used in this study, that is: "In vivo" tests were conducted on Indonesian pigs for a six month period; semi-precious alloys were implanted in the cheek and the maxillary and an amorphous one was placed between the palatal bone and fibromucosis.

The "in vitro" approach is a necessary screening test before use of new materials in man (1). An in vitro test was performed using human dental pulp maintained in a culture medium in such a way as to obtain the cellular response and evaluate the corrosion behaviour of the different tested alloys.

STUDY OF SEMI-PRECIOUS ALLOYS

In 1982 we presented 14 semi-precious Au-Pd-Ag-Cu alloys, with a gold content varying from 5 to 45wt%. Our research focussed on their structures before and after heat treatment and on their electrochemical responses. We arrived at the following conclusion: the gold content must be at least 30 wt% and the copper content must not exceed 10 wt% (2).

From 1983 to 1985, the following 4 alloys (wt%) were studied (Table 1).

TABLE 1

Number	Au	Pd	Ag	Cu
1	30	40	20	10
2	30	30	30	10
3	30	20	40	10
4	40	20	30	10

Particular attention was paid to their metallurgical properties (castability, heat treatments, microstructure) and electrochemical behaviour (after homogenisation, hardening, tempering). Particularly, the presence of silver-copper precipitation during tempering may lead to surface tarnishing and corrosion.

The final year was devoted to a biological study: "in vitro" using tissue culture, "in vivo" experiments using mini-pigs.

Tissue culture

The use of tissue cultures, developed in our faculty by Dumont and Magloire (4), is now standard practice in our center. Tissues of dental origin-pulpar or gingival-are cut into small pieces and put on inert glass plates (references) and on the metal specimens in Leighton tubes. EAGLE medium supplemented with foetal calf serum, penicillin and ascorbic acid are utilized. After 4 weeks, the cells are dehydrated, covered with a thin layer of gold and examined by scanning electron microscopy. For the 4 alloys, with a correct heat treatment(hardness), the cells growth on metal was compared with the same on the glass cover slip, and the following conclusions were drawn:

. the cell development is normal
. the growth of collagen fibres is identical with a satisfactory development of adhesive proteins
. the cells efficiently adhere to the metal

At this stage, these alloys, therefore, are well tolerated.

This is not the case for nickel-chromium samples used for comparison. The considerable disorder and absence of collagen fibres are sure signs of incompatibility between living cells and Ni-Cr alloys.

In vivo experiments

Our intention was to leave a crown and a few implants in an animal whose teeth and eating habits are similar to those of man. The Indonesian mini-pig, easy to rear and to handle, was chosen in preference to the dog or monkey.

. Preparation of the crowns: The crowns were made in a technical laboratory, in the alloy n°2 for the female, in the alloy n°3 for the male. They were fitted under general anesthesia, and in each pigs , three implants in the same alloy as that used in the crowns were installed, one in the cheek, one between the palate and the mucous fibre, and one deep in the alveolus. The male also had a strip of amorphous iron placed at the opening of the aveolus.

. First removal of samples: Three weeks later the general condition of the mouth was checked and everything found to be satisfactory. At the request of the specialist of the toxicology centre in Lille, Dr. Hildebrand, samples were taken, of urine, blood, bristle and nails, and kept at low temperature until they were submitted to a detailed analysis for metallic ions.

. Second removal of samples and extraction of the implants : Six months
later, the animals were killed and the implants observed. The crowns,
which had been present through the fifth month, had disappeared. Out
of the seven implants, only three remained and were recovered after
localization of the plates by X rays and needles and by parts of the
cheek being dissected step by step:

- a n° 2 alloy in the cheek of the female
- a n° 3 alloy, found, not at the bottom, but at the edge of the alveolus
of the male
- the amorphous alloy strip, which had been perfectly tolerated.

We noted that the n° 3 alloy darkened over a part of the surface, which
was in contact with the underlying tissues.

Analysis

Metallographic examinations after six months: alloy n° 2 (homogenised at
800° C, tempered 48 hrs at 500°C) under the scanning electron microscope
remained homogenious. For the alloy n° 3 (homogenised at 800°, tempered
one week at 500°C) we observed the appearance of superficial dark spots of
non adhering reaction products. These spots are so thin that they are not
detected by X-ray microprobe analysis and are probably composed of silver-
sulphur compounds, as the same appearance of dark surface sulphuration is
seen in the chemical conditions of the TOCILLO test in sodium sulfide
solution (5) (Figure 1).

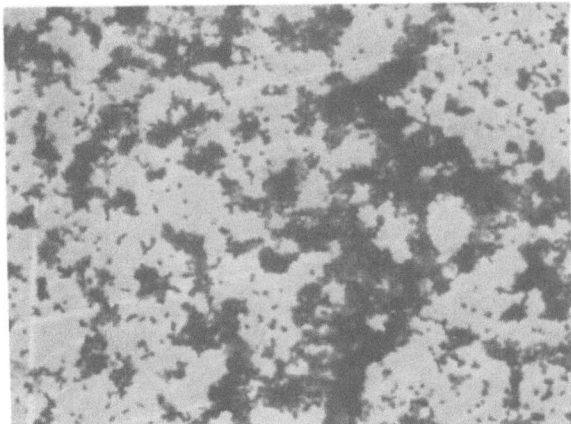

Figure 1 : X 800

Histological study

The samples were prepared as for a scanning electron microscope examination.
Fixed in glutaraldehyde then cacodylate buffer, the bone parts were demine-
ralized with HNO_3 (10 %, decreasing to 2 % over a fortnight). Finally they
were dehydrated using alcohol and embedded in paraffin. The coloring was
obtained by using Masson's trichrome.

Cheek sample of alloy n° 2 is observed at a place where the site of the implant is clearly visible (Figure 2). At a higher magnification, we observe an accumulation of contact cells and numerous collagen fibres (coloured in blue) (Figure 3). The fibroblasts appear in their normal spindle shapes (in red). A few sparse round red inflammatory cells are visible. The implant is well tolerated but tends to be rejected.

Seen in Figure 4 is an alveolar sample in contact with the alloy n° 3: As with the sample from male, the bone tissue maintains its Hawers system intact. The osteocytes are visible and correctly in place. No trace of the dark surface film which has disappeared during the preparation is visible.

In short, these two semi-precious alloys are bio-compatible, but the semi-wild animals used in the experiment have a high power of expulsion of foreign bodies.

Table 2
Copper (mg/l)

♂		3 weeks	6 months	evolution
	Blood with citrate	1.8	1.98	=
	Plasma with heparin	1.7	5.4	↗
	urine	15.6	10.7	↘
♀	Blood with citrate	0.8	0.86	=
	Plasma with heparin	2.7	{7 / 3.8	↗
	Urine	74.7	?	

Copper and silver contents

The levels of silver and copper ions in the blood and urine were measured after 3 weeks and 6 months. The technique used is atom absorption spectro-photometry in a graphite oven, establishing concentration levels by the method of measured doses of additives. Out of the 46 series of measurements, made by Dr. Mercier in Lille, we report results for 24, concerning total citrated blood, heparin plasma, and urine for both animals. The results are given in the Tables 2 and 3.

Figure 2. X 100

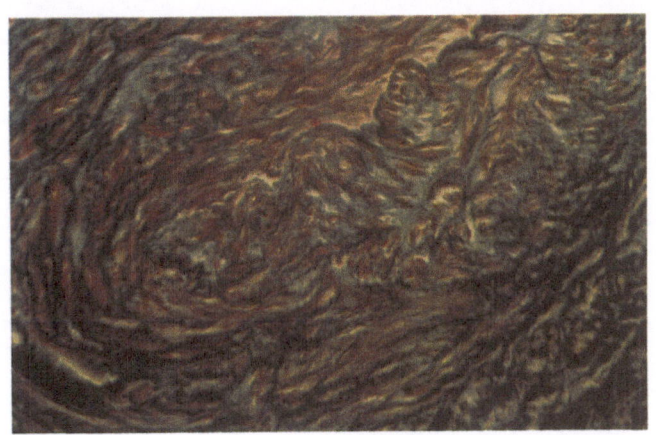

Figure 3. X 250

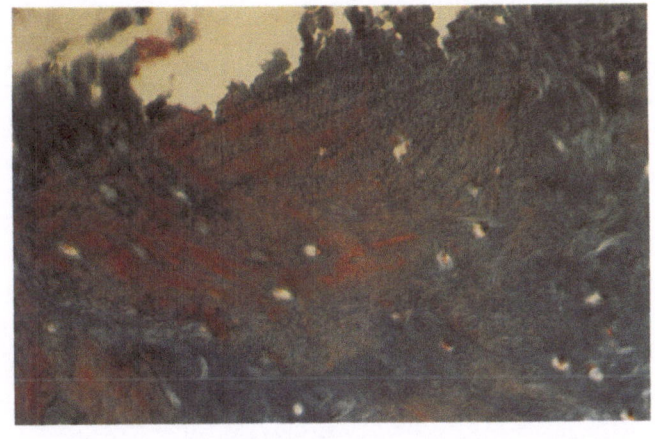

Figure 4. X 100

As an explanation of the copper levels, higher in the male and constant in the citrated blood, increasing in the heparin plasma, decreasing in the urine, one plausible hypothesis is that the kidneys might retain copper ions As silver levels exhibited a sharp drop after six months in both animals in the citrated blood, only increase being in the urine, it is reasonable to postulate that the kidneys are no obstacle to the passage of silver ions.

In short, as in the case of corrosion, the less noble ions, in solution or in aggregated particles, are the first to be eliminated.

Table 3
Silver (g/l)

		3 weeks	6 months	evolution
♂	Blood with citrate	41.7	6.4	↘
	Plasma with heparin	17.5	16.7	=
	urine	0.4	3.8	↗
♀	Blood with citrate	26.6	{0.5 / 4.4}	↘
	Plasma with heparin	3.1	{1.1 / 3.2}	↘ →
	Urine	1.4	?	=

STUDY OF AN AMORPHOUS ALLOY

The amorphous alloy we have chosen was prepared by Pont à Mousson Research Centre by the high speed quenching method from the liquid state. Ribbon-like specimens contain 87 wt% Fe, 4.6 wt% P, 2.7 wt% C, and 0.15 wt% Si (Fe76Cr4C11P8 wt%).

Qualitative analyses of the two sides of the ribbon were performed by AUGER and ESCA spectroscopy which give the evolution of elements and compounds as a function of depth. Metrological roughness measures of the surfaces were made too. The behaviour of this alloy was analyzed using CAMBRIDGE scanning electron microscopy and CAMEBAX electron microprobe analysis. These observations were performed before and after exposure to pulp cultures. The pulps used were derived from sound human teeth, cut into small pieces, and put on glass coverslips controls (controls) and on the amorphous ribbon (treated) in Leighton tubes. EAGLE medium supplemented with foetal calf serum, penicillin and ascorbic acid were utilized (4).

Initial state of amorphous surfaces

The manufacture of metal glasses gives ribbons with too different sides, distinct in composition and roughness, and called "bright" and "dull" faces. Preliminary analysis by Auger spectroscopy concerning the bright and dull sides is given in Figures 5 and 6.

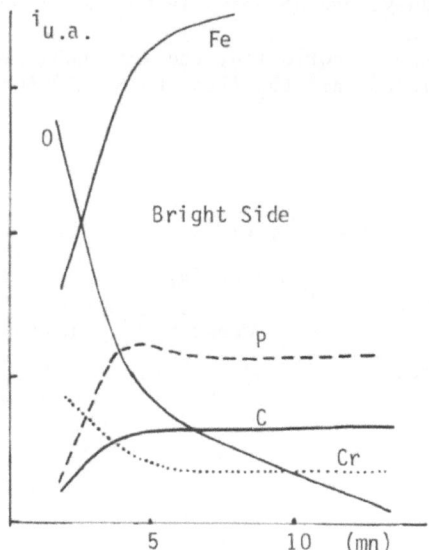

Figure 5.

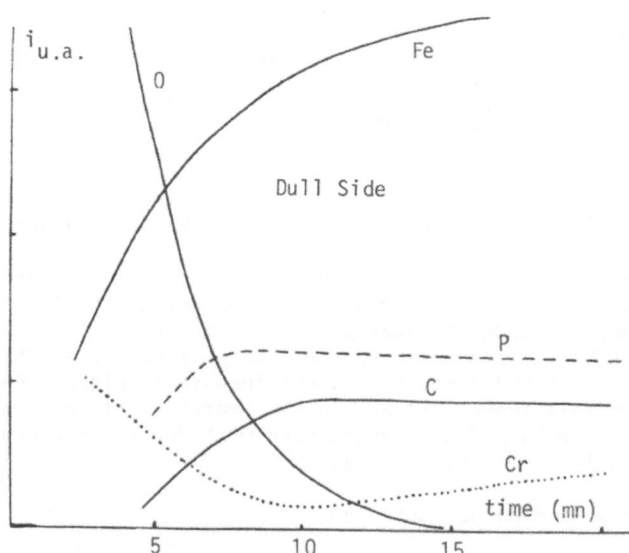

Figure 6.

A significative difference exists between the two sides, the dull side being more contaminated by oxygen in the surface layers (50-100nm). By ESCA, we have confirmed the presence of oxides. On the other hand, the dull side exhibits some topological peaks and breaks, while the bright one has a wavering profile (Figure 7).

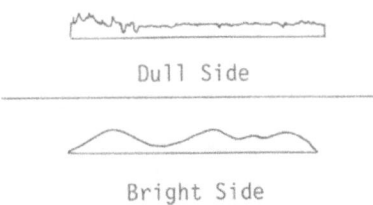

Dull Side

Bright Side

Figure 7.

Results after exposure to cultures

The scanning electron microscopy showed a modified aspect of the dull side of the ribbon after culture: gas bubbles which appeared before exposure to pulp culture were changed into cavities (Figures 8-9). No modifications appeared on the bright side. (Figure 10).

The same observations were done for the ribbon placed "in vivo" as an implant for the mini-pig in the first part of this study.

Figure 8 : dull side before culture
X 400

Figure 9 : Dull side after exposure to pulp culture
X 400

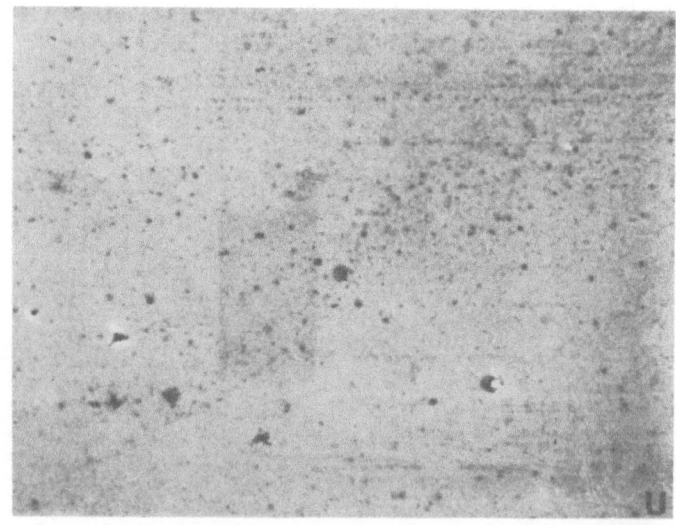

Figure 10 : Bright side before and after exposure to pulp culture
X 400

ESCA analysis permitted the detection of difference between compounds present on the surfaces (Figure 11). On both sides, after exposure to pulp culture, Na, N, Ca ions appear. The level of Na is higher on the bright side than on the other.

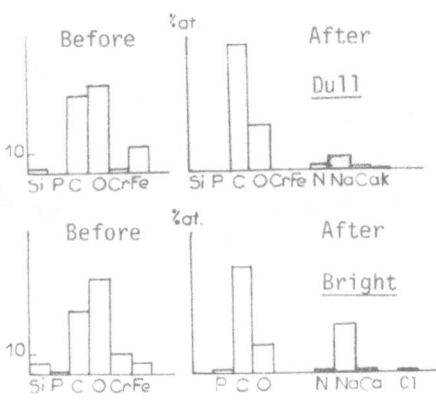

Figure 11

K and Cl are detected on both faces but to a lesser extent. The level of C is highly increased and that of O is decreased. Iron and Chromium ions are no longer detectable.

Cellular evolution

It should be pointed out that pulpal explants adhere faster on the bright face than on the dull face and than on glass coverslips. With the SEM, as shown in Figure 12, controls or treated cells exhibit a very spready aspect with numerous extends. At a ultrastructural level (TEM), cultures reveal dense inclusions into cells, grown on the bright face of the ribbon (Figure 13). The electron microprobe analysis (Figure 14) of such inclusion shows that they contain iron. Other detected elements come from the slide preparation and from the grid.

Apart from these inclusions, cytoplasmic ponctual random analysis of iron reveals that treated cells contain five to ten time more of this element than controls (Table 4).

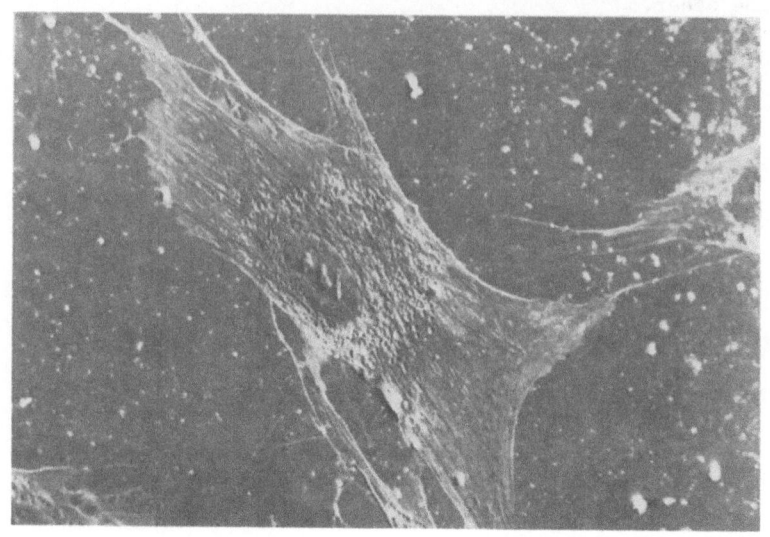

Figure 12 : X 900

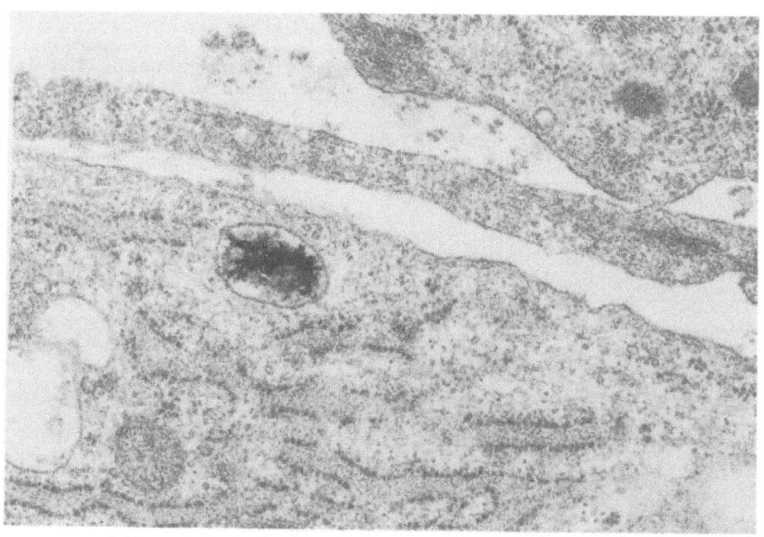

Figure 13 : X 45000

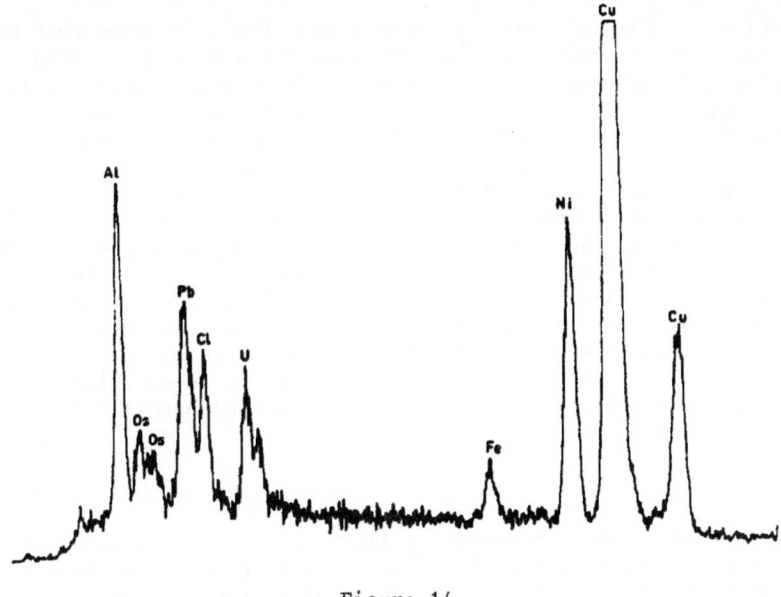

Figure 14

TABLE 4 : Impulses into each cell /20s

	mean value (10 spots)	± S.D
Controls	4.75	2.7
Bright side	47.2 22.8	12.8 6.2
Dull side	35.2 25.2	6.2 5.1

Discussion

Our interest in the amorphous alloys is focused on its surface aspect and composition, which differ with the analysed side. Our results show that the adhesion process may occur on both sides but the bright one is more favorable. In both cases too, fibroblasts grown on the material present the same morphological aspect, which differs from that of the reference. Differences of ions and compounds of the two surfaces have been established, before and exposure to pulp culture. Some elements are added, in varying quantities.

The topology of the dull face is changed too. Thus, it seems that cells and/or the culture medium may interact with the alloy. Also, iron seems to be of prime importance; this element is detected in intercellular organized inclusions, or in scattered form, probably binded to intercytoplasmic proteins. But this increased level of iron may be also related to an increase of cellular activity, by mitochondria for example. Nevertheless, these findings lead us to believe that a close correlation may exist between the presence of iron and a corrosion process of the amorphous alloy, either directly or through the culture medium. In conclusion, cells may adhere and grow on a metallic glass, but the compounds as well as the surface roughness analysis during the experiment exhibit a close inter-relation between this material and the culture itself, or the culture medium. The behaviour of the cells may depend on the composition and the roughness of the surface, and in the same way, the alloy may be modified by its micro-environment. Chromium content certainly has a real beneficial effect for the electrochemical passivity.

CONCLUSION

In the two cases, semi-precious alloys and amorphous material, favorable results concerning the cellular response were obtained. Even if the amorphous ribbon has a quite corrosion proof surface, a corrosion process appears. Similar results have been obtained previously using an in vitro implant method. In the same way, electrochemical studies of corrosion rate of the semi precious alloy led to the same conclusion as after the implantation process.

So, electrochemical, in vitro cell material interaction and in vivo studies give complementary indications and lead us to believe that the two types of materials concerned are biocompatible.

We must underline that the in vitro model used in this work permits control of a number of parameters and so, to keep a good homogeneity all along the experiments. Such a type of investigation must be developed in the future.

ACKNOWLEDGMENTS

The study has been realized with the help of Centre de Recherche de Pont à Mousson (M. de Guillebon) which supplied the amorphous ribbons.

REFERENCES

1. A. Pizzoferrato, A. Vespucci, G. Ciapetti, S. Stea Biocompatibility testing of prosthetics implant materials by cell cultures. BioMaterials 6, 346 (1985)

2. J. Brugiard, H. Mazille, Evolution of corrosion of some low gold alloys. Proc. of VIe Conf. of European Prosthodontic Assoc. BELGRADE sept. 1982, vol. 6, 44 (1983)

3. R. Charnay, P. Guiraldenq, J. Brugirard, H. Mazille, X. Heberard Influence de la structure et de la composition d'alliages semi-précieux sur leur comportement in vitro et in vivo - Metaux et Corrosion N°705, 161 (1984)

4. H. Magloire, J. Dumont Etude ultrastructurale de cellules pulpaires humaines cultivées in vitro. J. Biol. Bucc. 4, 3 (1976)

5. J.J. Tucillo, J.P. Nielsen Observation of the onset of sulfide tarnish on gold base alloys. J. Prosth Dent. 25, 629 (1971).

NICKEL AND CHROMIUM IN THE HUMAN BODY

A REVIEW AND A RESEARCH PROPOSAL

Jef A. Helsen

Department of Metallurgy and Materials Engineering
de Croylaan 2
3030 Leuven (Belgium)

INTRODUCTION

Claudius Galen, or Galenus in the original latin version of his name, is, as we all now, a very famous physician of Mysian origin and practicing in Rome in the second century, now some 1900 years ago. He was certainly not aware of the importance of trace elements in the bio-chemical factory of our body. But it might be possible that his dietary prescriptions for prevention of cancer resulted in a, in some way, con-trolled supply of trace elements. The food to be avoided included on one hand f.i. pickled meat. We can now accept that pickled food may contain substantially increased concentrations of some elements. On the other hand, he is in favour of sea food, the consumption of which we know is favourable for a regular supply of selenium and the anti-carcinogenic properties of this element are well established to day. The use of light wine is favoured by him over rich, heavy wines. If wines are analyzed, it is observed that the concentrations in nickel and chromium vary widely according to kind and origin : an Elzasser wine may show an average chromium content of 120 ppm, while a Beaujolais contains 10 ppm of chromium[1,2]. These 'gastronomic' comments illustrate that the uptake as well as the subsequent biological action of trace elements in biological media exhibits a very great specificity.

This unusual introduction is a 'paraphrase' for the hypothesis that biocompatibility studies of implant materials are bound to the specific environment, in which they finally will be used, because of the great specificity of action exhibited by the biological environment of these materials. This statement imposes severe restrictions on the conditions used in in vivo and in vitro compatibility studies.

The following points will be discussed in this communication : mean concentrations of nickel and chromium and confidence of the analytical determination; nickel and its biological incidence; chromium and its biological incidence; a research proposal : (a) epidemiological study, (b) physiological response of the body to the implant and its corrosion products, (c) search of representative in vitro test media.

Chromium is considered to be an essential trace element as established by Schwarz and Merz in 1959, nickel is suspected to be essential by authors like Nielsen and others since 1974[3]. The concentration of these elements is low with an order of magnitude of a few ng/ml or less in human serum and a few mg/g dry weight of muscle. In Table 1 some values for nickel in human serum are collated[4,5]. A drastic decrease is observed during the successive years due to improved experimental conditions. The values of Versieck are the lowest observed. The low average values and the numerous sources of contamination of the analyzed specimens makes the obtained accuracy questionable. An interlaboratory comparison of the total analysis of animal muscle has been done by the International Atomic Energy Agency in Vienna in 1980[6] and for animal bone in 1982[7]. Figure 1 and 2 show the values for chromium and nickel in muscle, the same for bone in Figure 3 and 4.

As can be seen no reference value or even recommended value for nickel is given and only a recommended value is given for chromium, although the values for chromium are not too bad. Parr[8] described in 1977 'in extenso' the difficulties related to the analysis of chromium by INAA. He finishes by quoting his report as 'a catalogue of failure'. And as we can see the IAEA reports of 1980 and 1982 do not contradict his statement ! If we scan at random the current literature (e.g. for nickel and chromium[9], for nickel[10]) for other critical analyses of reference samples in general the figures for chromium are conforting to some extent, what they are not for nickel (Table 2). Even the reference method recommended by the IUPAC[11] allows a rather large standard deviation (Table 3 and 4).

The examples shown indicate that care has to be taken in interpreting absolute low values for both elements.

Chromium is generally determined by INAA or electrothermal atomization in AAS. Nickel cannot be determined in a suitable way in low concentrations by INAA but is mainly determined either by electrothermal atomization in AAS or by the very promising differential pulse polarography of the dimethylglyoxime complex of nickel. Both methods are used for our own experiments[12,13]. A few values are given in Table 5. Chromium determination is performed by GFAAS and will also be done by INAA in collaboration with the University by Ghent.

Table 1 Metals in human serum. Concentration in ng/ml.

Year	mean	Nickel mean	Cobalt s.d.	Chromium mean	s.d.	range
1962	20	–	–	54	–	–
1970[a]	41	–	–	28	–	–
1970[b]	2.6	–	–	–	–	–
1975[c]	–	0.100	0.042	0.148	0.086	0.029–0.327

[a] J.M.H. Howard[5]
[b] F.W. Sunderman[5]
[c] J. Versieck[4]

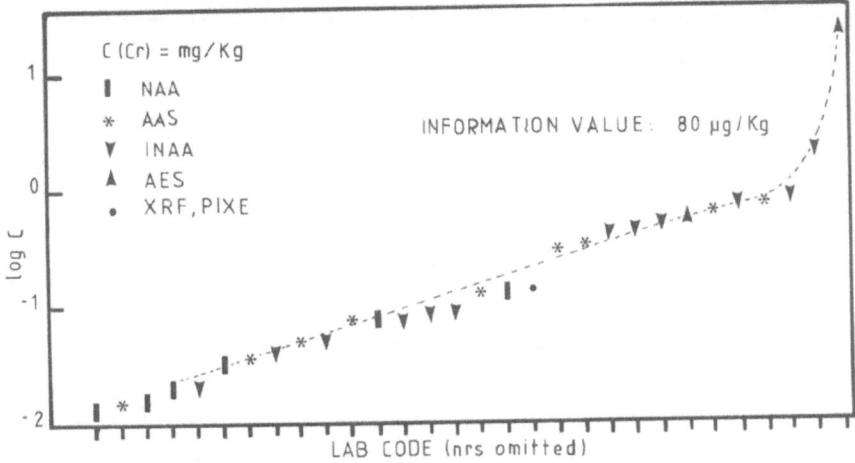

Fig. 1 Chromium in animal muscle. In ordinate the participating lab numbers (from Parr[6], Fig. 12).

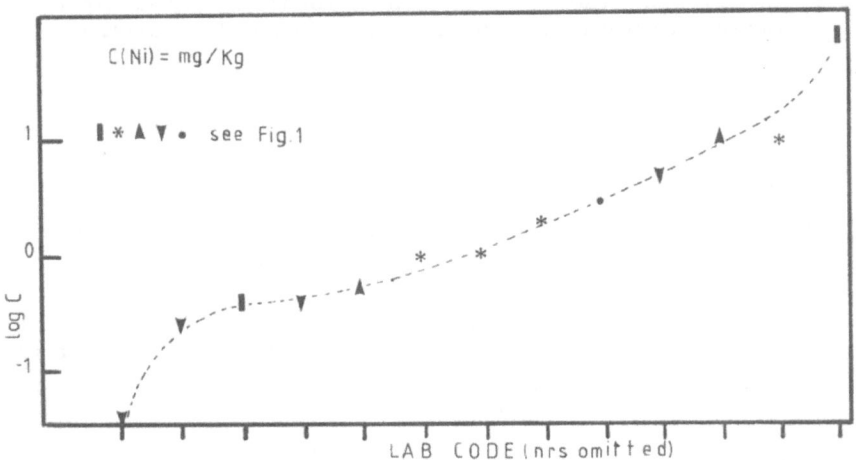

Fig. 2 Nickel in animal muscle. In ordinate the participating lab numbers (from Parr[6], Fig. 25).

Table 2 Nickel and chromium in different matrices. Concentration in mg/g.

SRM	Chromium[a]			Nickel[b]		
	A	B	C	A	B	C
1571 orchard leaves	2.6 0.3	2.6	2.8	1.3 0.2	1.4	1.5
1575 pine needles	2.6 0.3	2.8	2.8	(3.5)	4.2	3.7
1573 tomato leaves	4.5 0.5	4.3	4.3	–	–	–
1570 sinach	4.6 0.3	4.4	4.3	(6.0)	8.2	8.0

[a]According to Hoenig et al.[9]
[b]According to Clare and Robinson[10]
A : NBS proposed value; B : AAS; PE-400 + HGA-500; C : AAS; IL-751 + IL-555.

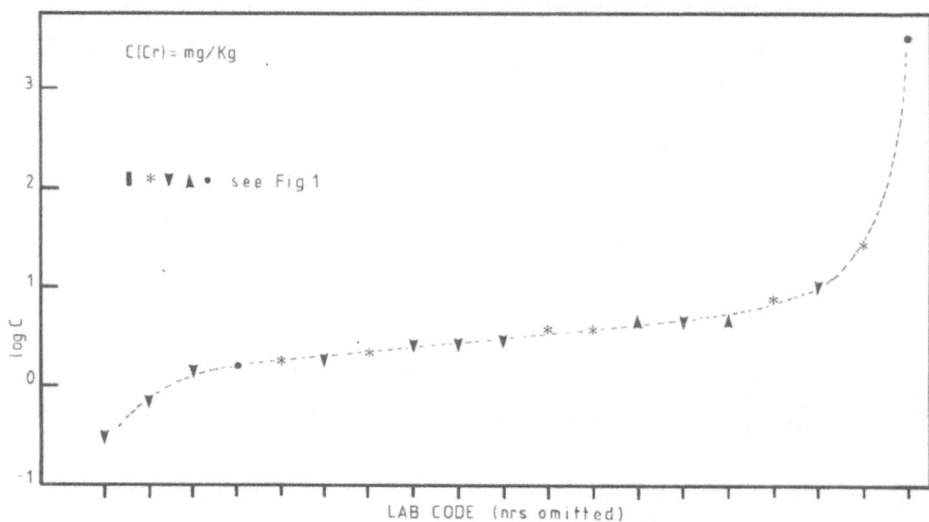

Fig. 3 Chromium in animal bone. In ordinate the participating lab numbers (from Parr[7], figures not numbered in his text).

Table 3 Interlaboratory comparison of nickel analyses in serum by electrothermal atomic absorption with 7 labs participating.

Sample	Concentration in μg/l[a]		
	mean	s.d.	range
Std. no. 1	3.4	0.8	1.8- 4.9
Std. no. 2	11.6	2.8	5.6-14.3
Pooled serum	12.7	3.0	7.2-16.2

[a]From Brown et al.[11]

Table 4 Precision of nickel analyses in serum with 7 labs participating.

	coefficient of variation[a]		
	mean	s.d.	range
within-run (4-6 replicates)	10.6	5.5	2.7-18.1
run-to-run (3 runs)	12.4	7.0	2.3-21.3

[a]From Brown et al.[11]

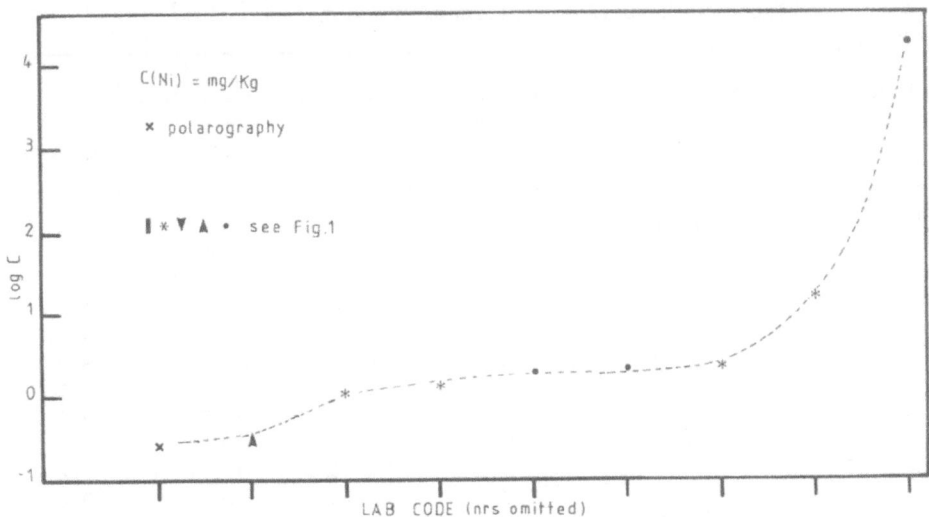

Fig. 4 Nickel in animal bone. In ordinate the participating lab numbers
(from Parr[7], figures not numbered in his text).

Table 5 Nickel determination by GFAAS and by DPP.

Sample	ref. value µg/g(a)	GFAAS ng/ml	DPP ng/ml	µg/g[d]
Std. no. 1[a]	14.85	–	–	14.1
Std. no. 2[b]	1.5	–	–	0.2
muscle (remote area)[a]	–	–	–	0.2
id. (near implant)[a]	–	–	–	5.0
urine A[c]	–	9.0	7.9	–
B	–	9.4	7.1	–
C	–	27.4	35.1	–

[a]From Helsen et al.[12]
[b]From Parr[7]
[c]From Nieboer et al.[13]
[d]On freeze-dried weight

Within the same laboratory, however, the precision of the determination
remains acceptable i.e. within 15 % or less, exceptionally 30 % as indicated by the IAEA intercomparison. Table 6 represents the values for
chromium in animal muscle taken from the IAEA intercomparison. I suspect
that the larger deviations observed in this table can be attributed to
uncorrect experimental conditions during the analysis. For nickel,
chromium and a number of other elements extreme care is indicated from
the spot were the sample is taken over the sample transport and pre-

Table 6 Results of intercomparison for chromium[a]

method	estim. lab. error %	no. of determ.	lab mean	lab. s.d. abs.	lab. s.d. rel %
A[b]	12.1	2	0.01	0.00	0.6
A	8.7	3	0.12	0.02	15.9
A	31.2	2	0.02	0.00	31.9
A	8.4	6	0.02	0.01	35.6
A*	0.0	3	0.07	0.09	126.6
B	15.0	3	0.28	0.02	5.4
B*	0.0	6	0.04	0.02	42.7
B	11.0	4	0.31	0.03	9.5
B*	14.7	5	0.05	0.01	14.6
B	28.3	4	0.71	0.06	9.2
B	0.0	2	0.01	0.00	0.0
B	0.0	5	0.12	0.03	23.4
C	0.0	3	0.47	0.08	16.9
C	34.4	3	0.08	0.01	6.9
C	10.0	6	0.80	0.09	11.2
C	24.8	2	0.42	0.19	44.9
D	1.9	3	0.50	0.24	47.6

[a] Parr[6], Table 12
A : activation analysis (non specified); B : atomic absorption (* : electrothermal); C : INAA; D : emission spectroscopy.

paration till the introduction in the measuring device.

Nevertheless, the table learns that concentrations measured within
the same laboratory have at least reliable relative value if counter-
indications can reasonnably be excluded. The table showing the coefficient
of variation of within-run and run-to-run data of the IUPAC nickel determination is also conforting in the sense that within the same laboratory
reliable relative values can be obtained. That is an important point
because of the hypotheses in our research proposal we want to formulate
further on is based the assumption that the wide dispersion of values
observed in some experiments is not due to an unfortunate combination of
random errors or runaways but represents true deviations.

NICKEL AND ITS BIOLOGICAL INCIDENCE

Toxicity of nickel for mammals is low, ranking in this respect with
chromium, barium and silver. Most of the toxicity studies are related
to workers in the nickel industry and acute intoxication is mainly provoked by nickelcarbonyl. But that is not our concern here.

Contact with nickel-plated articles and other nickel containing
products of daily use causes dermatitis. According to patch testing with
5 % nickelsulfate solution, 10 to 15 % of the population may be sensitive
and women represent a majority within this group. Important to note is
that cobalt may provoke cross-sensitization. And cobalt is a common
element in an number of implant alloys (see Marshall[14], p. 283).

Carcinogenicity is the other main concern for the use of nickel in the body. The effect of nickel subsulfide is in this respect beyond discussion. The activity of soluble divalent nickel, however, is more speculative. Kasprzak and collaborators performed extensive studies on rats with nickel hydroxide and sulfate. With nickel hydroxide haematuria and tumours developed but sulfate did not have a visible effect (see Brown and Sunderman[15], p. 59-60; rem : Brown and Sunderman[15] is often used as a reference, even where more recent references do exist. The more recent data do not change the viewpoint of this communication).

Whatever the reason for either allergy or cancer may be, nickel can only become a problem when it is extracted from the alloy or say oxidized. The biological oxidation cannot be simulated in vitro by contact with solutions used in ordinnary corrosion work. Andersen et al.[15] (p. 77-80, Fig. 3) deomonstrated clearly the different dissolution rates of nickel powder in Fe-citrate solutions, water, 0.9 % NaCl aqueous solution and whole blood (Fig. 5). Once dissolved in body fluids it is distributed between chelating agents of different strength. Jones and collaborators[15] (p. 77-76) simulated by computer the distribution of low molecular weight complexes in blood plasma (table 7). Nickel forms very stable complexes with sulfides and of course sulfhydryl groups. This is the reason why high nickel alloys are strongly corroded in biological environment. The cisteine-histidine complex seems to be a very stable one. A matter of fact complexation is an important phenomenon in the dissolving process because the redox potential is lowered, what displaces the thermodynamic equilibrium towards the oxidized form Ni(II). It is self-evident that addition of other ligands to blood provokes a redistribution of nickel. These complexes are the vehicles of nickel through the body where in turn they are distributed between different organs according to their specific affinity. The computer models used by Jones and colleagues provides the theoretical fundament to predict the efficiency of ligands injected for removal of metal ions from plasma.

Effects other than carcinogenity and allergy are related to coronary problems. Chronic exposure to nickel may lead to coronary vasoconstriction in dogs as mentioned by Kovach and colleagues[15] (p. 157-144).

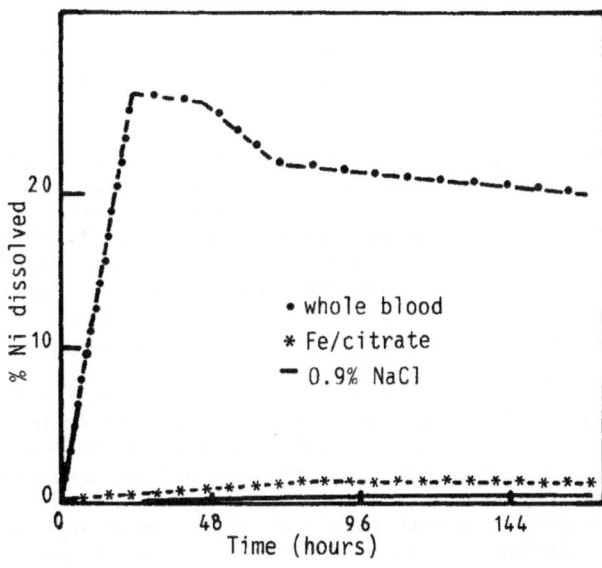

Fig. 5 Dissolution of nickel in Fe/citrate, whole blood, NaCl 0.9 % (from Andersen et al.[15], Fig. 3).

Table 7 Distribution of nickel(II) between ligands (predominating complexes in normal blood plasma).

Complex formed[a]	percentage of LMW metal
$NiCysHis^-$	44.2
$Ni(His)_2$	37.4
$Ni(Cys)_2^{2-}$	8.3
$NiHis^+$	3.3

[a]From Jones et al.[15]

Changes in nickel concentration in serum after infarct are reported but the differences are rather small and I do not know whether the obtained differences are sufficiently significant.

CHROMIUM AND ITS BIOLOGICAL INCIDENCE

In a remarkable review Mertz[16] summarized the biological effects of chromium up to 1969. The suspected cancerogenicity kept the interest in this element vivid. In this communication just a few facts will be mentioned which may be of interest to be monitored when chromium containing implants are used.

Chromium is considered as an essential element. As given in Table 1, the mean concentration in human serum amounts up to 0.148 ng/ml, the average value is somewhat higher for men than for woman : 0.164 ± 0.107 compared with 0.132 ± 0.059 for women. The values as obtained by Versieck[4], are much lower than obtained till 1974 and are generally accepted as the most reliable. Muscle tissue contains less than a microgram per gram freeze dried tissue.

The toxicology of chromium is not well understood. The acute minimal letal dose reported after subcutaneous injection of chromic acid is 330 mg/kg body weight for dogs or 75 mg for intravenous injection of sodium chromate.

When chromium is presented to cells, only chromate is absorbed, while the cell membrane is a barrier for non-compexed Cr(III). There is much debate about the oxidation state of chromium in living cells. In biocompatiblity of chromium-bearing implants, however, the discussion about the oxidation state is only relevant if the dissolved $Cr(III)_2$oxide from the passivating film can be oxidized biochemically. Rogers[17] discusses the in vivo production of hexavalent chromium by oxygen and its partially reduced intermediates. Once oxidized, it may enter the cell.

The hexavalentstate is suspected to be cancerogenic but one has to be more specific. Compounds of very low solubility do not consistently induce tumours. Only the medium solubility products as calciumchromate are consistently cancerogenic[18]. This compound is also known to induce DNA-strand breakage and -repair activity depending on concentration[19]. Highly soluble Cr(VI) products do not show tumour-inducing abilities. However, many cellular compounds are able to reduce Cr(VI) and trivalent chromium exhibits mutagenic and cytogenetic activity. As reducting agents are suggested : cystine, cysteine, lactic acid, haemoglobin[16]. Gluthatione is probably present in the redox pathway of Cr(VI)[20]. Merritt and Brown[21] found inflammatory and giant cell response from corroding stainless steel implants in rabbits and guinea pigs and their studies indicate the production of Cr(VI) after corrosion in serum. The life-time of Cr(VI) in the living cell, however, is small and remains questionable[22].

A well-established effect of chromium is the tanning activity of skin which is attributed to the trivalent form. The mean molecular weight of collagen can be increased from 65.000 tot 88.000-95.000 by crosslinking on addition of chromium(III)[23]. But this property is explained by the formation of glycidal, an epoxyaldehyde which is cancerogenic in rats and mice[18] and so, the over-all uncertainty remains !

The activity of different enzymes is stimulated by chromium. Well-known is the effect on phosphoglucomutase and the glucose tolerance factor. Cr(II) seems to be the active oxidation state. To my knowledge not very much advance is made in the study of this effect which is closely linked to the action of insulin[24,25].

A number of enzymatic reactions are also inhibited by chromium. An example is the enzyme β-glucoronidase[16]. Chromium-protoporphyrin is, probably along with other metalloporphyrins, a hyperbilirubinemia suppressor[26].

Special attention deserves the influence on sperm. Chromium deficiency in the diet of male rats leads to decreased sperm count with respect to chromium-supplemented controls. Accumulation of chromium after long term dietary doses was observed in bone, spleen and in testes[16,27,28].

Exciting new findings have been communicated at the PAC CHEM '84 by Shoji Okada from the Shizuoka College of Pharmacy, Shizuoka, Japan as reported in Chem. & Eng. News (14 jan. 1985, p. 47). Cr(III) is an integrating part of a monomeric protein with a molecular weight of about 70.000 in experiments with rats that have had part of their liver removed, it has been found that Cr(III) builds up in the nucleoli of the remaining liver cells and this build-up seems to precede the early stage of liver regeneration.

This brief review on chromium and its biological effects is given in order to suggest a number of reactions which are implicated when chromium bearing implants are used and are eventually accessible to monitoring.

A RESEARCH PROPOSAL

The research proposal of which the main points have been given earlier, is the result of a combination of experience and experimental work in the field. Whatever has been said about analytical determination, the biological incidence of nickel and chromium demonstrates the multi-dimensional character of the potential effect of corroding implants. The large domain of cellular response has not been treated but will be included in the epidemiological study.

When corrosion specialists got involved in the biomedical research, the classical techniques in classical conditions were applied to test, eventually select the candidate high strength alloys for implant devices[29]. It took quite a while for the non-biochemically trained corrosion scientists to realize the partial uselessness of in vitro testing for in vivo use by application of classical corrosion testing. The next step forward was the use of isotonic solutions as Ringer, Hanks and so. A definite advantage was the use of more complex biochemical substances as amino acids and proteins. A number of authors[30,31,32] published results based on this philosophy. All along the years this work was going on, in vivo testing was performed in many laboratories, very often by implanting in muscle, in the peritoneal cavity of rats and in bone[33,34,35,36,37,38,39]. But the combined results of this research studies do not permit, at least to my feeling, to complete the whole picture of biocompatibility.

The other, say ultimte way of testing is the clinical experience. There is ample report of clinical observations of failed or non-satisfactory prosthetic surgery and search for links between failure and physiological response. These reports often concern a restricted number of case studies or restricted to one hospital and thus often to one area with specific industrial, urban,... environments, thus not necessary a representative sample of the population. The size of the sample is often small to very small[40]. That was the immediate motive for the first point of the research porposal I want to present. This project is set up in multidisciplinary collaboration with the Department of Orthopedic Surgery at Pellenberg, the Laboratory of Immunology and the Department of Metallurgy in the framework of the joint research group for biomedical engineering at the University of Leuven (Belgium).

Epidemiology

A databank is being organized in order to collect information as complete as possible from different hospitals resulting in a representative sample of patients with metal prostheses. The data will be treated statiscally.

Physiological response of the body to the implant and its corrosion products

In the first chapter some data on nickel and chromium determinations have been shown. Neither precision nor accuracy of the determination of low levels is really satisfactory : lack of sensitivity, sample heterogeneity or contamination can be mentioned ax explanation or excuse for it. For the within-run precision and accuracy high relative errors are reported but not errors of decades. It is often observed that results from a given laboratory contian a set of values with very few runaways.

In 1977 we reported a set of data, all rather consistent : 5 to 6 microgram/gram freeze dried tissue near the implant and 1 or less milligram/gram for tissue sampled far away from the implant. But a clavicle bone from which a pin had to be removed, contained 325 microgram/gram Ni and 250 microgram/gram Cr[34,35]. Is the latter value a runaway ?

Castleman et al.[41] reported data on biocompatibility of nitinol. If we look at their data for nickel in bone and muscle of dogs exposed to a nitinol implant ant the corresponding controls, some values show excessively high concentrations for the exposed dogs. Castleman tends to quote these values as anomalous, or runaways. But is it true ? Because of the frenquency with which 'anomalous' values occur in reported data and, as mentioned before, the relatively good within-laboratory consistency of analytical data, it is hard to believe that they are all just analytical accidents.

Based on the said observations, one can hypothesize that these apparently anomalous values could have a physiological origin : either these patients or test animals accumulate for one reason or another the metals released from the implants, or a specific compound provokes an enhanced corrosion of the implant. The experimental verification of this hypothesis is in execution.

As already said, corroded implant surfaces have the aspect of corrosion by sulfur compounds rather than by chlorides. Fig. 6 demonstrates how sulfur compounds can act on stainless steel : intercrystalline attack with sulfur concentrated on grainboundaries followed by sequestration of grains. Chloride attack is of cours not excluded but to our opinion not the major cause of corrosion initiation. Protein adsorption may be considered here as a first step to this process. Bagnall and Arundel[42] published a method for prediction protein adsorption on implant surfaces and it is straightforward to apply these findings on corrosion initiation. But it is not necessarly the only step. As stated before some chemical compounds may interfere in a more specific way.

Corrosion follows a different route, however, when it takes place on e.g. a bone plate at the place of the healing fracture as was observed on several occasions. De Clercq and collaborators[43] are testing fibre reinforced bioglass dental root implants. A crack in the glass surface on a spot where intense bone neoformation is going on, entails intense corrosion of the embodied fibre, while fibres exposed to tissue at a remote place of bone formation do not corrode. Observations of this kind forced us to concentrate our efforts on testing experimental conditions simulating the complex processes of bone neoformation. The equilibria as given by Neuman and Neuman[44].

* Ca-proteinates $\rightleftharpoons$ Ca^{+++} $\rightleftharpoons$ CaHPO$_4$ aq. $\rightleftharpoons$ colloidal phase

$\qquad\qquad\quad \updownarrow$

calcium citrate complex, calcium malate complex, etc.

* H$_2$PO$_4^-$ $\rightleftharpoons$ HPO$_4^{--}$ $\rightleftharpoons$ CaHPO$_4$ aq. $\rightleftharpoons$ colloidal calcium salt

are examples of reactions that allow to explain why at the place of neoformation enchanced corrosion is to be expected. The incorporation of the dynamic of the system into the simulation is under study.

Another series of observations concern the presence of silicon often detected in the corrosion products. It is rather strange the find some accumulation of silicon in a yet unknown form on an implant. In case studies of corroded waterpipes silicates are very often detected in the corrosion products. As a consequence of these observations the hypothesis was advanced that nascent silicate precipitates may be the onset of corrosion. This is explained by the exchange properties of the precipitated mixed silicates and the membrane activity of the silicate sheet. Since a few years a detailed study of this phenomenon is underway and, although not yet proved, no experimental facts were encountered contradicting the hypothesis[45]. As a consequence of this research a systematic study of presence and origin of precipitating silicates on implants will be undertaken.

CONCLUSION

The aim of this cummunication was not an exhaustive treatment of all effects and biological and biochemical incidences of the elements nickel and chromium as important constitutive elements of may implant alloys. The main point was to juxtapose some facts and observations reported in the current professional literature and our own observations. From this confrontation if was tried to broaden the scope of the biocompatibility

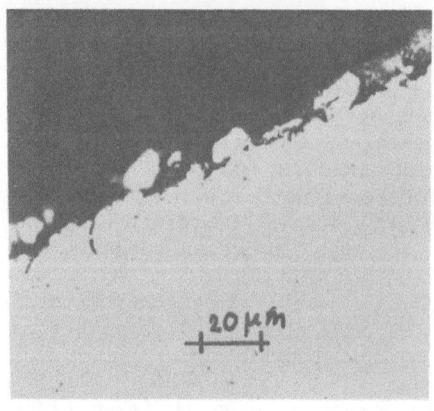 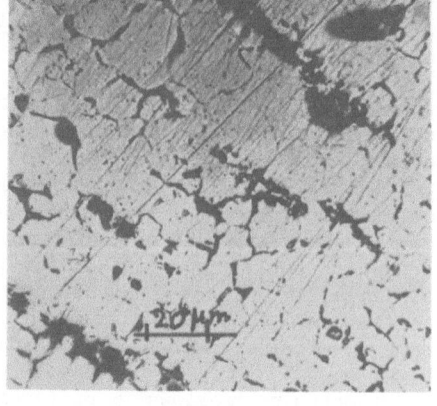

a   b

Fig. 6 Sulfur attack of an implant surface (AISI 316).
        (a) surface view; (b) cross section (photo obtained by the
        courtesy of Prof. M. Brabers, Dept. Metallurgy, Leuven).

studies in our laboratory. It aims to find out whether or not these exist
a link between failure of a prosthesis and a number of body response to
ions dissolving from a corroding implant. Subsequently, a hypothesis is
advanced on the possible link between accumulation of the ions nickel
and/or cobalt and chromium, allergic or other reactions and enhanced
corrosion. A research route is presented which is followed to establish
the biocompatibility of high alloy steel (and to be extended to other
alloys as well) in an environment of bone neoformation. Finally it was
mentioned that a study of the potential influence of silicates on the
corrosion behaviour of implants is underway.

REFERENCES

1. G.N. Schrauzer, The role of trace elements in the etiology of cancer,
      in : "Trace Element Analytical Chemistry", Walter de Gruyter,
      Berlin (1980).
2.   B. Medina et P. Sudrand, Teneur des vins en chrome et en nickel.
      Cause d'enrichissement, Connaissance Vigne et Vin, 14.79 (1980).
3. G.N. Schrauzer, "Trace metals in health and disease", N. Karash, ed.,
      Raven Press, New York (1979).
4. J. Versieck, Onderzoekingen over spoorelementen bij de mens, Ph.D.
      Thesis, Univeristy of Ghent, Belgium (1974).
5. F.W. Sunderman, S. Nomoto, A.M. Pradhan, H. Levine, S.H. Bernstein,
      R. Hirsch, Increased concentrations of serum nickel after acute
      myocardial infraction, New Engl. J. Med. 183:896 (1970).
6. R.M. Parr, Intercomparison of minor and trace elements in IAEA animal
      muscle, International Atomic Agency, Progress Report No. 2, Vienna
      (1980).
7. R.M. Parr, Intercomparison of minor and trace elements in IAEA animal
      bone (H-5), International Atomic Agency, Progress Report No. 1,
      Vienna (1982).
8. R.M. Parr, Problems of chromium analysis in biological materials :
      an international perspective with special reference to result for
      analytical quality control samples, J. Radioanal. Chem. 39:421
      (1977).

9. M. Hoenig, C. Lima et S. Dupire, Validité des déterminations par spectrométrie d'absorption atomique avec atomisation électro-thermique du cadmium, cobalt, chrome, nickel et plomb, Analysis 10:132 (1982).

10. C.E. Clare and M.F. Robinson, Copper, manganese, zinc, nickel, cadmium and lead in human foetal tissues, Br. J. Nutr. 39:639 (1978).

11. S.S. Brown, Shozo Nomoto, M. Stoeppler and F.W. Sunderman Jr., IUPAC reference method for anlysis of nickel in serum and urine by electrothermal atomic absorption spectrometry, Pure Appl. Chem. 53 : 774 (1981).

12. J.A. Helsen and P. Hermans, A polarographic method for determination of nickel in biological fluids and muscle tissue, in : "Biomaterials and Biomechanics", P. Ducheyne, G. Van der Perre and A.E. Aubert, eds., Elsevier, Amsterdam (1984).

13. E. Nieboer, P. Lavoie and D. Padova, A progress report on the DMG-sensitized polarographic method for nickel and a simple regime for controlling random background nickel contamination, in : "Nickel Toxicology", Academic Press, London (1980).

14. S. Marshall, "Toxis Metals", Noyes Data Corporation, New Jersey (1976).

15. S. Brown and F.W. Sunderman Jr., "Nickel Toxicology", Academic Press, London (1980).

16. W. Mertz, Chromium occurence and function in biological systems, Physiol. Rev. 49:163 (1969).

17. G.T. Rogers, In vivo production of hexavalent chromium, Biomaterials 5:244 (1984).

18. R.B. Hayes, Carcinogenic effects of chromium, in : "Biological and Environmental Aspects of Chromium", Langård, ed., Elsevier (1982).

19. S.H. Robison, O. Cantoni and M. Costa, Analysis of metal-induced, DNA lesions and DNA-repair replication in mammalian cells, Mutation Res. 131:173 (1984).

20. H.J. Wiegand, H. Ottenwaelder and H.M. Bolt, The reduction of chromium (VI) to chromium(III) by glutathione : an intracellular redox pathway in the metabolism of the carcinogen chromate, Toxicology 33:341 (1984).

21. K. Merritt and S.A. Brown, Effect of valence of chromium on biological responses, in : "Biomaterials and Biomechanics", P. Ducheyne, G. Van der Perre and A.E. Aubert, eds., Elsevier, Amsterdam (1984).

22. D. Beyersmann, A. Koster, B. Buttner and P. Flessel, Model reactions of chromium compounds with mammalian and bacterial cells, Toxicol. Environ. Chem. 8:303 (1984).

23. K.H. Gustavson, The effect of tanning agents on the shrinkage temperature of collagen, in : "The Chemistry and Reactivity of Collagen", Academic Press, New York (1956).

24. M.F. McCarty, Chromium and insulin. Comments, Am. J. Clin. Nutr. 36:384 (1982).

25. R.W. Tuman, Biological effects of glucose tolerance factor and inorganic chromium(III) in normal and genetically diabetic mice, diss. Abstr. Int. B 36:2789 (1985).

26. G.S. Drummond and Attalah Kappas, Suppression of hyperbilirubinemia in the neonate by chromium-protoporphyrin, J. Exp. Med. 156:1878 (1982).

27. R.A. Anderson and M.M. Polansky, Dietary chromium deficiency effect on sperm count and fertility in rats, Biomed. Trace Elem. Res. 3:1 (1981).

28. J. Krüger, Fundamental aspects of the corrosion of metallic implants, in : "Corrosion and Degradation of Implant Materials", B.C. Syrett and A. Acharya, eds., ASTM, Philadelphia (1978).

29. G.CF. Clark and D.F. Williams, The effect of proteins on metallic corrosion, J. Biomed. Mater. Res. 16:125 (1982).

30. D.F. Williams, I.N.Askill and R. Smith, Protein adsorption and desorption phenomena on clean metal surfaces, J. Biomed. Mater. Res. 19:313 (1985).

31. J.L. Woodman, J. Black and J.A. Jiminez, Isolation of serum protein organometallic corrosion products from 316 LSS and HS-21 in vitro and in vivo, J. Biomed. Mater. Res. 18:99 (1984).

32. A.B. Ferguson, P.G. Laing and E.S. Hodge, The ionization of metal implants in living tissues, J. Bone Joint Surg. 42A:77 (1960).

33. A.B. Ferguson, A. Yoshihiko, P.G. Laing and E.S. Hodge, Trace metal ion concentration in the liver, kidney, spleen, and lung of normal rabbits, J. Bone Joint Surg. 44:317 (1962).

34. J.A. Helsen and B. Bloch, Chemical compatibility of surgical implants, Acta Orthop. Belg. 43:62 (1977).

35. B. Bloch and J.A. Helsen, Chemical compatibility in total joint replacement, Acta Orthop. Begl. 43:66 (1977).

36. H.G. French, S.D. Cook and R.J. Haddad Jr., Correlation of tissue reaction to corrosion in osteosynthetic devices, J. Biomed. Mater. Res. 18:817 (1984).

37. J.C. Keller, G.W. Marshall and E.J. Kaminski, An in vivo method for the biological evaluation of metal implants, J. Biomed. Mater. Res. 18:829 (1984).

38. P. Ducheyne, G.W. Willems, M. Martens and J.A. Helsen, in vivo metal-ion release from porous titanium-fiber material, J. Biomed. Mater. Res. 18:293 (1984).

39. E.M. Evans, M.A.R. Freeman, A.J. Miller and B. Vernon-Robberts, Metal sensitivity as a cause of bone necrosis and loosening of the prothesis in total joint replacement, J. Bone Joint Surg. 56B:626 (1971).

40. A.H. Burstein and T.M. Wright, Neck fractures of femoral prostheses, J. Bone Joint Surg. 67A:497 (1985).

41. L.S. Castleman, S.M. Motzkin, F.P. Alicandri and V.L. Bonawit, Biocompatibility of nitinol alloys as an implant materials, J. Biomed. Mater. Res. 10:695 (1976).

42. R.D. Bagnall and P.A. Arundel, A method for the prediction of protein adsorption on implant surfaces, J. Biomed. Mater. Res. 17:459 (1983).

43. M. De Clerq, E. Shepers, P. Ducheyne, Microchemical analysis of bulk and fibre reinforced bioglass dental root implants in dogs, Proc. Int. Cong. on Tissue Integration in Oral and Maxillo-facial Reconstruction, May 1985, Brussels (to be published by Excerpta Medica, Amsterdam).

44. W.F. Neuman and M.W. Neuman, The nature of the mineral phase of bone, Chem. Rev. 53:1 (1953).

45. J.A. Helsen, P. Huybrechts and M.J. Brabers, Corrosion initiation of steel in ptabel water by silicates, Proc. Int. Cong. on Metallic Corrosion, 2:283 (1984).

ORTHOPAEDIC IMPLANTS AND CORROSION PRODUCTS

ULTRASTRUCTURAL AND ANALYTICAL STUDIES OF 65 PATIENTS

H.F. Hildebrand[1], P. Ostapczuk[3], J.F. Mercier[1], M. Stoeppler[3],
B. Roumazeille[2], and J. Decoulx[2]

[1] Institut de Médecine du Travail   [2] Service de Traumatologie
1, Place de Verdun F-59045 Lille Cédex, France
[3] Institut für Angewandte Physikalische Chemie
Kernforschungsanlage, Jülich, West Germany

INTRODUCTION

Cobalt, chromium and nickel containing alloys are frequently used
for orthopaedic implants and prostheses in spite of the increasing
number of reports dealing with reactions of intolerance to these metals.
Any implant introduced into the human organism undergoes at least two
strokes : physico-mechanical owing to the environment of the site of
implantation, and electrochemical due to the presence of chloride ions
in the biological milieu. There exist an interaction between the implant
and the tissue leading to a tissue response expecially in the vicinity
of the implanted material.

This reaction can be modified by different other factors such as a
local infection, a preexisting evolutionary pathology, local tissular
traumatism from the surgical intervention, movement of the implant like
a static foreign body. They may appear as inflammatory granuloma
mobilising phagocytic cells and representing in this way a fundamental
mecanism of non-specific defense. These cells are able to phagocyte and
to convey mineral or organic particles deriving from mechanical
degradation or from electrochemical corrosion.

In addition, histological observations reveal an increased number
of lymphocytes in the vicinity of the implants. The presence of
lymphocytes let also suggest immunological process. Indeed, clinical and
scientific research have evidenced a relation between the use of these
alloys and the increase of sensitization to the metals contained in
(11,14,19).

In the present work, the metabolism of the diffusing metal ions is
investigated in 65 patients after removal of orthopaedic implants :
osteosynthesis plates, nail-plates, centromedullary nails and total hip
replacements.

MATERIAL AND METHODS

Following biological materials were collected from patients :
blood, urine, hair, finger- and toenails and adjacent tissues close to
the implants.

Ultrastructural studies were carried out at described previously
(6) : tissue samples were stained with 2,5% glutaraldehyde followed by
1 % OsO$_4$ in the Millonig buffer system. Ultrathin sections were placed
on Pioloform-F coated grids and contrasted with lead-citrate and uranylacetate. Observations were made with a Philips EM300 electron
microscope.

Elementary determination of precipitates in the tissues was carried
out by X-ray Energy Dispersive Spectrometry (EDS) on unstained ultrathin
sections placed on Pioloform-F coated titanium or gold grids. These
analyses were performed with a Philips EM301 Electron Microscope coupled
to a Link-System microprobe analyser.

The Ni and Cr contents in body fluids and tissue samples were
determined by Electrothermal Atomic Absorption Spectrometry (ETAAS) on a
Perkin-Elmer Zeeman 5000 Spectrometer (16,17,20). Co and additional Ni
analyses were carried out by Adsorption Square Wave Voltametry (ASWV)
(12).

RESULTS AND DISCUSSION

Surface corrosion of implants

All orthopaedic explants - centromedullary nails, osteosynthesis
plates and total hip prostheses - displayed without any exception
corrosion areas. These areas were observed on any place of the material
and were generally characterized as pittings and crevasses (fig. 1). On
osteosynthesis plates, we found additional galvanic corrosion areas in
the holes and on the head of screws.

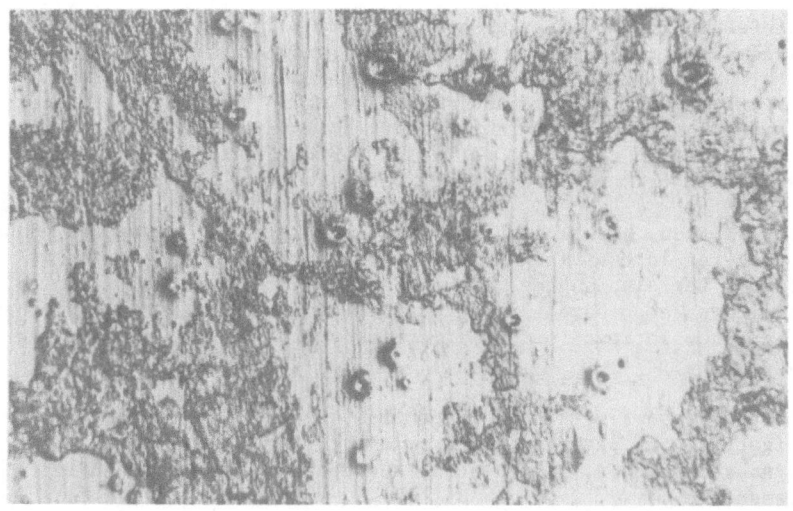

Fig. 1. Corrosion area of a stainless-steel osteosynthesis plate

Analyses of biological fluids were made from all patients. Normal values are represented in table I. It was impossible, however, to perform analyses of the other biological samples from all these persons, since several patients did not accept the removal of their implant.

Table I : Normal Nickel, Cobalt and Chromium Levels
in Human Biological Samples

Biological Samples	Ni	Co	Cr
total blood (μg/l)	< 1.80	< 0.11	< 3.00
plasma (μg/l)	< 0.90	< 0.10	< 1.40
urine (μg/g creat.)	< 3.30	< 0.36	< 3.00
hair (mg/kg)	1.25	0.08	/
nails (mg/kg)	1.38	0.08	/

The Ni-concentrations in the body fluids are represented in fig. 2. It becomes obvious that, in most patients, the Ni-level in total blood (fig. 2a) and in urine (fig. 2c) is significantly higher than the upper normal level. The increase of Ni-concentrations is even more evident in plasma (fig. 2b), where we detected only in one or two patients nearly normal values.

The increase of Co-concentrations in total blood (fig. 3a) and in plasma (fig. 3b) is yet more obvious, and one recognize that only very few patients can be placed within the normal level bracket. In urine, however, Co-concentrations are not so high (fig. 3c). These results are very interesting since only two patients had implants made from Co-Cr-alloys. The cobalt seems to be quite well eliminated in the urine, where we observed very high levels of 4.3 and 2.9 µg/g creatinine respectively (fig. 3c). All other patients had stainless-steel implants, which contain very low amounts of cobalt.

Except in very few cases, Cr-concentrations are generally normal in total blood (fig. 4a) and in urine (fig. 4c), and only in plasma (fig. 4b) we note significantly higher Cr-levels then the upper normal values. This phenomenon may be explained by the lower normal value in plasma than in total blood or in urine (cf. also table I). When one compare, however, the absolute values of the three body-fluids, no remarkable difference can be stated.

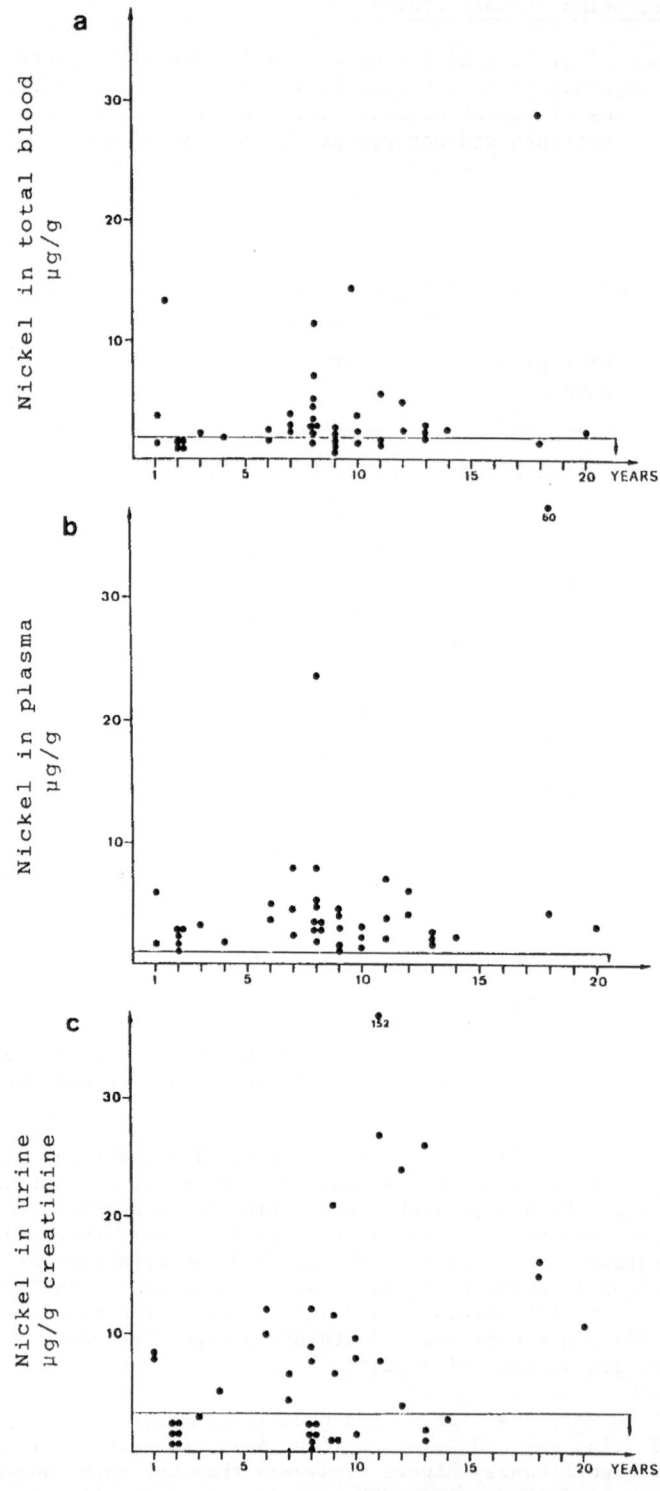

Fig 2. Nickel contents in total blood (a), plasma (b) and
urine (c) from patients with osteosynthesis plates and
centromedullary nails
Lines with arrow indicate maximal normal values

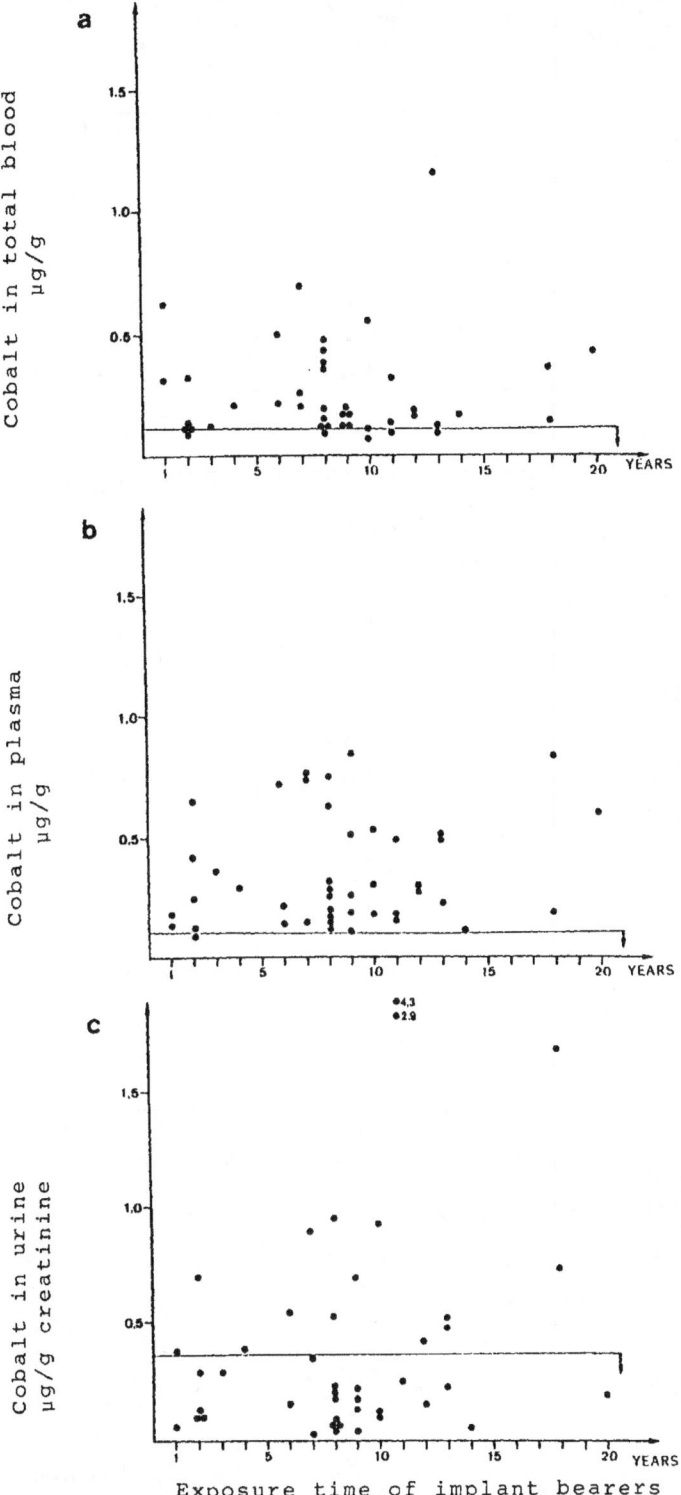

Fig 3. Cobalt contents in total blood (a), plasma (b) and urine (c) from patients with osteosynthesis plates and centromedullary nails

Lines with arrow indicate maximal normal values

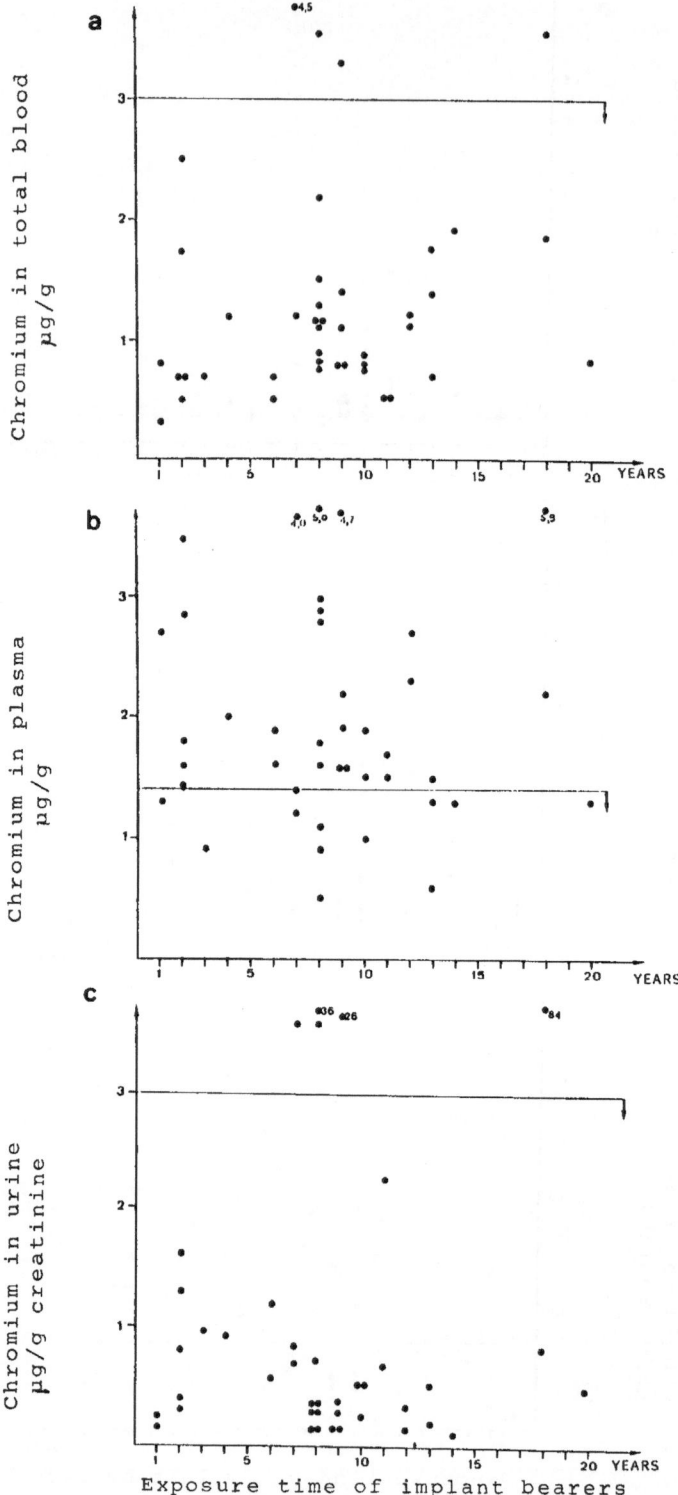

Fig 4. Chromium contents in total blood (a), plasma (b) and
urine (c) from patients with osteosynthesis plates and
centromedullary nails
Lines with arrow indicate maximal normal values

Some extremely high values for Ni, Co or Cr were detected for some patients. These values can not be explained by outer contamination due to storage containers or analytical instruments, since these analyses were repeated critically several times, in both laboratories and from different storage vials. The results were always identical. These high levels must be due either to a different physiological behaviour of some patients or to differences of the physico-chemical quality of the implant. In order to clear up this latter argument, metallurgical analyses of the removal material are now undertaken.

We avoided in the previous figures to present the Ni, Co and Cr concentrations in body fluids from patients with total hip replacements. These values were generally so high (table II), that they would have vitiated an eventual correlation between body-burden and exposure time. At present similar studies on metal-ions release from patients with total hip joints are quite rare and not so wide-spreading than the present investigation. Elevated Nickel-values have been described previously by Dobbs and Minski (4) and high Cobalt-levels were mentioned by Coleman et al. (1,2) who measured 6.5 µg/l in blood a few months after surgery. The high concentrations should be considered either as a result of corrosion and consequently the quality of the alloy or as the consequence of clinical complications such as loosing related to allergic reactions and/or to infections. In two cases, surgical failure may be the reason of the wear of prostheses (patients 243 and 276).

Ultrastructural studies

Some morphological changes of the implant surrounding tissue can be observed. Rather often, we find granulomata the dark colour of which indicates the presence of metallic precipitates or wear particles. The connective tissue displays an increase of cell density and high amounts of metallic intra- and extracellular precipitates. The intracellular deposits are generally observed in macrophages, fibroblasts and histiocytes. The granulomous tissue is characterized by multinuclear giant cells containing a lot of electron dense particles as well (fig. 5). Ultrastructural examination reveals quite different aspects of these structures varying from the patient, the implant type and the examined tissue. So, wo observe amorphous precipitates besides crystalline particles and granular organic deposits next to membranous structures (fig. 6).

The elementary composition of the particles and precipitates varies in function of their morphological differences and, of course, of the initial composition of the implant-alloy, like we can see in the next paragraph.

X-ray microprobe analyses

The various precipitates in tissue samples next to osteosynthesis plates, as shown in fig. 6, were analysed by EDS. In extracellular crystalline areas (fig. 7), Cr, Fe and Ni are found to be the main components. In intracellular deposits, we recover especially Cr and Fe and other additional elements such as Cl, Ca, S and in particular P. A large homogenous crystalline structure (fig. 8) contains mainly Cr and P. This association suggests the existence of a Cr-P-complex, which can be considered as a Cr-metabolite.

More organic structures display the same association of alloy-originating and supplementary elements. Membranous structures contain mainly Cr (fig. 9), whereas granular structures display essentially Fe (fig. 10).

Table II : Distribution of Ni, Co and Cr in urin, plasma and blood from patients with total hip replacement (SS : Stainless Steel)

Patient number	Exposure time (years) Alloy	Urin (µg/g creatinine)			Plasma (ng/g)			Blood (ng/g)			Observations
		Ni	Co	Cr	Ni	Co	Cr	Ni	Co	Cr	
243	2 Ni-Cr-Mo	36.75	4.70	45.82	18.40	1.52	13.66	83.30	0.76	6.50	Loosening allergy to Cr
265	3 SS	10.99	0.28	0.23	5.70	0.12	1.79	5.44	0.28	0.98	Second replacement Allergy to Ba
264	8 Ni-Cr-Mo	49.70	6.40	/	/	/	1.47	/	/	/	
208	12 Vitallium	6.30	2.63	112.04	5.00	1.54	8.43	2.59	0.33	27.08	Loosening
269	15 SS	3.95	0.53	1.75	4.70	0.40	2.58	0.54	0.18	2.88	Loosening
271	2 Co-Cr	9.2	62.1	39.3	10.6	41.6	33.2	15.0	37.1	22.3	Loosening allergy to Co
272	7 SS	1.86	0.36	0.90	9.5	0.08	0.50	20.45	0.75	1.80	Loosening allergy to Ni
276	2 Francobal	11.2	63.7	27.5	11.1	47.4	17.5	5.12	45.6	16.70	Third replacement Allergy to Co and Ni
Normal	0	< 3.30	< 0.36	< 3.00	< 0.90	< 0.10	< 1.40	< 1.80	< 0.11	< 3.00	

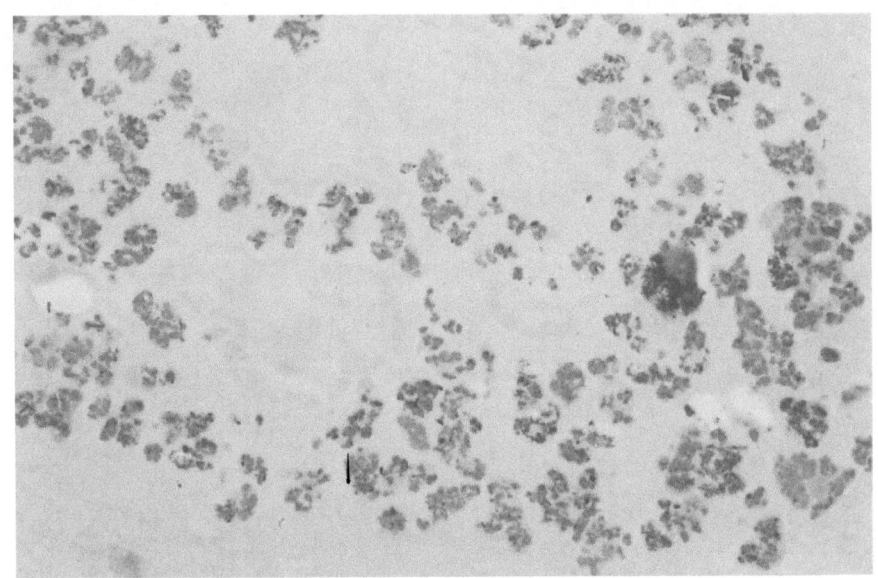

Fig 5. Intracellular deposits in a multinuclear giant cell of granulomous tissue adjacent to a total hip replacement (Patient Nr 208, exposure time: 12 years, prosthesis made from Vitallium/Vitallium). X 6,000

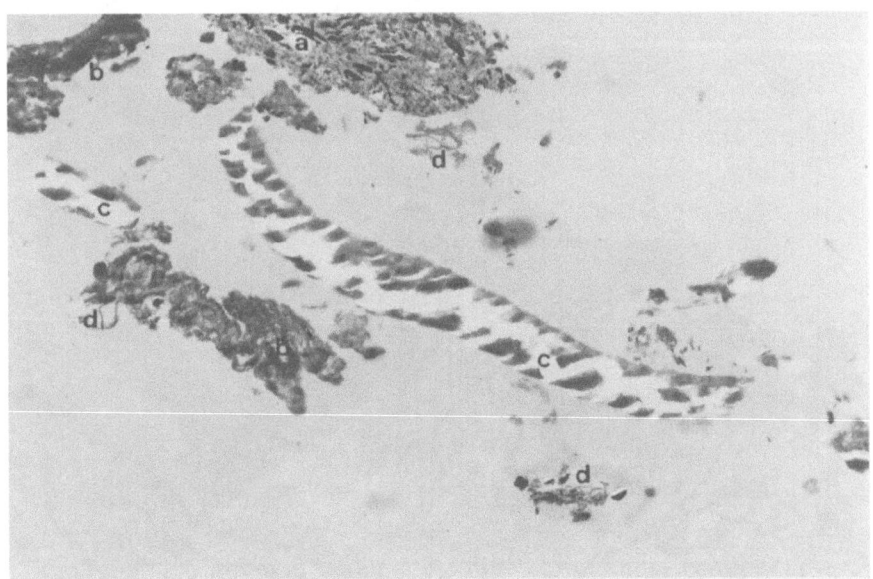

Fig 6. Extra- and intracellular deposits in connective tissue under a osteosynthesis plate. Note the variation of precipitates: extracellular crystalline particles (a) and granular organic deposits (b), intracellular amorphous precipitates (c) and membranous structures (d). (Patient Nr 254, exposure time: 8 years, stainless steel plate) X 6,000

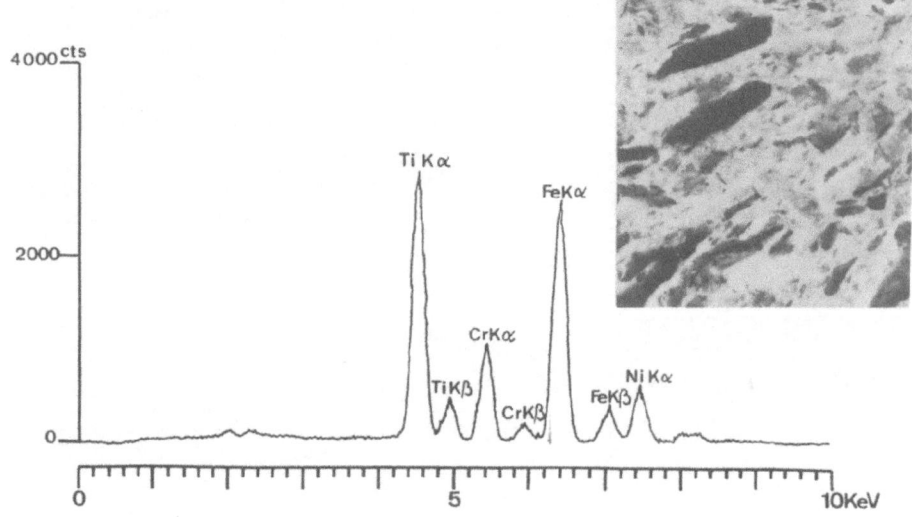

Fig 7. EDS-analysis of extracellular crystalline particles in connective tissue under a osteosynthesis plate (Fig 6, area a). These particles essentially contain Fe, Cr and Ni in similar ratio than the alloy itself. Analysed area: $4\mu m^2$

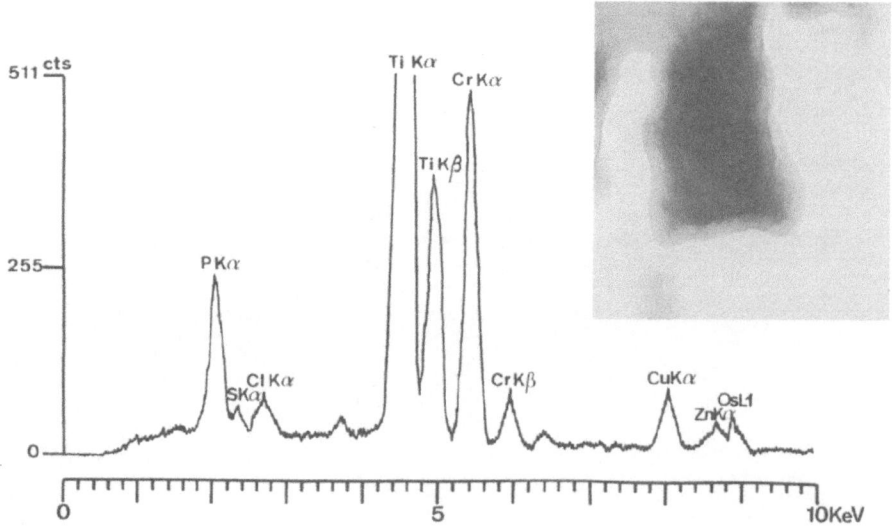

Fig 8. EDS-analysis of intracellular amorphous precipitate in connective tissue under a osteosynthesis plate (Fig 6, area c). This precipitate contains only Cr originating from the alloy. Furthermore, the analysis reveals an important peak of P. Analysed area: $4\mu m^2$

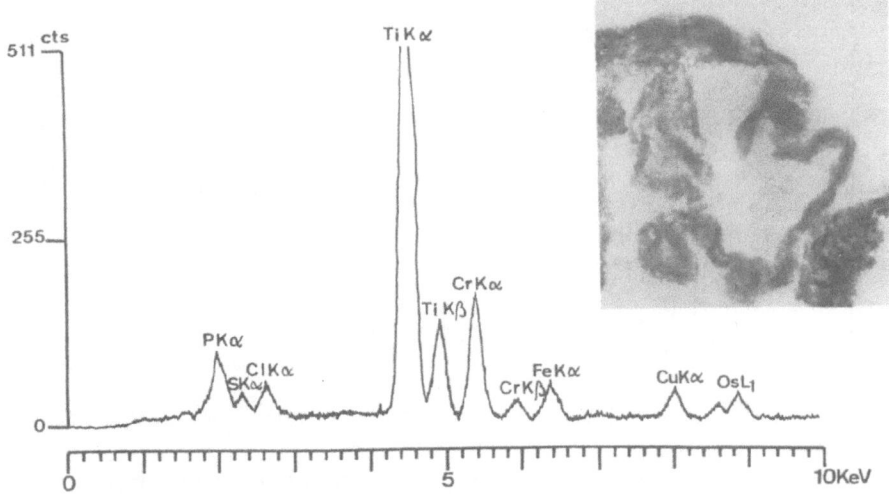

Fig 9. EDS-analyses of intracellular membranous structures in connective tissue under a osteosynthesis plate (Fig 6, area d). Cr is the main component of these structures, very little Fe is present. Also here we detect an important amount of P. Analysed area: $1.5\mu m^2$

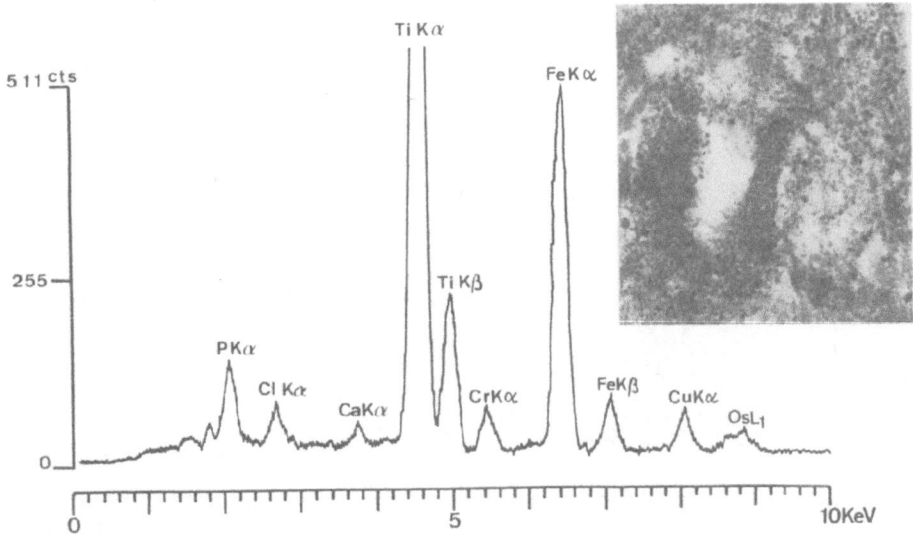

Fig 10. EDS-analyses of extracellular granular deposits in connective tissue under a osteosynthesis plate (Fig 6, area b). Fe and very little Cr are detected as well as an important amount of P. Analysed area: $4\mu m^2$

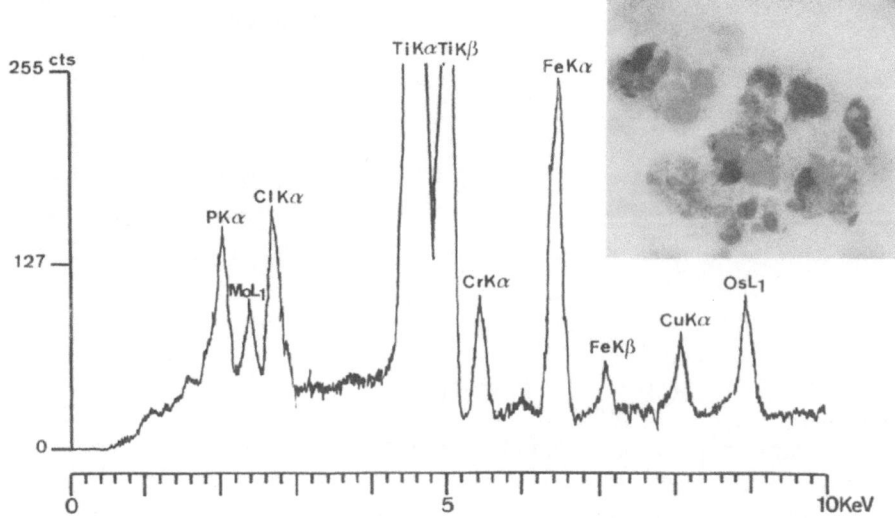

Fig 11. EDS-analysis of globular particles in a multinuclear
giant cell next to a total hip replacement (patient
Nr 208). Cr and Mo of the Vitallium alloy are recovered
but no Co, even in traces. Note the presence of Fe and
especially of P and Cl. Analysed area: $4\mu m^2$

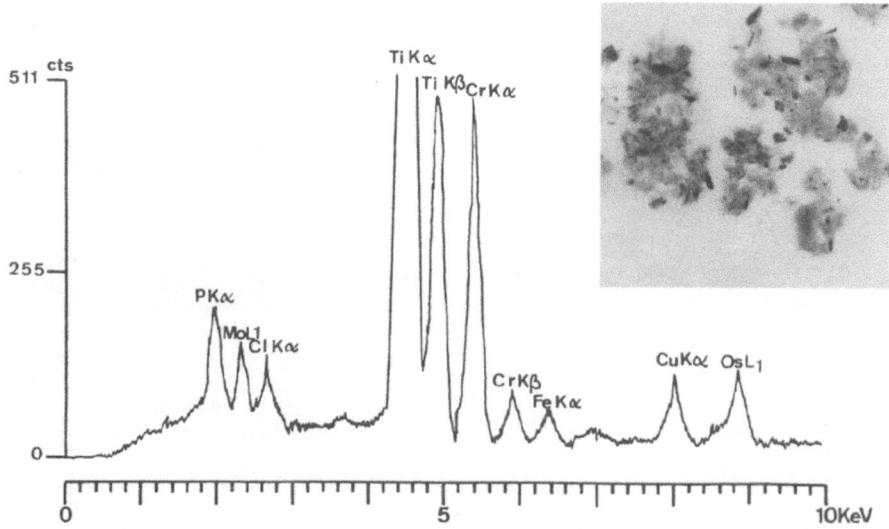

Fig 12. EDS-analyses of crystalline particles in a macrophage
next to a total hip replacement (patient Nr 208).
These particles mainly contain Cr, Mo and always P and
Cl. Also here we cannot recover Co, essential component
of the Vitallium alloy. Analysed area: $9\mu m^2$

We noted the presence of two different types of precipitates in connective tissue surrounding total hip prosthesis from a Co-Cr-alloy (Vitallium). The amorphous structures in lysosomes of multinucleated giant cells mainly contain Fe (fig. 11). Crystalline structures in mononucleated macrophages, however, display a high amount of Cr (fig. 12). Both types of particles contain alloy-originating, Mo and additional important amounts of P and Cl. Co never could be detected in all these precipitates.

Tissue analyses

In order to estimate the possible storage of corrosion products in implant surrounding tissues, we performed quantitative analyses from different tissue samples obtained during removal of orthopaedic implants and prostheses. Since we could not provide our own reference material, we give in table III the normal values of Co, Cr and Ni as found in adequate tissue by Michel and coll. (8,10) after instrumental neutron activation analysis. These values must be considered as indicative only, since we did not always dispose of the same tissues than these authors. Especially no normal values are known for bone and skin. Table III, however, shows that there are no remarkable distribution differences between various analysed tissues.

Table III : Medium values of Co, Cr and Ni distribution in normal human tissues
All values are given in µg/g dry weight

TISSUE	Co	Cr	Ni	Ref.
articular capsule	0.05	0.20	0.20	8
fascia lata	0.20	0.50	0.50	8
connective tissue	/	0.50	0.50	10

In patients with centromedullary nails (table IV), we observe generally a remarkable increase of Ni and Cr distribution, especially in connective tissue next to the nails. In addition, the concentration increases from the eye of the nail to its point and becomes extremely high in long-term implant bearers. The distribution of Ni and Cr in indefined tissue inside the nails is lower but also increases with duration of exposure.

The Co-distribution, even much lower than that of Ni and Cr, follows the same criteria : more important at the nail point than at its eye, higher in the tissue outside the nail than inside, and increasing with the duration of exposure. The low Co-concentrations are directly related to the low percentage of Co contained in the stainless-steel.

The same differences between Co and Ni-Cr-concentrations can be observed in tissues in the vicinity of stainless-steel osteosynthesis plates (table V). There is, however, an extremely high Co-concentration up to 3 orders of magnitude over the normal values in the connective tissue taken on a Vitallium plate (patient 231).

Table IV : Typical distribution of Ni, Co and Cr in tissue samples
from patients with centromedullary nails

Patient number	Exposure time (Y)	Tissue Sample	Ni	Co (μg/g dry weight)	Cr
237	8	C.T. outside nail : point	331.10	3.45	798.90
241	8	C.T. outside nail : point	138.70	1.50	270.80
235	9	Inside nail (undefined)	3.92	0.67	4.71
242	12	Inside nail (undefined)	53.90	1.20	99.20
245	14	C.T. outside : eye	0.06	0.01	68.90
		C.T. outside : middle	38.90	0.63	92.20
		C.T. outside : point	1,204.80	7.41	2,551.00

Stainless Steel nails of the 316L-type
C.T. : Connective Tissue

Table V : Typical distribution of Ni, Co and Cr in tissue samples from patients with osteosynthesis plates

Patient number	Exposure time	Tissue Sample	Ni	Co	Cr
				(μg/g dry weight)	
230	6 Y	C.T. on screw	815.50	1.30	> 5,000.00
		Bone on plate	4.05	0.36	7.28
248	6 Y	C.T. under fibula plate	123.70	1.88	134.00
		C.T. under femur plate	308.00	3.42	481.20
		Bone on femur plate	8.31	0.02	58.20
		Skin of thigh	0.42	0.02	0.24
254	8 Y	C.T. under plate	159.40	10.70	245.80
		Bone on plate	9.34	0.85	24.70
		Bone under plate	399.20	8.92	507.20
		C.T. distant 2cm	100.80	3.54	67.40
231	11 Y	C.T. on plate	3.53	140.10	95.80

Stainless Steel plates, except patient 231, who had two Vitallium plates
C.T. : Connective Tissue

The Ni and Cr distribution differs quite strongly from one tissue to another. Connective tissue generally contains very high concentrations, especially samples from the vicinity of a screw. We also observe an enrichment in bone, but much more in bone under the plates than in new-formed bone which is usually found on the plates after several years of exposure. The storage in bone under a plate is difficult to explain otherwise than by a faster biological transport in bone on the plate being in intime relation to other tissues and the vascular system.

Table V also shows that the metal concentrations in the tissues fall with the distance from the implant. In patient 254, we observe a significant decrease of the concentration of the three metal ions in connective tissue distant 2 cm from the plate. In patient 248, we find absolutely normal values of all elements in the skin of the high distant 5 to 6 cm from the implant.

Some other interesting results were obtained on tissues from the vicinity of total hip prostheses (table VI). We here present a 2 years exposure of a patient (243) with a stainless-steel prosthesis and a polyethylen cup. The patient suffered from loosening of his prosthesis and an intensive allergy to Cr (+++). The metal concentrations in the tissues are not very elevated. We note also here a higher concentration in connective tissue (granuloma) than in bone. No correlation seems to exist with the Cr-allergy. It is interesting, however, to compare the tissue values with those found in the body fluids (table II), where we observe very high concentrations.

Another patient (208), with a metal/metal Vitallium prosthesis had very high Co-concentrations in the different tissues, especially the peace of bone grown into the hole of the pin of the prosthesis, and in the Fascia lata. This patient had, except for Cr, rather low concentrations in the body fluids (see table II).

Elevated contents of Ni, Co and Cr have also been observed in hair and in finger and toe nails. Nevertheless, it appears that, due to endogenous influences, these samples may not be significant enough and further analyses have been given up in the frame of this investigation. Whole blood, plasma and (if properly collected) urine are the most significant indicators for excessive internal exposure.

A general statement is that, in the majority of cases where we observed high metal concentrations in the contact-tissues, we find lower concentrations in body fluids. The situation is reverse in patients with high concentrations in body fluids. In the case of infection, loosening and allergic reactions, we generally found elevated amounts in body fluids and in adjacent tissue as well (13).

A special interest should be given to the Co-distribution. Co is never detected in the precipitates by EDS analyses, but is generally revealed in body fluids sometimes in a more than ten-fold concentration in patients with stainless-steel implants, and even in a two to three hundred-fold concentration in patients with Co-Cr prostheses. Very high Co-contents could also be revealed by analyses of whole contact tissue.

The absence of Co in the extra- or intracellular precipitates can only be explained by a faster biological transport of Co relative to Cr and Ni or by another metabolic pathway. Co may be bound to whole tissue in an unprecipitated form which is detectable by the analyses of very small areas.

Table VI : Typical distribution of Ni, Co and Cr in tissue samples
from patients with total hip replacement

Patient number	Exposure time (Y)	Tissue Sample	Ni	Co	Cr
				(µg/g dry weight)	
243	2	Granuloma	4.95	4.35	6.62
		Bone	1.52	1.21	1.43
208	12	C.T. Inside Pin Sheath	0.26	2.00	1.37
		Bone in Pin Hole	0.74	59.50	32.50
		Fascia Lata	3.34	117.70	83.70

243 : Stainless-Steel / polyethylen, patient allergy to Cr
208 : Vitallium / Vitallium
C.T. : Connective Tissue

An excessive supply of Co to the whole body is of particular importance with regard to its role as an essential trace elements and because of the danger of Co poisoning. Smith and Carson (15) have published a large review on these items, and it is clear-cut that Co is not only bound to Vitamin B_{12} coenzyme, but has a lot of other possibilities to interact with cellular constituents. By reacting with the thiol groups of aminoacids, proteins and other coenzymes and cofactors, Co inactivates several biochemical pathways, especially those pathways associated with the production of cellular energy. According to several studies (3,5,18,21), Co inactivates dihydrolipoic acid, thus limiting the activity of the Krebs cycle, the production of ATP, and energy metabolism in general. On the other hand, Co has a profound action by repression on the synthesis of δ-aminolevulinic acid (ALA) synthetase, and by this way, it reduces total microsomal heme, inhibits heme synthesis and induces the enzyme heme oxygenase (9).

The biological transport of Ni seems to be slower than that of Co but faster than that of Cr. This may be the reason why Ni is found only in some precipitates, whereas Cr is recovered nearly in all intracellular deposits. This also may be the reason for a significantly higher amount of Ni in body fluids. We remind that Cr-values in body fluids are quite proximate to normal values. Thus the Cr is sequestered in the cell and, by this way, less bio-available than Ni and especially Co.

Another debatable point is the origin of Fe in the precipitates. If these are observed close to stainless-steel implants, they may be one of the corrosion products. This explanation is not so plausible when Fe-containing precipitates are recovered in tissue proximate to Co-Cr-alloys. This Fe probably comes from hemoglobin degraded in a hematoma arising during the introduction of an implant and not during its removal. The excised tissue samples are stained immediately in glutaraldehyde, so that there is no possibility and time for Fe to enter the tissue and the cells and to be transformed into crystal-like structures in the lysosomes.

Another important event in the metabolization of the corrosion products is the association of the metal ions with phosphorus. It seems to be evident that the phosphorus revealed by EDS-analyses is extracted from bone and consequently has formed complexes with Cr or Ni ions, especially since both metals have a high affinity to phosphate groups.

Other experiments with Ni_3S_2 in culture systems with human lung cells (L132) (7) where no exogenous P is present, however, show that this Ni-compound is transformed structurally to small tiny particles and chemically to a compound containing mainly Ni and P without any trace of preexisting sulfur. These particles have a high affinity to cellular, cytoplasmic and nuclear membranes and to euchromatin. These specific sites of fixation suggest the formation of a Ni-complex with the phosphate groups of membranous phospholipids and DNA as well.

These experiments clearly show that the detected P in the precipitates of corrosion products may also have an endogenous origin. This means that Cr and Ni may sequester P of organic structures and alter consequently energy metabolism and other cellular functions.

CONCLUSIONS

Orthopaedic implants of Co-Cr-Ni alloys can corrode in biological milieu and the corrosion product is recovered in biological fluids as well as in implant surrounding tissue.

The metabolism and clearance of the metal ions is different. Ni and especially Cr are observed in form of extra- and intracellular precipitates, whereas Co seems to be bound to whole tissue in an unprecipitated form.

The presence of other elements such as S, Cl, Ca and especially P suggests the metabolization of the corrosion products and the existence of metallo-organic metabolites.

In contrast to occupational exposure, orthopaedic implants are an internal - quite distinct - source of contamination by allergenic and toxic elements. Sometimes, alarming high concentrations of these elements are recovered in tissues and body fluids, so that more intensive monitoring of patients with such implants is strongly to be recommended.

REFERENCES

1. COLEMAN R.F., HERRINGTON J. & SCALES J.T.
   The concentration of wear products in the body following total joint replacement.
   Phys. Med. Biol., 1972, 17, 744.

2. COLEMAN R.F., HERRINGTON J. & SCALES J.T.
   Concentration of wear products in hair, blood and urine after total hip replacement.
   Br. Med. J., 1973, 1, 527-529.

3. DINGLE J.T., HEATH J.C., WEBB M. & DANIEL M.
   The biological action of cobalt and other metals. II. The mechanism of the respiratory inhibition produced by cobalt in mammalian tissues.
   Biochem. Biophys. Acta, 1962, 65, 34-46.

4. DOBBS H.S. & MINSKI M.J.
   Metal ion release after total hip replacement.
   Biomaterials, 1980, 1, 193-198.

5. GOLDWASSER E., JACOBSON L.O., FRIED W. & PLZAK L.
   Mechanism of the erythropoietic effect of cobalt.
   Science, 1958, 125, 1085-1086.

6. HILDEBRAND H.F., KRIVOSIK I., GRANDIER-VAZEILLE X., TETAERT D. & BISERTE G.
   Perineal rhabdomyosarcoma of a new-born child pathological and biochemical studies with emphasis on contractile proteins.
   J. Clin. Pathol., 1980, 33, 823-829.

7. HILDEBRAND H.F., COLLYN-D'HOOGHE M. & HERLANT-PEERS M.C.
   Incorporation of Ni_3S_2 and NiS into human embryonic pulmonary cells in culture.
   In : "Progress in Nickel-Toxicology" S.S. Brown & F.W. Sunderman Jr, edts. Blackwell Scientific Publications, London, 1985, 61-64.

8. HOFMANN J., WIEHL N., MICHEL R., LOER F. & ZILKENS J.
   Neutron activation studies of the in-body corrosion of hip-joint prostheses made of Co-Cr-alloys.
   J. Radioanal. Chem., 1982, 70, 85-107.

9. MAINES M.D. & KAPPAS A.
Metals as regulators of heme metabolism.
Science, 1977, 198, 1215-1221.

10. MICHEL R. & ZILKENS J.
Untersuchungen zum Verhalten von Metallspuren im umgebenden Gewebe
von AO-Winkelplatten mit Hilfe der Neutronenaktivierungsanalyse.
Z. Orthop., 1978, 116, 666.

11. MUNRO-ASHMAN D. & MILLER A.J.
Rejection of metal to metal prosthesis and skin sensitivity to
cobalt.
Contact Dermatitis, 1976, 2, 65-67.

12. OSTAPCZUK P., FRONING M., STOEPPLER M. & NURNBERG H.W.
Square wave voltametry : a new approach for the sensitive
determination of nickel and cobalt in human samples.
In : "Progress in Nickel Toxicology" S.S. Brown & F.W. Sunderman Jr,
edts. Blackwell Scient. Publ., London, 1985, 129-132.

13. OSTAPCZUK P., HILDEBRAND H.F., MERCIER J.F., RACZEK U. &
STOEPPLER M.
Elevated nickel, cobalt and chromium levels in body fluids, tissue
and other specimens of patients with implants from Ni-Co-Cr-alloys.
In : "Heavy Metals in the Environment". T.D. Lekkas, edt. CEP
Consultants Ltd, Edinburgh, U.K., 1985, vol. 2, 92-94.

14. ROSTOKER G., ROBIN J., BINET O. & PAUPE J.
Dermatoses d'intolérance aux métaux des matériaux d'ostéosynthèse et
des prothèses (Nickel-Chrome-Cobalt).
Ann. Dermatol. Vénérol., 1986, 113, 1097-1108.

15. SMITH I.C. & CARSON B.L., edts.
Trace metals in the environment (vol.6, Cobalt).
Ann Arbor Science Publisher, Ann Arbor, 1981.

16. STOEPPLER M.
Analysis of nickel in biological materials and natural waters.
In : "Nickel in the Environment". Nriagn J.O., edt. J. Wiley, New
York, 1980, 661-821.

17. SUNDERMAN F.W., Jr
Analytical biochemistry of nickel.
Pure Appl. Chem., 1980, 52, 527-544.

18. TAYLOR A. & MARKS V.
Cobalt : a review.
J. Human Nutr., 1978, 32, 165-177.

19. TILSEY D.A. & ROTSTEIN H.
Sensitivity caused by internal exposure to nickel, chrome and
cobalt.
Contact Dermatitis, 1980, 6, 175-178.

20. VEILLON C., PATTERSON K.Y. & BRYDEN N.A.
Chromium in urine as measured by atomic absorption spectrometry.
Clinical Chemistry, 1982, 28, 2309-2311.

21. WIBERG G.S., MUNRO I.C., MERANGER J.C., MORRISON A.B. & GRICE H.C.
    Factors affecting the cardiotoxic potential of cobalt.
    Clin. Toxicol., 1969, 2, 257-271.

ACKNOWLEDGEMENT

    This work was supported by grants from the association
"Prothésor" and from the "Institut National de la Santé et de
la Recherche Médicale (INSERM)": PRC Investigations Cliniques
Nr 82/849.

HISTOCHEMICAL, ELECTRON MICROSCOPIC AND MICROANALYTIC INVESTIGATIONS OF TISSUE SURROUNDING NI-CR-ALLENTHESIS IN MAXILLO-FACIAL SURGERY

Eckhard Dielert and Eberhard Fischer-Brandies

Klinik und Poliklinik für Kieferchirurgie
der Universität München
D-8000 München 2

INTRODUCTION

Special metal devices can be suited for a temporary alloplasty and for the fixation and stabilization of fractured bone structures. Due to their good electroconductivity, however, metals tend to corrode by electrochemical processes. Special alloys are manufactured for surgical purposes. Because of their composition and homogeneity, these alloys are claimed to be electroenergetically inactive and chemically resistant. However, this holds true only if the passive surface layer is intact. Because the surface layer in steel alloys is only 1/200 000 mm thick, the question arises, whether it can act as a barrier under biological conditions. Repeatedly, metallotic damage to the surrounding tissue and fracture of the plates have been reported (Frank and Zitter, 1971; Schuster, 1970, 1975; Münzenberg et al., 1972). The question arises, to what extent corrosive processes account for these problems.

MATERIAL AND METHODS

In 4 patients mandibular reconstruction plates had to be removed because of soft tissue necrosis with exposure of the allenthesis (n=2) and osteolysis in the screw area (n=2). The plates were in position for 6 - 14 months. During surgical reintervention biopsies were retrieved from the surrounding tissue, 4 samples for histochemical and 12 samples for electron-microscopical investigations from each patient. Specimen were taken with a certain distance to the screws to exclude influences due to crevice and abrasion corrosion, and well away from the area of plate exposure. The adjacent soft tissue did not exhibit any metallotic discolouring. The preparation procedure of the tissue samples for histochemical, electron microscopic and microanalytic investigation are described in detail elsewhere (Dielert et al., 1981). The chemical composition of the plate alloy (X5 Cr Ni Mo 18/12; DIN 17006) is Fe 62, Cr 17.5, Ni 12.0, Mo 2.5, Mn < 3, Si < 3, C 0.03 weight percent.

HISTOCHEMISTRY

Ferric ions can be identified with potassium ferrocyanide and ferrous ions with potassium ferricyanide by the resulting blue color. With light microscopy small non-translucent particles are seen with a blue reaction in the vicinity (fig. 1). These pictures, however, cannot answer the question whether the iron determined by the blue reaction originates from hematoma or from the allenthesis.

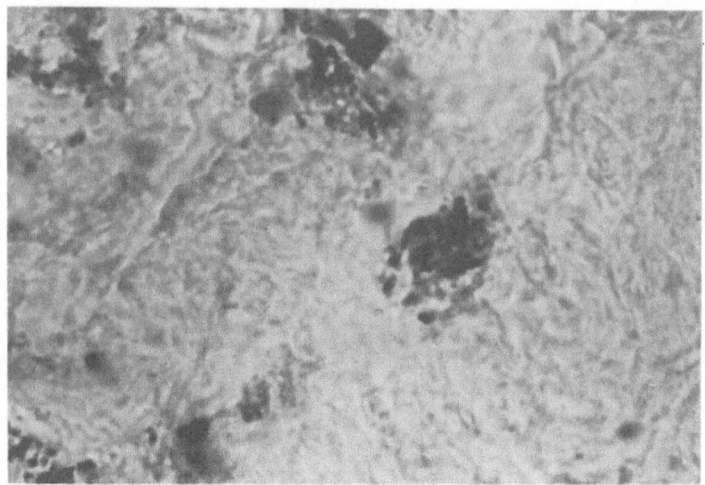

Fig. 1. Accumulated enclosures in the connective tissue surrounding the allenthesis with Berlin blue reaction.
x600

TRANSMISSION ELECTRON MICROSCOPY

The next step to answer this question consisted in transmission electron microscopic investigations. The particles varied from 0.01 to 20 microns in size (fig. 2). Their shape was often comparable to that of the grains found in plate surface micrographs. Foreign material could also be detected lying inside histiocytes and fibrocytes (fig. 3).

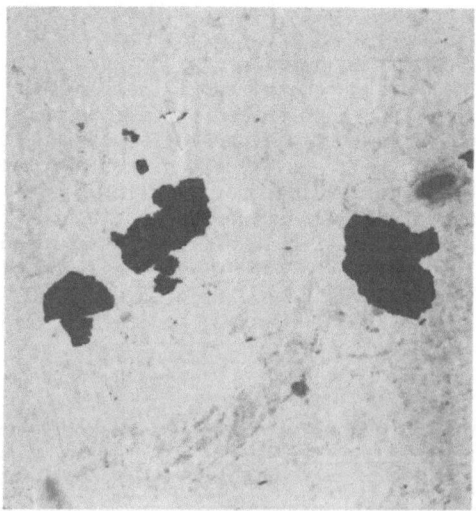

Fig. 2.

Sharp-edged particles, diameter up to 20 µm, near the implant.
TEM.

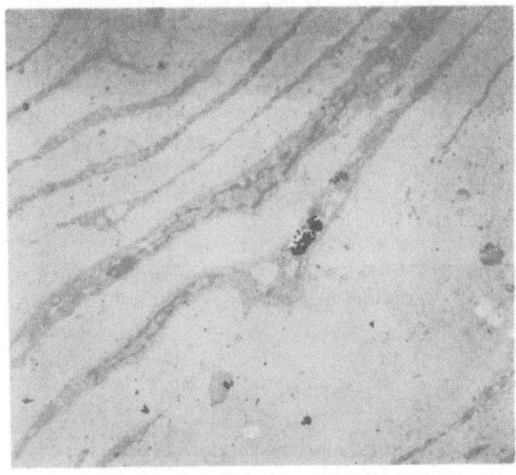

Fig. 3.

Foreign bodies lying in-
side a longitudinally
cut fibrocyte and some
smaller extracellular
particles.
TEM x 7400.

MICROANALYSIS

After a special preparation procedure these TEM specimens were used for microanalysis; with x-ray spectrometry, amounts of 10^{-10} to 10^{-14} grammes can be detected; characteristic spectra for the different elements are obtained. The point analysis of 50 particles localized by TEM revealed mean values of 62 % for Fe, 18.57 % for Cr and 12.45 % for Ni (fig. 4).

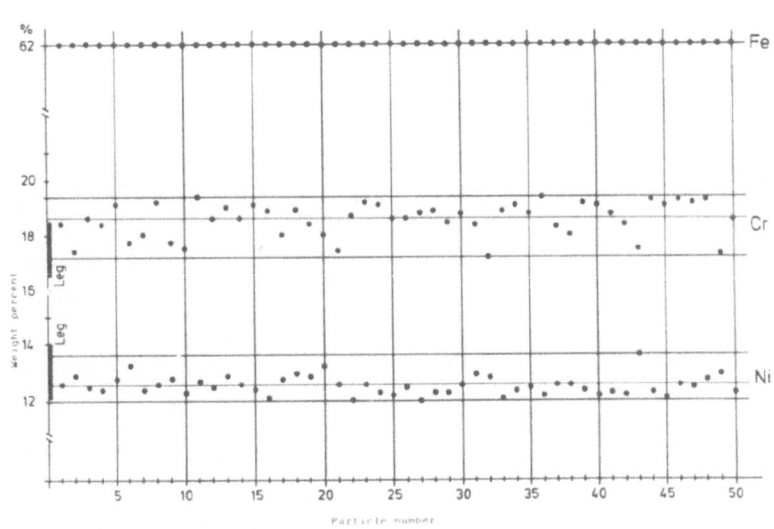

Fig. 4. Element concentration (Fe, Cr, Ni) in weight percent
        revealed by x-ray spectrometry (point analysis) of
        50 crystalline particles. The bars on the vertical
        axis represent the values given by the manufacturer.

The standard deviation amounted to 0.62 % (Cr) and 0.35 % (Ni). When comparing these values to the chemical composition of the plate alloy (s. above) a correspondence becomes evident.

The next step consisted in the determination of the structure of the same particles by electron diffraction microprobe analysis. The principle consists in the deflection of an electron beam at the crystal lattice according to Bragg. Depending on the method used ring or point diagrams are obtained. The analysis of the characteristic maxima of deflection (fig. 5) leads to a surface-centered structure with a lattice constant of 3,60 ± 0,1 Å. Thus the particles have to be considered to be polycrystalline; their structure corresponds to the Cr-Ni-Mo-steel of the reconstruction plates, which is cubic surface centered and has a lattice constant of 3.59 Å.

DISCUSSION

The incrustations found in the tissue surrounding reconstruction plates exhibit a positive reaction to iron. In TEM the particles have a shape similar to the grains in the plate surface micrographs. X-ray spectrometry reveals a chemical composition corresponding to that of the alloy itself. The lattice structure and constant as determined by electron diffraction analysis is also identical. Thus the particles can be considered as fine crystalline corrosion products of the austhenitic Cr Ni steel.

Transcrystalline corrosion of the particles initially measuring 10-80 microns leads to microcrystals which are found between the cells. The attack along the grain boundaries is stimulated by chromium carbides concentrated at the surface of the single grain. The result is a relative reduction of Cr in the surrounding. Therefore the formation of a correct passivation layer is not possible (Frank and Zitter, 1971). In the body fluids a stable anode will form, which leads to corrosion processes. The metallic impregnation does not result from a solution of single components of the alloy, but from inter- and transcrystalline decay of the material.

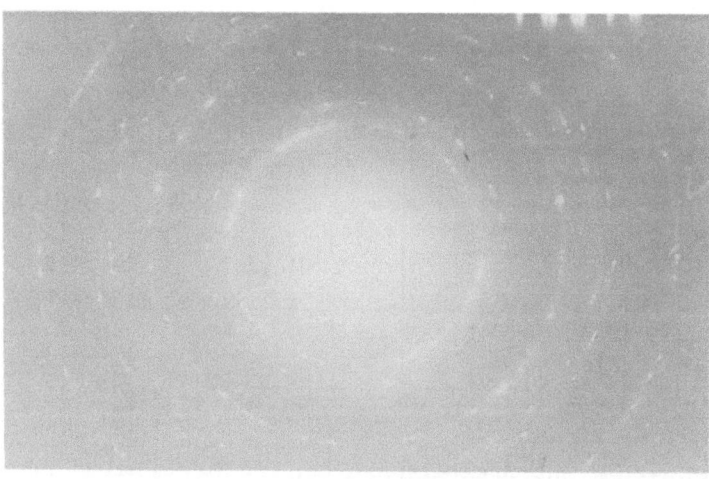

Fig. 5. Ring diagram of a crystalline particle.

The chemical and electrochemical interactions between the implant and its bed can negatively influence the therapeutical effect of the allentheses. As an alloy, the elements Cr and Ni (Ferguson et al., 1960; Contzen et al., 1967; Hulliger et al., 1967; Riede et al., 1974) do not exhibit toxicity. With further corrosive decay, however, these elements are set free. Inhibition of growth and calcification is described. Dependent on the specific toxicity, local inflammation may occur. Furthermore, changes in the homeostasis of the trace elements are possible.

CONCLUSIONS

The investigations lead to the following conclusions:

1. Microanalysis can contribute to knowledge of metallosis.

2. Even with modern implant alloys for surgical purposes an impregnation of the surrounding tissue with corrosion products can be observed.

3. Not the single components of the alloy are set free primarily, but the metal structure itself decays.

4. The electrochemical attack leads to an inter- and transcrystalline corrosion.

5. The metal particles in the adjacent connective tissue result in further galvanic elements.

REFERENCES

Contzen, H., Straumann, F., and Paschke, E.,1967, "Grundlagen der Alloplastik mit Metallen und Kunststoffen," Thieme, Stuttgart.

Dielert, E., Schinko, I., and Hamm, G., 1981, Zum Einsatz des Elektronenstrahls beim Identifizieren korrosiver Zerfallsprodukte von Osteosyntheseplatten in biologischem Material, Dtsch. Z. Mund-Kiefer-Gesichts-Chir., 5:159.

Ferguson, A., Laing, P., and Hodge, E., 1960, The joinisation of metal implant in living tissues, J. Bone Jt. Surg., 42:77.

Frank, E., and Zitter, H., 1971, "Metallische Implantate in der Knochenchirurgie," Springer, Wien New York.

Hulliger, L., Pohler, O., and Straumann, F., 1967, Wachstum von Kaninchenfibrozyten in Gegenwart gelöster Metallchloride, Z. ges. exp. Med., 144:145.

Hulliger, L., Pohler, O., and Straumann, F., 1967, Einfluß einiger Metalle und Legierungen auf das Wachstum von Kaninchenfibrozyten in Gewebekulturen, Z. ges. exp. Med., 144:145.

Münzenberg, K. J., Flajs, G., Roggatz, J., and Süßenbach, F., 1972, Korrosionsbedingte Oberflächenschäden am lebenden Gewebe, insbesondere am Knochen, im rasterelektronenmikroskopischen Bild, Z. Orthop., 110:336.

Riede, U. N., Rüedi, Th., Rohner, Y. L. E., Perren, S., and Guggenheim, R., 1974, Quantitative und morphologische Erfassung der Gewebereaktion auf Metallimplantate (Osteosynthesematerial). Teil I Eine morphometrische, histo-

logische, mikroanalytische und rasterelektronenmikro-
skopische Studie am Schafsknochen, <u>Arch. Orthop. Unfall-
Chir.</u>, 78:199.

Riede, U. N., Rüedi, Th., and Limacher, F., 1974, Quantitative
und morphologische Erfassung der Gewebereaktion auf
Metallimplantate (Osteosynthesematerial). Teil II
Untersuchungen am Menschen, <u>Arch. Orthop. Unfall-
Chir.</u>, 78:199.

Schuster, J., 1970, Metallurgische Untersuchungen von Osteosynthesematerial aus V4A-Stahl, <u>Mschr. Unfallheilk.</u>,
73:13.

Schuster, J., 1975, "Die Metallose," Prakt. Chir. H. 90,
Enke, Stuttgart.

MEASUREMENT OF NICKEL AND CHROMIUM AT THE SITE OF METALLIC DENTAL IMPLANTS

A. Garuet[2], M. Simonoff[1], B. Berdeu[1], Y. Llabador[1],
F.X. Michelet[3], and P.F. Caitucoli[2]

1) Chimie Nucléaire, C.E.N.B.G. 33170 Gradignan
2) UER d'Odontologie 14 Cours de la Marne 33000 Bordeaux
3) Chirurgie maxillofaciale et Stomatologie CHR Bordeaux
   Place Amélie Rabat-Léon 33000 Bordeaux

Since 1945 alloys of nickel-chromium have been widely used in Odontology thereby permitting the development of metallic dental implants.

The choice of metals for such implants is dictated by the requirements of resistance to wear, and, most importantly, biocompatibility. While these alloys fulfill the second requirement, their long-term biocompatability and resistance to corrosion in situ have been brought into question in the light of reports citing adverse effects following implantation.

Both chromium and nickel have been implicated as carcinogens in the industrial environment. However, the carcinogenicity of chromium is associated with the hexavalent ion and that of nickel with volatile or readily absorbable forms. In fact, the hexavalent chromium ion is rapidly reduced in vivo to the trivalent ion and it may be that this, in fact, is the actual carcinogen (1). At present, this is an open question.

In principle, the problems associated with these implants are physical, where there is mechanical failure, and corrosion with the release of soluble salts of the metals into the surrounding tissue, and possibly into the circulation. Of course, the corrosion of itself can be a factor in the failure of the implant. In the case of both nickel and chromium very little is known concerning the effects of the deposition of these metals in the tissues. It is assumed that they will be, at least at first, complexed with low molecular weight substances such as organic acids and amino acids. Subsequent combination with proteins would result in accumulation at the site of the implant while the low molecular weight complexes could be transported to other tissues and excreted. In spite of the many thousands of prostheses that have been implanted, our knowledge concerning the deposition of corrosion products and their biological activity remains fragmentary. The very few reports of malignant tumours associated with chronic metal implants would suggest that their carcinogenetic potential is very low. It must be borne in mind, however, that the majority of the limb prostheses have been implanted in elderly subjects and carcinogenesis may not become apparent without a long incubation period.

Nickel is the most allergenic of the metals, and chromium the second (2) although in the past chromium allergy has been associated with industrial exposure to hexavalent salts. This may be due to the very low tissue penetrating potential of the trivalent chromium ion (3). There have been reports of the failure of implants as the result of allergic reactions (4). The evidence for allergic manifestations with dental implants is contradictory, varying from severe local and generalised manifestations (5) to no apparent effects (6).

Because the forms of nickel and chromium associated with industrial toxicity are those to which the skin and mucous membranes are most permeable, the potential toxicity of the "non-penetrable" forms of the metals has not received great consideration. Evidently, this must be investigated further.

The inevitable surgical correction poses considerable removal of tissue. There are large anatomical lesions and the sequelse cause hollowing as well as scar tissue which compromises the retention of subsequently istalled prosthesis.

The appearance of allergic phenomena and those of rejection directed our etiological investigation towards the importance of the metals themselves and the phenomena of deterioration of the alloys used. The liberation of metal ions in biological milieux due to corrosion from temperature and salinity is such as to produce cytotoxic and biomechanical problems.

EXPERIMENTAL

In principle the dental prosthesis is retained by the extra-gingival portion of the implant which lies in the buccal cavity while the base is juxta- or endo-osteal (figure 1).

Inserted surgically, they align single teeth as well as total prostheses in the case where a complete set has been replaced. In both cases the post which protrudes from the fibro-mucous serves to keep the internal and external aspects in line.

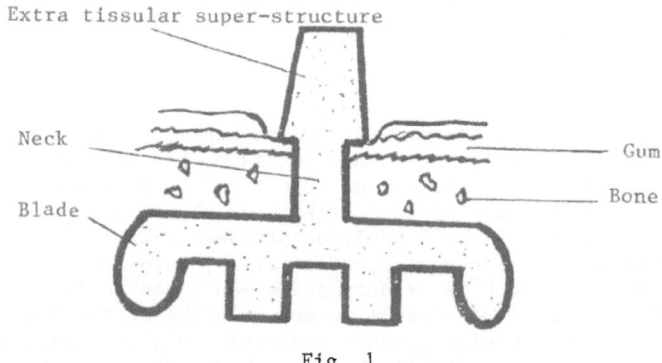

Fig. 1

These were not homogenous structures because they have two distinct part (figure 2) :

- a sub-mucal infrastructure which assures alignment and anchorage

- an extra-tissular superstructure traversing the periosteum and mucosa within the buccal cavity to attach any dental prosthesis
Composition has been measured elsewhere (7).

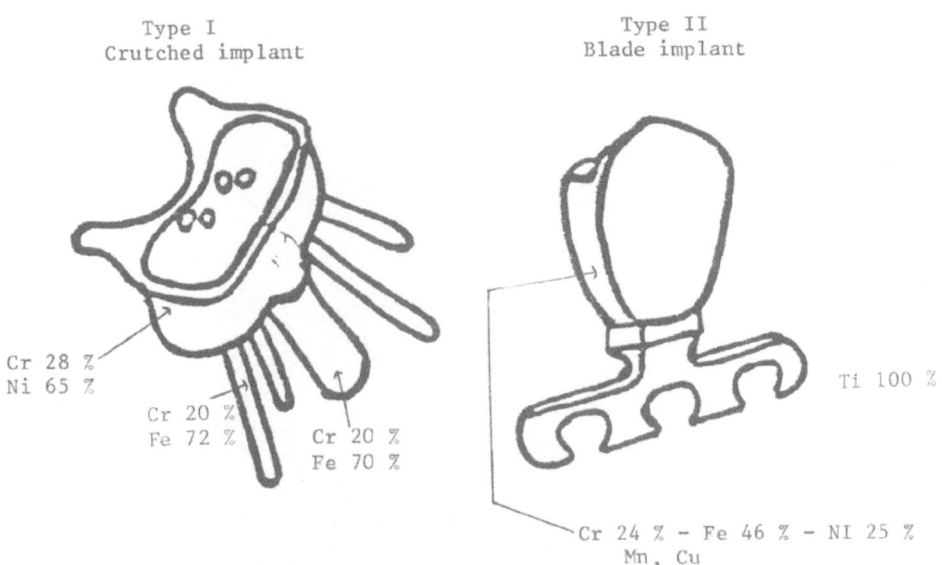

Type I
Crutched implant

Type II
Blade implant

Cr 28 %
Ni 65 %

Cr 20 %
Fe 72 %

Cr 20 %
Fe 70 %

Ti 100 %

Cr 24 % - Fe 46 % - NI 25 %
Mn, Cu

Fig. 2

CLINICAL STUDY OF FAILED IMPLANTS

Clinical and Histological Study of Failed Implants

Panoramic and retroalveolar radiographs showed bone formation, destruction or rarefaction according to the state of the patient and the date of installation (figure 3).

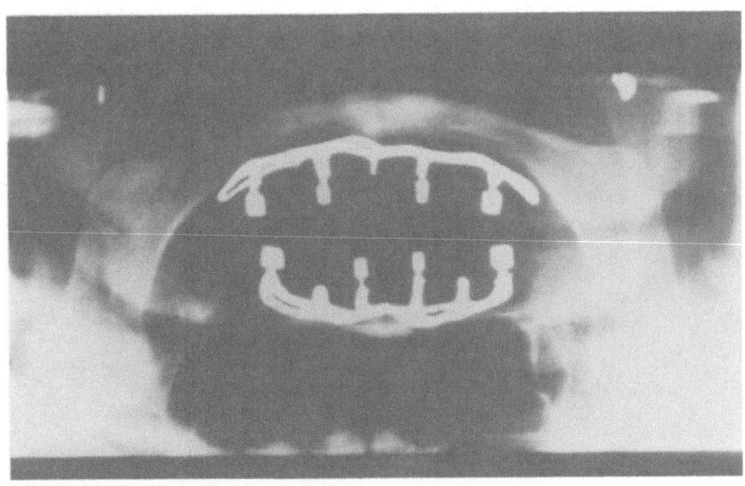

Fig. 3a

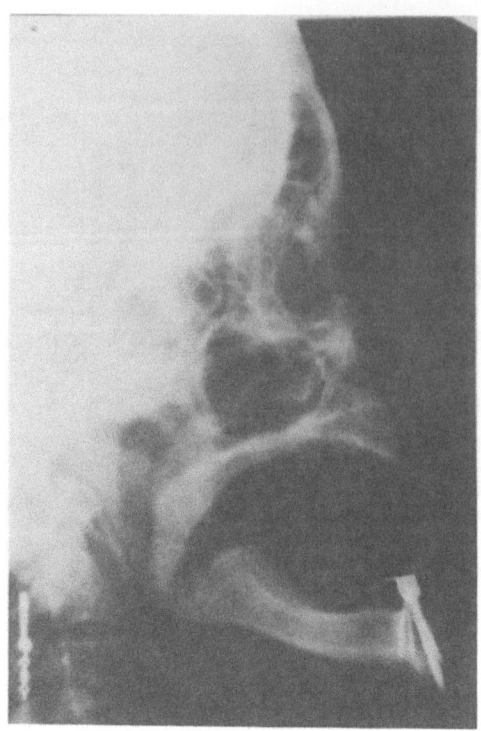

Fig. 3b

Frontal (3a) and lateral (3b) views of implants showing loss of
surrounding bone

Histological examination of neighbouring tissue showed changes in
macrophages and an inhibition of lymphocyte migration which favorised bone
resorption (figure 4). It was evident that such manifestations occured with
single implants as with larger replacements. These cellular abnormalities
are increased by mastication characterised by pulling structure, both the
implant and prosthesis.

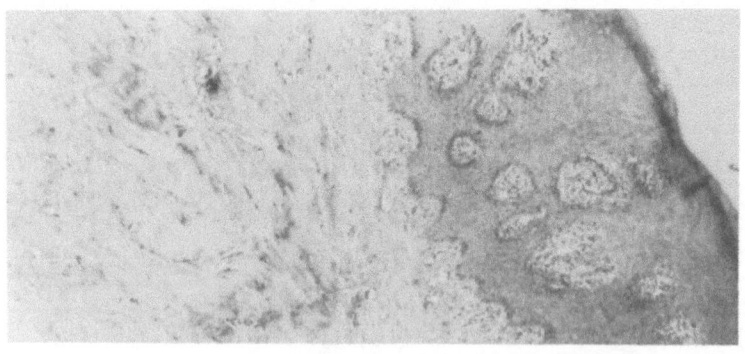

Fig. 4a

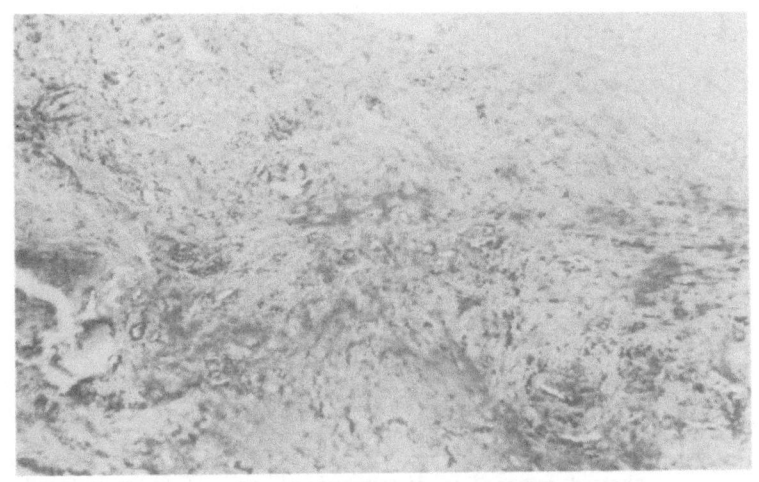

Fig. 4b

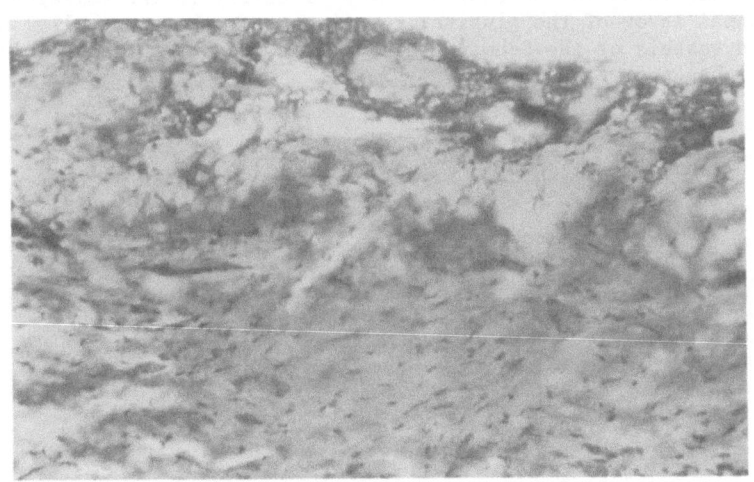

Fig. 4c

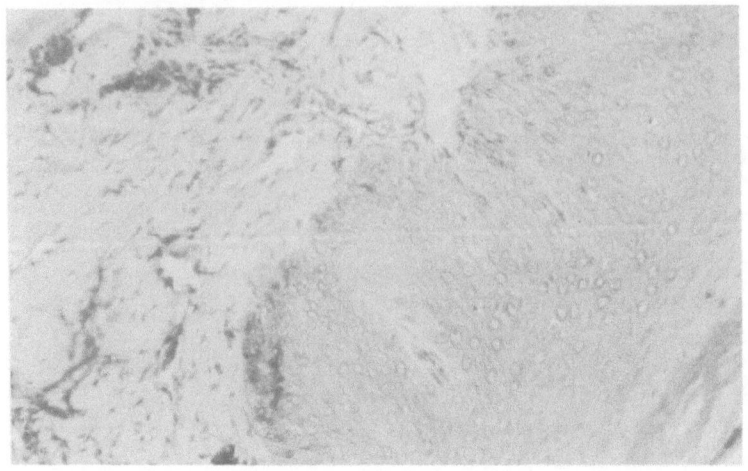

Fig. 4d

Fig. 4 : (4a) Normal gingival tissue
  (4b) Tissue removed from implantation site Type I implant
       showed extensive disorganisation and filsous replace-
       ment
  (4c) Control tissue
  (4d) Tissue removed from implantation site Type II implant
       with almost complete loss of normal structure

Demonstration of traces of metals

An ultrastructure study of the fragments of gum and tissue by electron
microscopy revealed the presence of metal particles (figure 5). They were
present essentially in the chorion extracellularly in the collagen fibres,
in the basal layer of the gingival keratinocytes, the endothelial cells of
the blood vessels or the Schwann cells ; also present intra-cellularly in
the macrophages or fibroblasts.

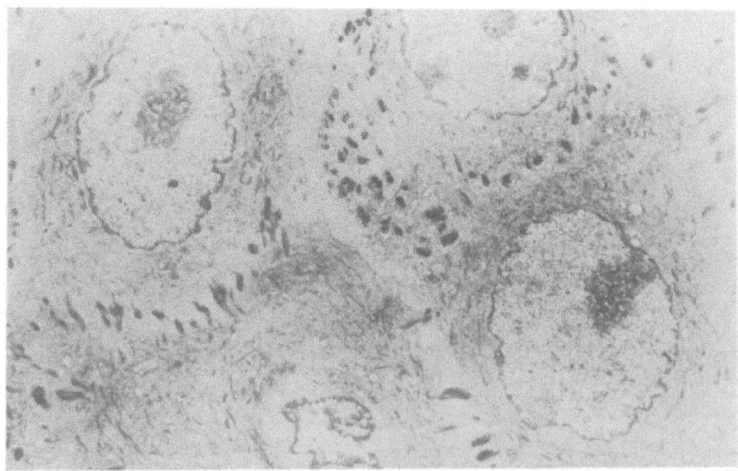

Fig. 5a

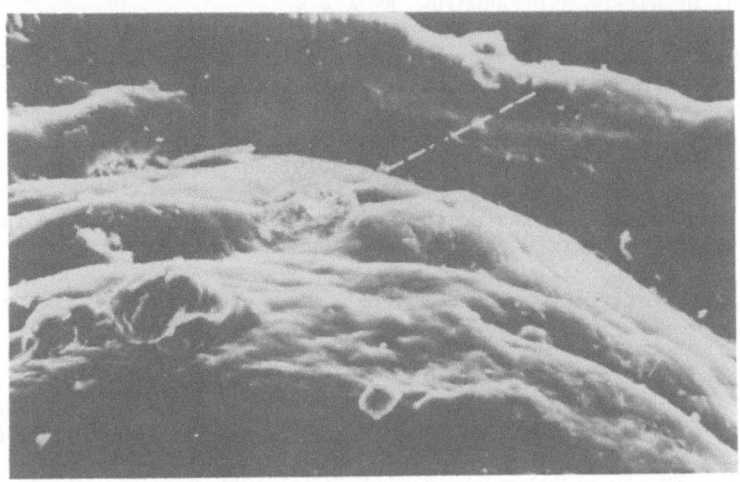

Fig. 5b

Fig. 5 : (5a) Electron micrograph of tissue from the immediate vicinity
of implant showing metal fragments (black granules)
(5b) Scanning electron micrograph demonstrating the presence
of metal fragment

Sample collection

Samples were collected under local aneasthesia at 9 sites for the type
I implants (figure 6), and at the centre post and extremity for Type II.
The sample refered as zero was the control taken at distance form the nine
other samples

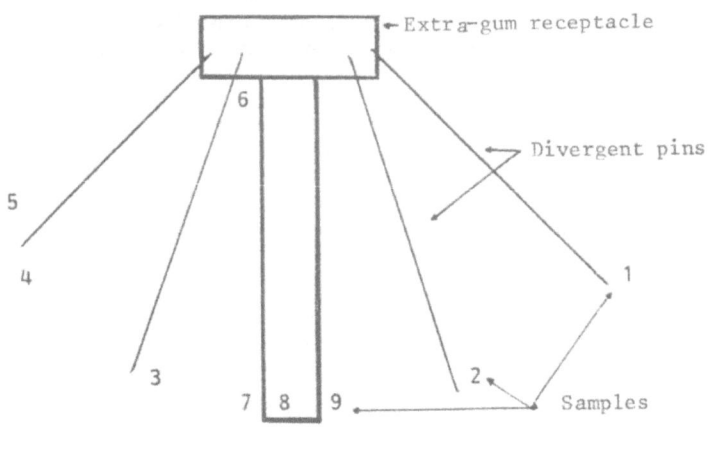

Fig. 6

MEASUREMENT OF NICKEL AND CHROMIUM BY A.A S.

Because of its sensitivy, we chose atomic absorption spectroscopy (A.A.S) with electrothermal volatilisation to measure these metals.

The apparatus was Varian 1475A with graphite furnace GTA 95. The use of autosamples enabled automatic sample dilution and standard preparation.

Sample prepration : The weighed tissue sample was dissolved in 0.5 mL of Suprapur (Merck) nitric acid and made up to 3 mL with water.

By experimentation we found the following programmes to give optimal results.

Nickel	Drying	75,90°	Chromium	Drying	75,90°
	Chaning	1200°		Chaning	1200°
	Atomisation	2400°		Atomisation	2500°

Argon gas was used throughout with stopped-flow during atomisation.

Under these conditions a linear response was obtained for 0-30 ppb nickel and 0-15 ppb chromium.

Samples were processed neat, 1 : 4, 1 : 10, 1 : 40, 1 : 44 as appropriate.

METAL DETERMINATION BY PIXE

To determine the content of metals in the samples by a multielement technique we used proton-induced-X-ray-emission (PIXE). In this technique, atoms irradiated by elementary particles absorb energy causing displacement of outer electrons, which return almost instantaneously to the basal state with emission of X-rays which are characteristic for the element. With this technique atoms heavier than 24 Daltons can be identified and quantitated.

The samples ashed, by low temperature ashing with an oxygen plasma, doped with internal standard (Yttrium) were dried on polycarbonate films and exposed to a beam of 3 MeV protons from the Van de Graaff accelerator at the Centre d'Etudes Nucléaires, Bordeaux. Detection and measurement was by silicon dètector. The PIXE set-up is presented in figure 7.

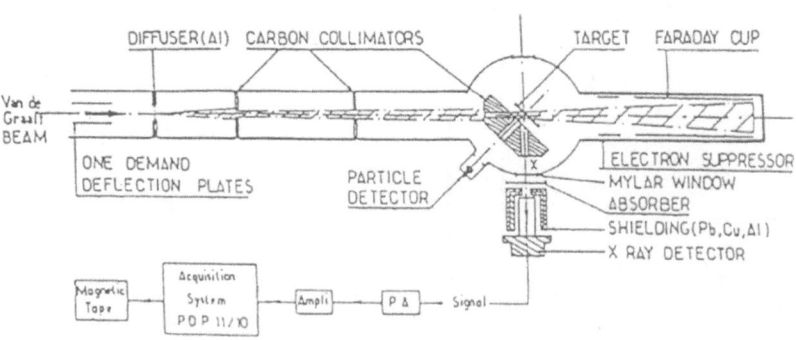

Fig. 7

RESULTS

- Atomic absorption spectroscopy
Nickel and chromium (Type I implant) (Table 1)

Table 1

Site	Nickel content (ppm)	chromium content (ppm)
0	Not detected	0.944
1	11.901	35.523
2	11.919	180.26
3	15.616	30.23
4	38.172	25.116
5	18.944	16.581
6	23.367	41.175
7	3.653	18.078
8	0.453	1.427
9	1.934	10.291

Nickel was not detected in the control sample, and was present at highest concentration around the side struts.

Chromium was present in all samples including the control at high level, although the control level was below that of all the test samples. As with nickel, the concentration was higher around the side struts than at the center post.

- PIXE analysis

Table II : B_1, B_2, B_3

	B_1 Biopsy type I (Cr-Ni-Fe) in ppm		B_2 Biopsy type I (Cr-Ni-Fe) in ppm	
	Control	Test	Control	Test
K	174	164	179	148
Ca	68	62	43	161
Ti	6,1	7,5	1,6	2,1
Cu	1,17	0,73	1,01	1,28
Zn	13	3,1	7,9	6
As	3,3	3,1	0,14	0,2
Br	-	-	0,42	2,72
Rb	1,4	-	1,83	-
Fe	80	19	29	38
Cr	0,5	-	2,6	0,3
Ni	0,05	0,19	0,44	1,06

(continued)

Table II (Continued)

Biopsy II (Ti-Fe-Ni-Cr) in ppm ; B_3

	Control	Test
K	732	2,241
Ca	2,549	5,642
Ti	19	139
Cu	1,31	–
Zn	162	337
As	3,88	17,4
Zr	10,6	20
Fe	290	306
Cr	39	1,7
Ni	10,5	19,3

All results are expressed as ppm metal (μg/g dry weight). The error is about 15 %

Between the test and the control samples for zinc and iron, there are differences probably due to decreased circulation. The large variations of ratio Ca/K between the different biopsies sign the different origins of the samples (tissular, osseous, periostic).

PIXE and AAS measurements show an increased nickel in the test near the implant of type I compared with the control biopsy. Such a migration (indepently of the process) is not seen for chromium in the samples measured py PIXE.

A high level of titanium (139 ppm) compared with the control (19 ppm) has been detected near the titanium blade of this implant (type II)

In order to compare the different samples we expressed the results in terms of concentration metal/concentration of potassium as index element. Both were in ppm.

Figure 7a and b show PIXE spectra from tissue of type I implants, Figure 7c and d show spectra from type II implants.

DISCUSSION

It is evident that the presence of these dental implants may be associated with severe tissue reaction leading to disorganisation and breakdown. How much of this is due to the physical force exerted by the prosthesis and how much is due to chemical irritation due to dissolution of the metals composing the implant is not clear.

Similarly, the presence of metal fragments in the tissue could be due to chemical erosion by biological fluids or, as has been reported, damage to the prosthesis occuring during the implantation, or from drilling with metal bits. Nickel and chromium were found to be present in the tissues both by PIXE and AAS.

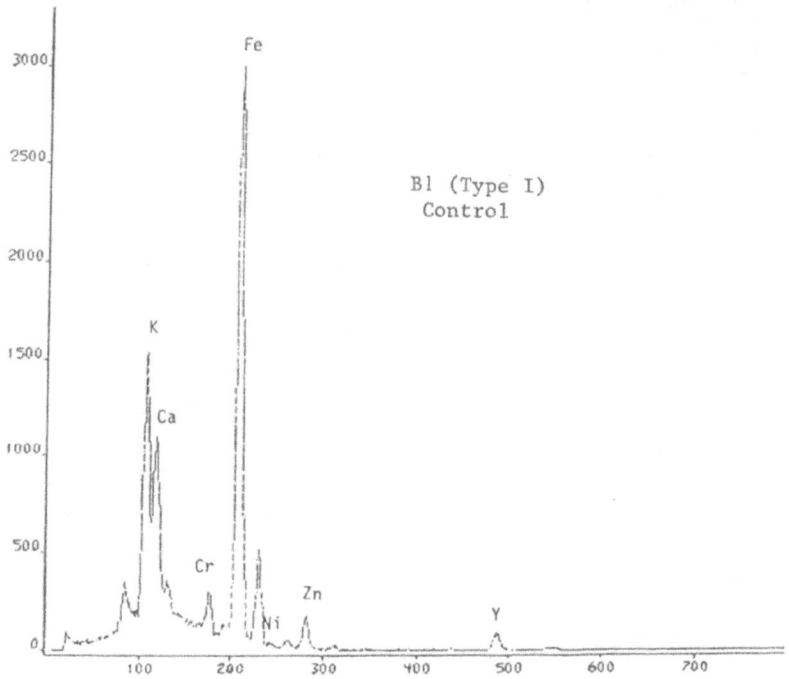

Fig. 7a

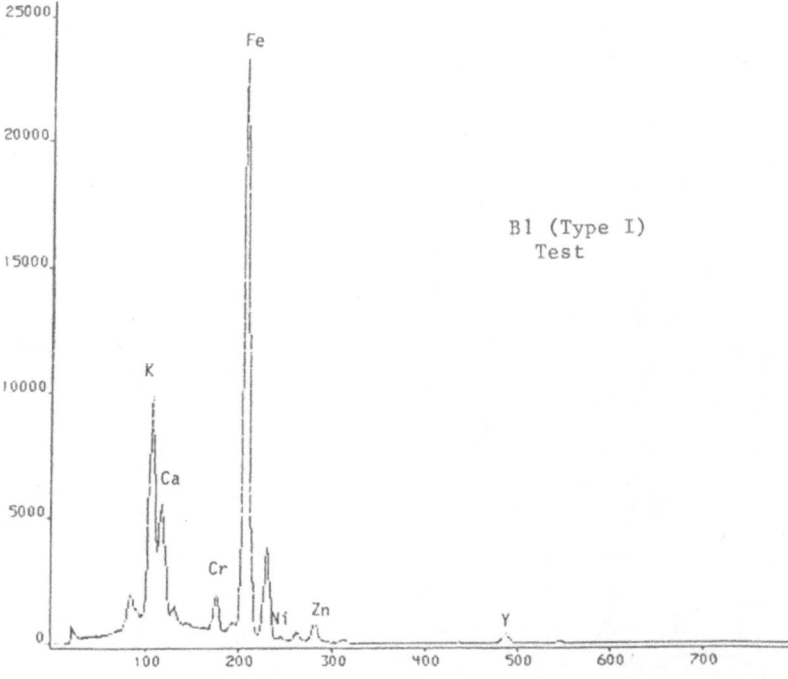

Fig. 7b

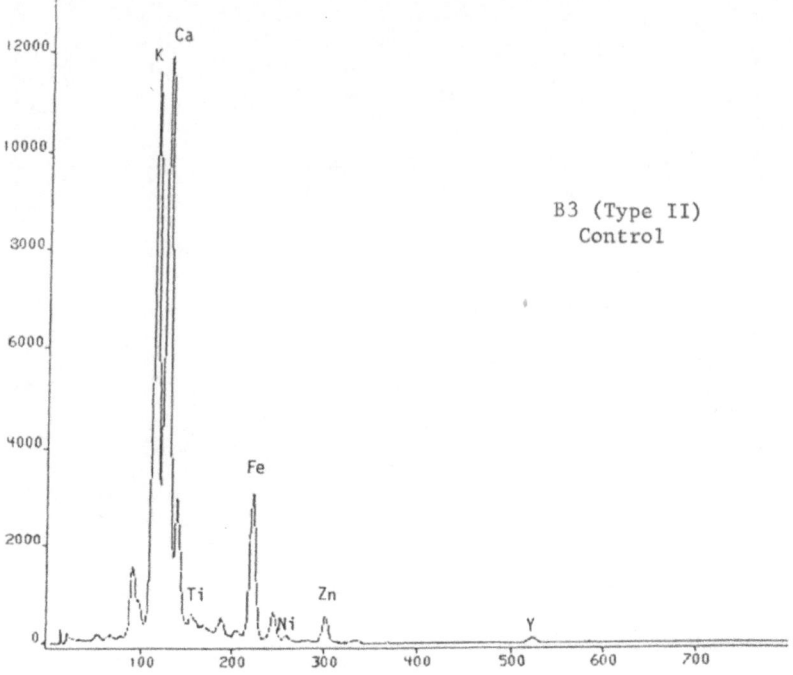

Fig. 7c

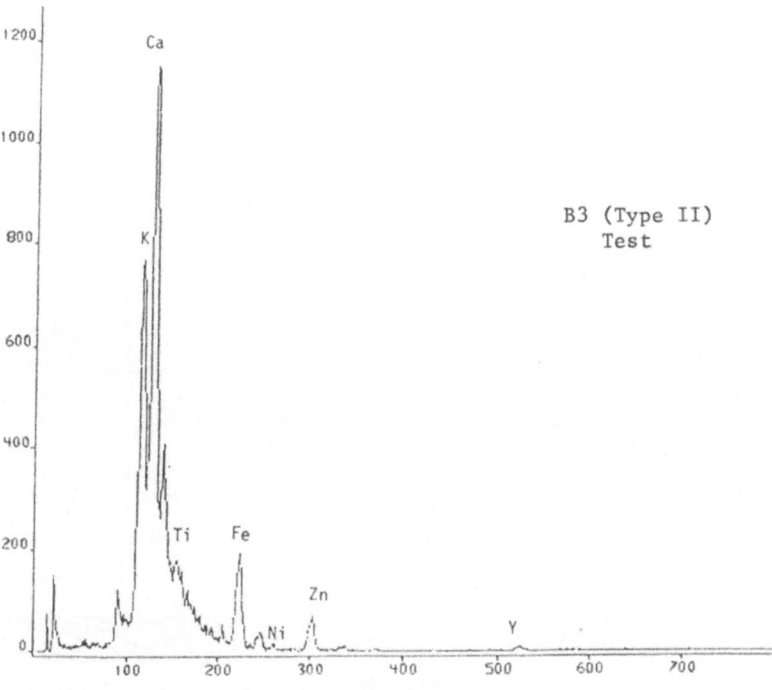

Fig. 7d

The nickel concentration, as measured by AAS, is about 2 ppm near
the central stem, and 20 ppm near the lateral pins. By PIXE, the nickel
increasing for the test compared with the control is in average, a factor
3 and these Ni concentrations are situated in the range 0.2 to 1 ppm. The
concordance between the results using the two techniques PIXE and AAS is
excellent for these biopsies coming from the area situated near the central
stem. The nickel concentrations measured by Dielert et al. (8) by emission
spectroscopy (ICP - OES) in mandible and periosteum of dogs implanted
with a Co - Cr - Mo - Ni alloy were situated in the range 0,5 - 1.5 ppm.

The large chromium variation and some very elevated Cr contents as
measured by PIXE and AAS suggest a contamination of the tissue from the
instruments during the biopsy.

As to the B3 Test, with its very high titanium concentration, it must
be presumably explained by the presence of a microfragment of titanium
comming from the titanium blade during the extraction.

The marked difference in chromium and nickel levels in the tissues
in contact with the different parts of the implant supports the hypothesis
that the prosthesis, in contact with biological fluids, loses metals by
electrolysis.

CONCLUSION

This preliminary investigation confirms that metallic dental implants
can cause severe tissue reaction probably by dissolution and liberation of
metal atoms into the surroundging tissue. In view of the very large number
of these implants in use, further research is imperative to evaluate both
short term and long term effects.

References

1) P.H. Connett, and K.E. Wetterhahn, Metabolism of the carcinogen chromate
   by cellular constituents, in Inorganic Elements in Biochemistry,
   M.J. Clarke et al. eds, Springer Verlag, Berlin P. 93 (1983)
2) L. Polak, Immunology of chromium, in Chromium Metabolism and Toxicity,
   ed. D. Burrows CRC Press Boca Raton, Fla, p. 51 (1983)
3) J.W.H. Mali, W.J. Van Kouten, and F.C.T. Van Neer, Some aspects of the
   behaviour of chromium compounds in the skin, J. Invest. Derm. 41 : 111
   (1963)
4) E.C. Brown, M.D. Lockshin, E.A. Salvati, and P.G. Bullough, Sensitivity
   to metal as a possible cause of sterile loosening after cobalt-chromium
   total trip replacement artheroplasty, J. Bone Joint Surg. 59A : 164 (1977)
5) A.A. Fisher, Safety of stainless steel in nickel sensitivity, J. Amer.
   Med. Ass. 221 : 1279 (1972)
6) G.D. Ovnitsky, and A.D. Ulyanov, Allergy to chromium using steel dental
   prosthesis, Stomatologia Moscow 55 : 60 (1976)
7) M. Simonoff, B. Berdeu, Y. Llabador, A. Garuet, P.F. Caitucoli,
   F.X. Michelet- Microanalysis of the composition of dental implants before
   and after implantation. NATO-OTAN Advanced Research Workshop on Biological
   incidences of Co-Cr-Ni alloys used in Orthopaedic Surgery and Stomatology,
   Bishenberg, France, 30 sept-4 oct. 1985.
8) E. Dielert, W. Winter and P. Schramel, Biochemical and histological investigations in the ambient tissue of metallic implants, Trace Element -
   Analytical Chemistry in Medicine and Biology, Vol. 2, 1983, ed. Walter de
   Gruyter & Co., Berlin-New York - Printed in Germany

IN VITRO EVALUATION OF THE BIOCOMPATIBILITY OF A NI-CR ALLOY (WIRON 88):

AN ULTRASTRUCTURAL STUDY

Patrick Exbrayat*, Marie-lise Couble, Daniel Hartmann[2], and
Henri Magloire

Faculté d'odontologie de Lyon, 69372 Lyon Cedex 2, France
[2]Institut Pasteur de Lyon, rue Pasteur, France

INTRODUCTION

Nickel-containing dental alloys are finding wider application in
dentistry. Epidemiologic and experimental studies raise the problem of
nickel or chromium carcinogenicity and allergic hypersensitivity. The
biocompatibility of Ni-Cr alloys have often been implicated in refering
to the toxicity of each metallic component, especially nickel compounds:
Ni_3S_2, $NiCl_2$, Ni(II), NiO, NiS (Hildebrand et al. 1981, Bearden and
Cooke 1980, Jacobsen 1977). Some authors have used animal experimentations:
Bergman et al. 1980 implanted subcutaneously in the neck region of mice
specimens of four Ni-Cr dental casting alloys and established that nickel
was released from the alloys and concentrated in the soft tissue capsule
and in some organs. Newman et al. 1981 showed that due to electrochemical
corrosion nickel is solubilized from two Ni-Cr dental casting alloys in
autoclaved human saliva. An other interesting study was performed by Woody
et al. 1977 who cultured mouse fibroblasts and HeLa cells on Ni-Cr dental
casting alloys and also with Ni-Cr powders. They found that cast metals
are passive and that metallic powders released toxic ions and induced cell
alterations and zone of lysis in the culture.
To complete these approaches, the aim of our work was to realize an
in vitro model as near as possible to the reality using human gingival
cells (fibroblasts and epithelial cells) grown on a Ni-Cr dental casting
alloy. The behavior of the cultured cells was appreciated through ultrastructural features and their ability to synthesize fibronectin, a glycoprotein involved in adhesion to substrata .

MATERIALS AND METHODS

Alloy

The non-precious dental casting alloy used in this study was the
Wiron 88 (BEGO, Bremen BRD). It is mainly used in the fabrication of
porcelain-fused-to-metal restorations for dental prostheses. Its chemical
composition as given by the manufacturers appears in table 1.
The alloy specimens were prepared by usual dental casting and polishing
techniques, in agreement with the recommendations of the manufacturers.
The culture's metallic supports were cast in the same dimensions as

Table 1. Chemical composition of the Wiron 88 (Wt%)

Ni	Cr	Mo	Si,Al,Ce,Mn,B	C
60-75%	16-25%	5-15%	0-5%	0,02-0,03%

the control glass supports : 10 mm x 30 mm.

The alloy sheets were cleaned by ultrasond and sterilized in alcohol before use.

Tissue culture

Cultured cells were obtained from biopsies of clinically healthy human gingiva upon patients without dental metallic prostheses. Samples were stored in physiologic serum, then cut in small pieces of about two mm^3 and put on alloy' sheets placed beforehand in Leighton tubes.

Cultures were suspended in Eagle basal medium (BME, Biomérieux, F) supplemented with 10% foetal calf serum (KC biological, Lenexa, Kansas) Penicillin, Streptomycin, sodium ascorbate and glutamin for 3 or 4 weeks. The medium was changed every three days.

Preparation for scanning electron microscopy

- fixation in 2% glutaraldehyde-0.1 M sodium cacodylate (pH 7.4) solution for 2 hours at 4°C
- dehydration with ethylic alcohol and freon
- cultures on its support were mounted on silver dag and DC-sputtered with 15 nm gold in a coating-unit
- examination with Cambridge S 600 microscope

Preparation for transmission electron microscopy

- fixation for 10 mn at room temperature in 2% glutaraldehyde-0.1 M sodium cacodylate solution (pH 7.4)
- three rinses in 0.2 M sodium cacodylate solution (pH 7.4) for 5 mn each
- postfixation for 30 mn in aqueous 2% OsO4 in 0.2 M sodium cacodylate solution (pH 7.4)
- dehydration in a graded ethanol series
- embedding in Epon 812
- ultrathin sections of the embedded cultures using an ultra microtome (Ultratome, LKB)
- staining with uranylacetate and lead citrate
- examination with either Philips EM 300 or Hitachi HU 12 A electron microscope

Intracellular immunoperoxidase labelling of fibronectin for transmission electron microscopy (according to K. HEDMAN 1980)

The cultures were rinsed twice in phosphate buffered saline (PBS) and fixed at 4°C for 30 mn in a 0.05% saponin- 4% paraformaldehyde- 0.1M sodium cacodylate (pH 7.4)solution. After washing, the cultures were maintained overnight at 4°C in a 0.05% saponin- bovine serum albumin- 0.1M glycin- PBS solution (DBS). The cultures were reacted for 2 hours at 4°C with anti-fibronectin antibodies diluted 1/20 in DBS and then washed again for 1 hour in DBS. The cultures were incubated for 2 hours at 4°C with goat anti-rabbit IgG peroxidase conjugate diluted 1/50 and

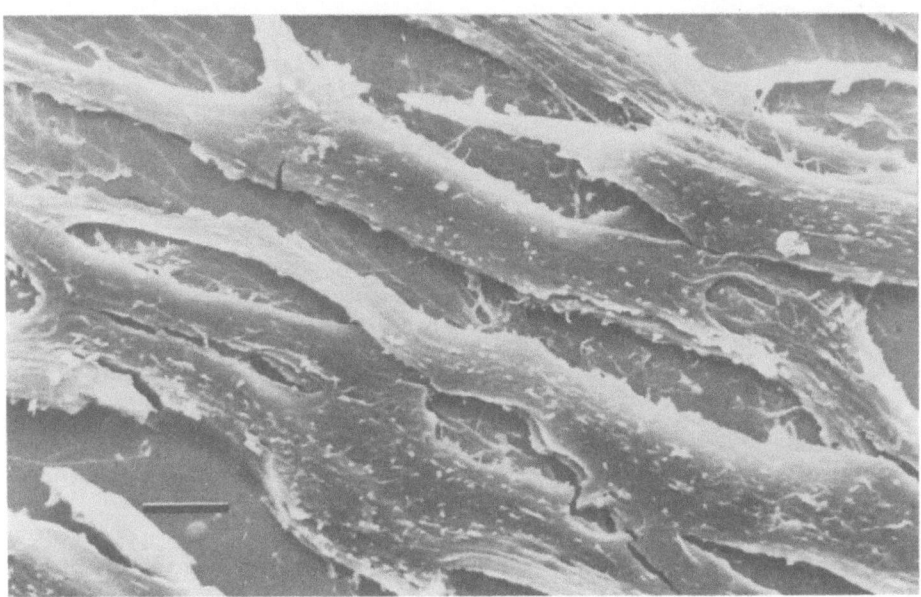

Fig. 1. Fibroblasts grown on alloy sheets for three weeks : many extensions assure adherence to the support.(bar 5 μm).

rinsed overnight. The cultures were treated for 30 mn at 37°C with PBS plus 0.05% DAB and then for 15 mn with PBS- 0.05% DAB- 0.01% H_2O_2 solution. After washing three times in PBS, the cultures were post-fixed for 30 mn in a 1% OsO4- 0.1M PBS solution (pH 7.4) and washed again. Finally, the cultures were dehydrated in ethanol and embedded in epon.

Control cultures were prepared as described above and incubated in the conjugate alone or in non immune rabbit serum instead of anti-fibronectin antibodies.

Thin sections of the exposed and control cultures were then observed with Philips EM 300 or Hitachi HU 12 A microscope.

RESULTS

Scanning electron microscope observations

The surface of the alloy sheets appeared uneven with many grooves and enhanced the inadequacies of polishing techniques.

The appearance of cultured gingival cells was similar on glass coverslips and on alloy sheets. In a parallel plane to that of cellular spreading the fibroblasts exhibited fusiform and flattened profiles with many even thin extensions (Fig. 1). They were particularly clear at the periphery of the culture and adhered both to glass coverslips and alloy sheets.

Transmission electron microscope observations

Ultrastructure of cultured gingival cells on Wiron 88

Fibroblasts

The ultrastructural morphology of cultured fibroblasts grown on the alloy appeared on Fig. 2 in cross section and presented different features:

• elliptic and offset nuclei
• the paranuclear cytoplasm was moderately electron dense but contained most of the organelles in a polarized arrangement

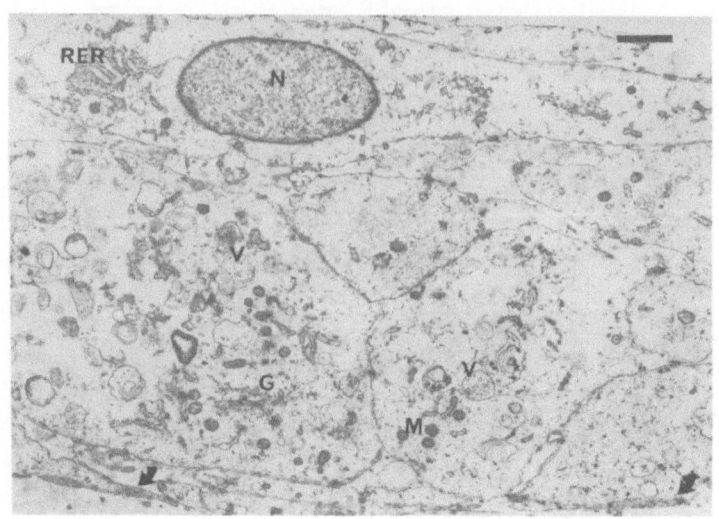

Fig. 2. Cross section of fibroblasts grown on alloy sheets for three weeks : the cytoplasm is slight dense, but containes the essential organelles (rough endoplasmic reticulum RER, mitochondria M, Golgi apparatus G) and many vesicules V . The plasma membrane in contact with the alloy presents a densification of granular material (arrows). (bar 1 µM).

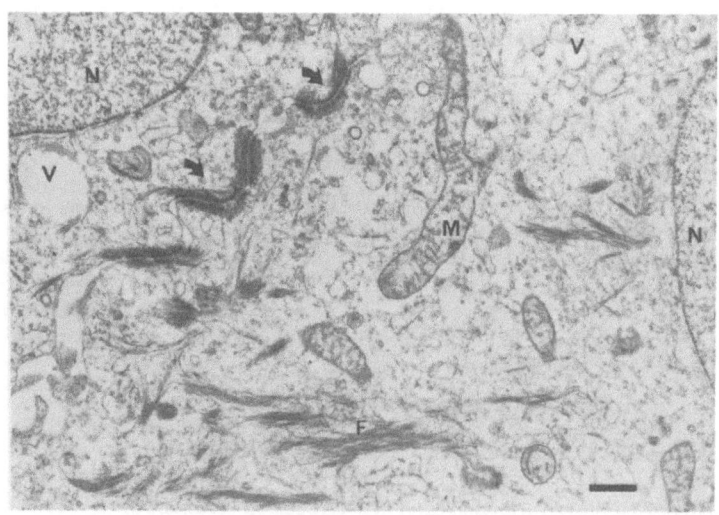

Fig. 3. Epithelial cells grown on alloy sheets for three weeks . We note the presence of two characteristic features: desmosomes (arrows) and intermediate filaments F . We observe mitochondria M and vacuoles V . (bar 0,5 µm).

- the rough endoplasmic reticulum comprised cisternae filled with finely granular material
  - mitochondria had high electron dense inner matrix
  - Golgi complex were well developed with arrays of packed cisternae numerous small vesicles and coated vesicles
  - many lysosomes and autophagosomes
  - we observed also intimate cell-to-cell contacts
  - electron dense, fine granular substances occured on cell surface in contact with the alloy

Epithelial cells

They presented numerous intracellular intermediate filament patches (probably keratin components) and desmosomes.

We observed a poorly developped Golgi apparatus, but a great number of vacuoles or lysosome-like bodies. Some mitochondria showed an inner matrix with a low electron density and few cristae (Fig. 3).

Extracellular matrix

The extracellular matrix included two types of filaments : collagen fibers with typically periodic striation and a great amount of fine micro-fibrils without cross banding.

Immunostaining of fibronectin

First, we observed fibronectin in human gingival cultures grown on glass coverslips at light microscopical level. The cytoplasm of both fibroblasts and epithelial cells, especially around the nuclei, was clearly stained and filled with peroxidase deposits. The extracellular localization of fibronectin was obvious on the surface and between the cells. Control specimens gave a negligible background.

Second, we had the aspects at ultrastructural level of human gingival cells grown on alloy sheets. Fibronectin was detected both intracellularly and extra cellularly in fibroblasts (Fig. 4,5) and in epithelial cells (Fig. 6). Peroxidase deposits were clearly seen into rough endoplasmic reticulum and in Golgi area. The rough endoplasmic reticulum was filled with fine granular material. Nuclei and mitochondria were unstained. We noted the very clear aspect of the cytoplasm following the Saponin treatment.

The detection of fibronectin confirms its synthesis by the fibroblasts in culture on alloy sheets.

Extracellularly, the peroxidase products were observed on microfibrils sometimes arranged in bundles and in close association with plasma membrane; they were heavily stained. In cross section, fibronectin appeared as dot-like material scattered between the cells. These structures formed a dense network. These microfilaments probably corresponded to those seen previously.

DISCUSSION

Our study assessed that human gingival cells grown on WIRON 88 compared to control cultures grown on glass coverslips are able to adhere and grow on this alloy.

The main points of the ultrastructure of cultured human gingival fibro-blasts described by Engel et al.(1980) have been found again in the tested fibroblasts. However, the cytoplasm of these cells grown on alloy sheets is less dense, even if it contained the most important organelles. The epithelial cells are not apparently modified by the contact of the casting alloy, exept for vacuoles. These findings are similar to those obtained by Woody et al.(1977) on four non-precious alloys.

The intracellular detection of fibronectin, the main glycoprotein mediating cell adhesion, lead us to believe that fibronectin is synthesized by gingival cells under our experimental conditions. This confirms the data of Grinnell and Field (1979) and Kleinman et al. (1981). Furthermore, on ultrastructural level, adhesion plaques were detected as specialization at

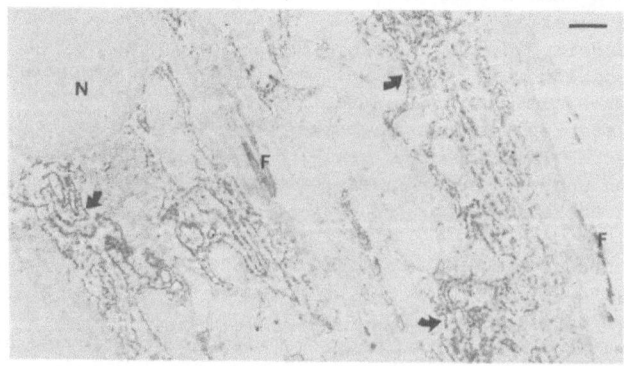

Fig. 4. Immunoperoxidase labelling of fibronectin in fibroblastic
cells grown on Wiron 88 for three weeks : the intracellular staining appears in the rough endoplasmic reticulum's cisternae(arrow)
and we note also extracellular staining of fibronectin F.(bar: 1μ).

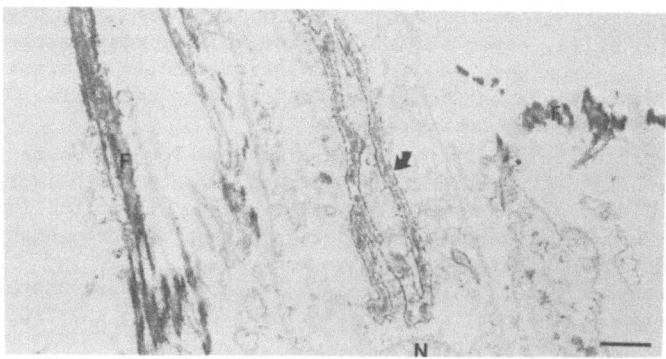

Fig. 5. Higher magnification of immunolabelling of fibronectin in
the rough endoplasmic reticulum (arrow) and in the extracellular
matrice (F). (bar: 1μ).

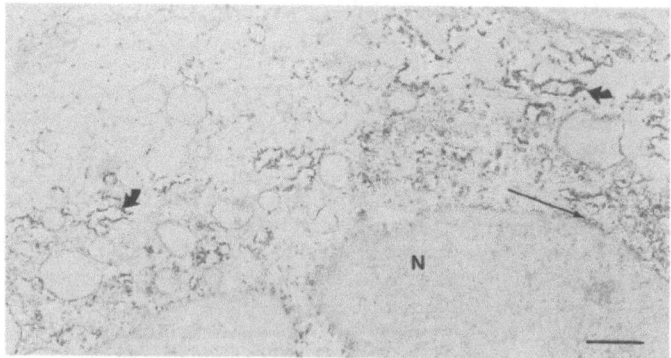

Fig. 6. Immunoperoxidase labelling of fibronectin in epithelial
cells grown on Wiron 88 for three weeks. The ultrastructure is
less well preserved, but we distinguish the intracellular staining in the rough endoplasmic reticulum's cisternae (large
arrows) and around the nuclear membrane (thin arrow).(bar:1μ).

cell-alloy contact. These presence was related previously when Grinnell and Field (1979) cultured human fibroblasts on glass or collagen gels and Magloire et al. (1986) on glass coverslips coated with cyanoacrylate films.

So cultured gingival cells may adhere to the Wiron 88 and the adhesion may be mediated both by fibronectin and adhesion plaques.

Various authors appreciate the biocompatibility of dental alloys through evaluation of the release of lactate dehydrogenase or lysosomal enzyme markers (Rae 1986), animal implantation (Niemi et al. 1985 , Bergman et al. 1980), counting of total cell number and of abnormal cell number (Evans and Thmas 1986), etc.

These different methods used associated with ultrastructural analysis lead us to believe that the appreciation of biocompatibility of dental casting alloys must cover a lot of markers of cellular activities enhancing the opinion of Pizzoferrato et al. (1985) claiming that cell cultures are a reliable and sensitive method for initial screening in testing the bio-compatibility .

ACKNOWLEDGEMENTS

This work was supported by CNRS (RCP 814).

REFERENCES

Bearden, L. J., and Cooke, F. W., 1980, Growth inhibition of cultured fibroblasts by cobalt and nickel, J.Biomed.Mater.Res.,14:289-309.

Bergman, M., Bergman, B., and Söremark, R., 1980, Tissue accumulation of nickel released due to electrochemical corrosion of non-precious dental casting alloys, J.Oral Rehab., 7:325-330.

Engel, D., Schroeder, H.E., Gay, R., and Clagett, J., 1980, Fine structure of cultured human gingival fibroblasts and demonstration of simultaneous synthesis of type I and III collagen, Archs.Oral Biol., 25: 283-296.

Evans, E. J., and Thomas, I. T., 1986, The in vitro toxicity of cobalt-chrome-molybdenum alloy and its constituent metals, Biomat., 7 (1):25-30.

Grinnell, F., and Field, M. K., 1979, Initial adhesion of human fibroblasts in serum free medium : possible rate of secreted fibronectin, Cell., 17:117-128.

Hedman, K., 1980, Intracellular localization of fibronectin using immunoperoxidase cytochemistry in light and electron microscopy, J.Histo chem.Cytochem., 28 (11):1233-1241.

Hildebrand, H. F., and Tetaert, D., Ni^3S^2-induced leiomyosarcomas in rabbit skeletal muscle : analysis of the tumoral myosin and its signi-ficance in the retro differentiation concept, Oncodev.Biol.Med., 2:101-108 (1981).

Jacobsen, N., 1977, Epithelial-like cells in culture derived from human gingiva : response to nickel, Scand.J.Dent.Res., 85:567-574.

Kleinman, H. K., Klebe, R. J., and Martin, G. R., 1981, Role of colla-genous matrices in the adhesion and growth of cells, J.Cell.Biol., 88: 473-485.

Magloire, H., Calle, A., Hartmann, D. J., Joffre, A., Serre, B., Grimaud, J. A., and Shue, F., 1986, Type I collagen production by human odontoblast-like cells in explants cultured on cyanoacrylate fibers : electron immunolocalization of fibronectin at cell-film interface, accepted in cell and tissue research.

Newman, S., Chamberlain, R. T., and Nunez, L. J., 1981, Nickel solu-bility from nickel-chromium dental casting alloys, J.Biomed.Mater.Res., 15:615-617.

Pizzoferrato, A., Vespucci, A., Ciapetti, G., and Stea, S., 1985, Biocompatibility testing of prosthetic implant materials by cell cultures, Biomat., 6:346-351.

Rae, T., 1986, The biological response to titanium and titanium-aluminium-vanadium alloy particles : tissue culture study, Biomat., 7:30-36.

Woody, R. D., Huget, E. F., and Horton, J. E., 1977, Apparent cyto-toxicity of base metal casting alloys, J.Dent.Res., 56 (7):739-743.

CHAIRMEN SUMMARY: TISSUE RESPONSE

J. Black and D.F. Williams

Many factors control the local tissue response to implanted materials. In the case of metals, the subject of this conference, the characteristics of the local response are determined by the rate of metal
particle and ion release and the "biological activity " of these degradation products.

Corrosion rates and machanisms are strongly influenced by the
presence of proteins. Furthermore, the biological activity of many metals
depends upon their ability to interact with proteins and form complexes,
either on the corroding metal surface, in solution or within cells.
However, the reactions with proteins are different at each location and,
as a result, elicit different tissue responses. For a given situation, the
local response is quite specific to the metal itself but may depend upon
concentration and time after implantation. Thus tha local response to
metallic implants may not correlate with toxicology , based upon ingestion
or injection in acute animal models or upon human epidemiological studies.
The situation is further complicated since the same celltype in tissue
culture, in various animal species and in humane may respond differently
to the same challenge. degradation products may circulate in the blood
and lymphatic systems, evoking systemic responses or they may be stored
and concentrated in tissues or organs distant from the original implant
site, causing a secondary local response, called a remote response.
Systemic and remote responses to metallic implants were briefly discussed
and might be considered more fully at a future conference.

However, the general experimental situtation is improving slowly.

Papers presented in the sessions on local tissue response reflect an increased sensitivity, on the part of the investigators, to the need to characterize the implant as to particle size and shape, chemical composition, microstructure, surface morphology, dose, carrier, etc... Increasing knowledge and control of these aspects of the implant leads to more reproduceable results and to a more fundamental understanding of biological response to metallic implants.

Finally, it is exciting to note the increased formation of multi-disciplinary research teams whose members are bringing the knowledge and procedures of many different sciences to beat simultaneously on a single problem in biomaterials : predicting the behaviour of the implant-tissue interface.

IMMUNOPATHOLOGY

FUNDAMENTAL ASPECTS OF CONTACT SENSITIVITY TO NICKEL

Claude Benezra and Maurice J.F. Leroy

Département de Chimie (UA CNRS 31) et Ecole Européenne des
Hautes Etudes des Industries Chimiques (URA CNRS 405)
Université Louis Pasteur, 1, Rue Blaise Pascal
67008 Strasbourg, France

INTRODUCTION

Contact sensitivity (or allergic contact dermatitis or delayed hypersensitivity to, <u>nickel</u> is an allergic reaction.

Allergy (also known as hypersensitivity) has been conveniently divided by Gell and Coombs into four different types I-IV. The first three types are "immediates" (types I-II) or semi-delayed (type III): they occur very soon (a few minutes to 2-4 hours) after the attack of an <u>antigen</u>.

<u>Type I or Anaphylaxis</u>: this allergy type includes urticaria, asthma, hay fever, anaphylactic shock (which may lead to death due to insect-wasps in particular-bites), to medications such as penicillin, etc. The allergen or antigen is generally a protein (from pollen for instance) or a polysaccharide. Reaction occurs between this antigen and antibodies, either induced or inherited (known as immunoglobulins) present in the blood.

<u>Type II or Cytotoxic reactions</u>: this includes the hemolytic disease of the newborn, accidents in blood transfusion due to the incompatibility of blood groups, etc. This is again a reaction between an antibody (immunoglobulin) and the attacking antigen (a protein usually).

<u>Type III (also known as Arthus phenomenon)</u>: this takes a few hours to show up. The clinical aspects is that of an oedema (swelling of the skin at the site of attack), an erythema (redness of the skin) or even necrosis, allergic vascularitis, glomerulonephritis, etc. Again the antigen is usually a high molecular weight compound.

Contact dermatitis due to nickel belongs to type IV allergy. <u>Type IV</u> includes such important immune phenomena as homograft rejection, autoimmune diseases, and allergic contact dermatitis (ACD). The aggressor is a small molecule (molecular weights usually do not exceed 1000 daltons) called <u>hapten</u> or incomplete antigen. In order to be recognized in the organism, it needs to attach to a carrier molecule (usually a protein), thus forming a complete antigen or allergen. The latter word, allergen, is very often used for small molecular weight compounds which should correctly be named haptens.

CELLULAR MECHANISM OF ALLERGIC CONTACT DERMATITIS

It is generally thought that ACD occurs in two distinct phases. In the <u>first contact</u> with the hapten, let us say here with nickel (metal or salt, this will be developed later on) nothing appears at the clinical level. This is the induction phase. What must probably occurs is the following[1].

The hapten penetrates into the skin, gets bound to a carrier protein molecule in the epidermis, thus forming the complete allergen. This allergen attaches itself to several skin cells and in particular to Langerhans cells, a dendritic cell equipped with HLA receptors (or MHC receptors) and in particular with the D_R (in man) or I_a (in guinea pigs and mice) antigen which enables this cell to deliver a message to another important cell, a T-lymphocyte. This close contact between the Langerhans cell and the T-lymphocyte, also called "flirt", seems a major event in the induction of sensitization (= induction of allergy).

In a second phase, also known as proliferation phase, the T-cells go to the closest node where they multiply through clonal selection into effector and memory cells[2]. The former will be responsible, through the release into the organism of chemical mediator, lymphokines, for the following clinical events (<u>not apparent in the first contact with the allergen (hapten)</u>): redness (= erythema), swelling (= oedema).

If there is no further contact with the hapten (nickel for instance), there will be no clinical evidence of active sensitization.

In turn if after a few days (the whole process described above takes less than a week), there is a second contact with the allergen (nickel) the same molecular and cellular events take place much faster, and a clinical lesion of the skin (eczema, the combination of erythema, and oedema) shows up. Fortunately, a third subpopulation of T-lymphocytes also appears after a few days: these are called suppressor cells. They are responsible for the decrease of the intensity of the skin reaction. Being sensitized or not is probably due to a delicate balance between suppressor and effector cells.

Cellular events are summarized in Figure 1.

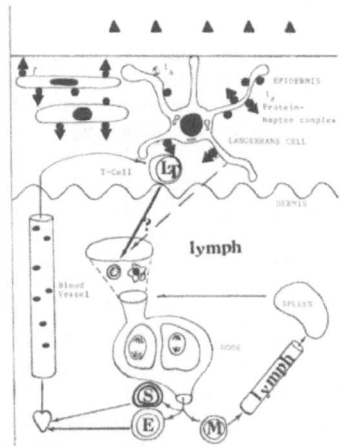

Figure 1.

Summary of cell events

I_a(mouse)=D_R(Man) Major Histocompatibility System antigen; L_T=T-Lymphocyte; S=Surpressor cells, M=Memory cells, E=Effector cells

Looking at the chemical structure of skin haptens a chemist is immediately struck by the fact that most of them are Electrophiles (i.e. are electron deficient). For instance, Nickel ions lack two electrons and have empty orbitals. All proteins, in turn possess electron-rich groups, called Nucleophiles (for instance, amino groups, NH_2, thiol groups, SH). An "autologous" protein (= a protein which is normally present in the skin) will not provoke any adverse reaction as it is considered by the organism as "normal" or "self". If this protein is modified, in turn, it will become foreign (or antigenic) to the body and will trigger the cellular events abovementioned. This is probably what happens with cutaneous haptens. A hapten (HAPT) reacts with a protein to give a complete antigen or allergen:

HAPT + Protein $\longrightarrow$ Protein-HAPT (the antigen)

Well-known exemples of cutaneous antigens include chromium salts (and in particular bichromates, $Cr_2O_7{-}{-}$), well-known sensitizers in cement workers, methylenelactones from Compositae plants, quinones from tropical woods or from primrose (Primula obconica), etc. (figure 2.)

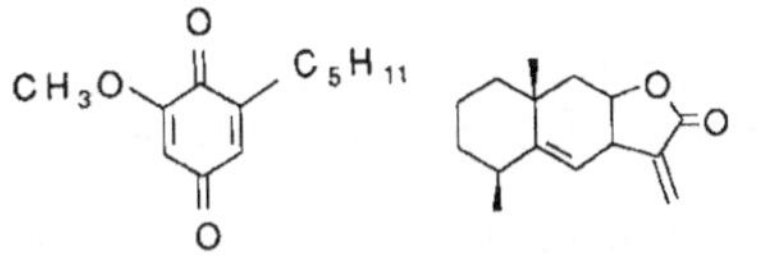

Primin (from Primula)   Alantolactone (from Inula)

Figure 2. Structures of Primin, a parabenzoquinone and of alantolactone, a sesquiterpene lactone.

Some sensitizers (a word used as synonymous for allergen or hapten) have no electrophilic properties. Take for instance eugenol, an antiseptic often used by dentists[3]: direct reaction with a nucleophilic group from a protein is not possible. Take also paraphenylenediamine or PPD, one of the most famous contact sensitizers an especially commun problem in hairdressers[3]. In order to become truly haptenic or electro-philic, they must undergo a chemical reaction (which can easily take place in vivo), for instance oxidation to ortho-quinones (figure 3.)

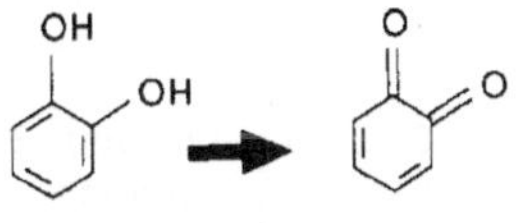

Figure 3. Oxidation of pyrocatechol into orthobenzoquinone.

Finally, there are some reactions of skin sensitizers which only occur in the presence of light. These compounds are called photosensitizers[2]. They involve the formation of one-electron species also called radicals[3]. For instance, the antiseptics used in soaps: tribromosalicylanilide, tetrachlorosalicylanilide, etc., and the neuroleptics such as promethazine, etc., probably react through this mechanism.

In the case of ACD due to nickel , the nickel can be reactive either in the metallic form (as nickel has empty d-orbitals) or as a salt, the oxidation into Ni^{++} salts supposedly occuring _in vivo_.

In a maximization study on 25 volunteers, it was demonstrated that 12 subjects could be sensitized to a nickel sulfate solution[6]. In the Kligman scale, nickel is therefore expected to be a moderate sensitizer. However, figures of ACD due to nickel are higher than predicted: for instance, as much as 10% of the female population of San Francisco is allergic to nickel (H.I. Maibach, personal communication, 1983). Why is nickel so allergenic ?

HOW CAN NICKEL GET INTO SOLUTION[7]

A) Metallic nickel must undergo oxidation to be transformed into ionic Ni^{++}. This corresponds to the reaction:

$$Ni \longrightarrow Ni^{++} + 2e^-$$

with an E° oxidation potential of + 0.25 volt. This indicates that nickel oxidation should be easy as nickel is a good reducing agent. What could be the oxidizing partner? Molecular oxygen (E° = 0.07 V), or proton (H^+, E° = + 0.25 V) are likely candidates.

A first oxidation step of Ni metal could lead to Ni^+ which would then disproportionate into Ni° (metal) and Ni^{++}.

The nickel ion Ni^{++} thus obtained can be complexed by organic ligands.

B) Metallic nickel itself can dissolve through complexation. Complexed nickel is then _easier_ to oxidize than free metal.

NICKEL AND COMPLEXATION

Nickel belongs to the first series of transition elements like zinc, iron, copper, manganese which all are present in substantial concentration in cells or biological fluids and which are not known for their sensitizing properties. What are the differences between nickel, a contact sensitizer and these elements ?

A) Ionic radius[8]

The ionic radius of Ni^{++} is inferior to the radii of all the other divalent ions: Fe^{++}, Zn^{++}, Cu^{++} and Mn^{++}. As a consequence, ligands may come closer to the ion and the complexation energy such as hydration, for instance, is larger for nickel.

B) Chemical affinity

Though the elements Fe, Co, Ni, Cu and Zn show the same affinity for oxygen, nitrogen or sulfur-based ligands (they are classified, in Pearson-Basolo HSAB scale between hard and soft elements), the geometry

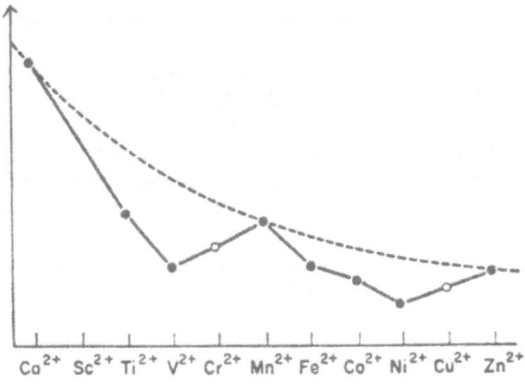

Figure 4. Ionic radii

of their complexes are different: Nickel prefers octahedral arrangement while all the others would rather form tetrahedral complexes.

A possible interpretation of the sensitizing properties of nickel: Since the energies involved in complexing nickel ions are stronger than the ones for Fe^{++}, Zn^{++}, Cu^{++} or MN^{++}, it is possible to imagine that Ni^{++} could displace the latter ion from complexes. Complete allergens are hapten-protein complexes. If nickel displaces the normal coordinated central ions in homologous proteins (which are therefore not antigenic), these proteins then become heterologous and therefore antigenic. The antigenic properties of a protein are probably related to its confor- mation (arrangement in space). First, as nickel "prefers" octahedral arrangement in complexes and since the other natural coordinated metal ions prefer other geometries, replacement of these by Ni^{++} ions will alter the conformation. Second, the small radius of Ni^{++} ions will change the metal-ligand bond lengths and again modify the conformation. Therefore the biological properties of proteins mostly due to their tertiary structure are changed.

CONCLUSION

In summary, the sensitizing capacity of nickel can be understood if one considers some of the fundamental properties of this metal :

- the facility of metallic nickel oxidation especially when comple- xed,
- the energy of complex formation which would allow nickel ion to replace other central metal ions in protein,
- change of the tertiary structure of a carrier protein due to intrinsic geometrical properties of nickel complexes.

REFERENCES

1. L. Polak, "Immunological aspects of contact sensitivity", S. Karger, Basel, 1980.

2. C. Benezra, G. Dupuis, "L"allergie de contact", La recherche, 1983, 14: 1062-1072.

3. J. Foussereau, C. Benezra, H.I. Maibach, "Occupational contact dermatitis - clinical and chemical aspects", Munksgaard, Copenhagen, 1982.

4. C. Benezra, G. Ducombs, Y. Sell, J. Foussereau, "Plant contact dermatitis", B.C. Decker, Toronto and C.V. Mosby, Saint Louis, 1985.

5. G. Dupuis, C. Benezra, "Allergic contact dermatitis to simple chemicals: a molecular approach", Marcel Dekker, New York, 1982.

6. A.M. Kligman, "The identification of contact allergens by human assay. II. The maximization test : aprocedure for screening and rating contact sensitizers", J. Invest. Derm., 1966, 47: 393-409.

7. J.E. Huheey, Inorganic Chemistry, Principles of structure and reactivy Harper and Row, New York, 1972.

8. F.A. Cotton and G. Wilkinson, Advanced inorganic chemistry John Wiley, New York, 1972.

TOTAL HIP ARTHROPLASTIES AND

ALLERGY TO METALS

Francisque Leynadier and Francois Langlais

Service de Médecine Interne et Centre d'Allergie
Hôpital Rothschild - 75571 Paris Cédex 12 - France
Service d'Orthopédie et Centre de Traumatologie
Hôpital de Pontchaillou - 35000 Rennes - France

It is well known that, in previously sensitized subjects, contact allergens or systemic exposure can trigger cutaneous and/or systemic hypersensitivity reactions. Hypersensitivity to nickel is common and its clinical aspects easy to recognize. Eczema to wrist watches, jean's buttons or necklaces are typical examples of nickel allergy (1, 11, 18, 19, 21).

Another classical example is eczema to earrings. Earpiercing seems to represent a major way to sensitize young women to nickel and probably, within the next years, young men who now pierce their earlobes! From an allergological point of view, the major problem for public health is poor quality jewelry and not orthopedic implants.

On the other hand, acute general eczema after exposure to nickel is less frequent and mainly due to foods contamination; e.g. an acute generalized eczema may appear within 6 to 24 hours in patients hypersensitive to nickel after eating meals cooked in a saucepan made of pure nickel. In some cases acute facial eczema may mimic, for the first hours, angioedema because of the volume of swelling.

Hypersensitivity to chromium is less frequent except occupational allergies (cement). The most common source is the leather of shoes.

In most cases, the diagnosis is easily confirmed by patch-tests which are simple to perform. A standard battery including about 30 allergens is attached to the back of the patient and the results are read 48 hours later : a positive patch test is defined by the presence of erythema and vesicles (19).

HYPERSENSITIVITY TO METALS IN ORTHOPEDIC SURGERY

1 - <u>Skin lesions induced by metals allergy</u>

The first case of hypersensitivity reaction to orthopedic implant was reported by Laugier and Foussereau (16). In their case report, the skin lesions appeared after the insertion of the implant and disappeared when the implant was removed. Some other cases have been then reported, but remain relatively uncommon (13, 20, 28).

In a prospective study conducted during 2 years at the Cleveland clinic, KUBBA et al. (13) found only 19 cases of cutaneous eruption of unknown origin after 800 joint replacements or internal fixations. However metal sensitivity was found and considered to be relevant in only 2 patients out of 19. They concluded that in an implant recipient in whom skin problems develop, the likelihood of these problems to be allergic in nature is low but higher :

- if the implant is of the static type,
- if there is a history of metal hypersensitivity
- if the skin eruption is located at the anatomic zone of the implant and
- if the eruption is eczematous and has developped and persisted late.

We never observed such cases in our allergic department. The only patient reffered for this diagnosis was a woman who developped skin lesions 6 to 8 months after total hips replacement. Unfortunately, a positive patch test to nickel was found. In fact these skin lesions were historical aspect of bromides induced by a drug (Bromoserine*) containing bromine. All skin lesions disappeared within a few months after discontinuing the drug.

This case demonstrates that skin lesions after orthopedic surgery are not necessarily metal-related, even if the patch tests are positive !

2 - <u>Aseptic loosening and allergy to metals</u>

Some years ago the main problem was the relationship between aseptic loosening of hip arthroplasties and sensitivity to metals (2, 3, 4, 5, 6, 8, 9, 12, 22, 23, 27).

Four questions should be asked :
2 - 1 : First question : what is the prevalence of nickel, chromium and cobalt sensitivity in the general population ?

On table I, are shown published results on nickel sensitivity in 3 different populations considered as representative of a "normal population" :
i) subjects programmed for an orthopedic implant and tested in prospective study;
ii) healthy volonteers;
iii) patients tested, for research purposes, in departments other than dermatological and allergological. Indeed, nickel sensitivity may be found in 13 % of patients tested in a dermatological clinic as reported by Baer et al. (1). Sensitivity to nickel is significantly more frequent in women (table I).

In study conducted in a gynecologic department, we patch tested 119 young women with nickel and copper. Positivity to nickel was found in about 6 % (7).

On the other hand, studies on chromium and cobalt sensitivity in a general population are less frequent. Deutmann et al. (5) found a sensitivity to chromium in 1,2 % of patients tested before total hip arthroplasty. Magnusson et al. (18) found 1,8 %. This sensitization is considered to occur more frequently in men than in women (11).

For cobalt, Deutmann et al. (5) found 3.5 % of sensitivity, Munro-Ashmann et al. (23) 1.8 % and Magnusson et al. (18) 1.8 %.

In summary, the incidence of sensitivity to nickel in a general population is less than 7 %, to cobalt less than 4 % and to chromium less than 2 %.

2 - 2 : Second question : what is the prevalence of allergy to metals in patients with total hip arthroplasties ?

TABLE I : NICKEL SENSITIVITY AMONG GENERAL POPULATION

AUTHOR	MEN	WOMEN	TOTAL
CARLSSON et al.	3.9 % (51)	12.4 % (89)	7.9 % (140)
DEUTMANN et al.	3.2 % (131)	4.9 % (142)	4 % (273)
DRY et al.	NT	6.7 % (119)	6.7 % (119)
PELTONEN et al.	0.8 % (478)	8 % (502)	4.5 % (980)
PRYTOWSKY et al.	0.9 % (460)	9 % (698)	5.8 % (1 158)
ROOKER er al.	3 % (33)	5.6 % (36)	4.4 % (69)
TOTAL	1.3 % (1 153)	8.3 % (1 586)	5.3 % (2 739)

NT : non tested

$X2 = 64$; $p < 0.001$

As shown table II, sensitivity to nickel is significantly higher in metal-metal prostheses, such as Mac-Kee Farrar's prostheses. Sensitivity to nickel in patients with metal-plastic prostheses gives approximatively the same percentage as in a control population (6.6 % versus 5.3 %). Three prospective studies (Carlsson et al., Deutmann et al. and Rooker et al.) have clearly demonstrated that sensitivity to nickel is not induced by the metal-plastic hip arthroplasty : the percentage of positive patients remain the same after metal-plastic arthroplasty.

TABLE II : METAL SENSITIVITY (Ni, Cr, and/or Co) AMONG PATIENTS WITH
JOINT REPLACEMENT ARTHROPLASTIES
(between parentheses : number of subjects)

AUTHOR	ARTHROPLASTY	
	METAL TO METAL	METAL TO PLASTIC
BENSON et al.	28 % (32)	2.6 % (39)
BROWN et al.	0 % (20)	NT
CARLSSON et al.	NT	9.7 % (134)
DEUTMANN et al.	9.5 % (21)	6 % (66)
DRY et al.	20 % (50)	8.7 % (23)
ELVES et al.	36.6 % (41)	NT
EVANS et al.	23.7 % (38)	NT
JONES et al.	16.2 % (37)	NT
ROOKER et al.	NT	1.8 % (54)
TOTAL	21.9 % (239)*	6.6 % (316)

NT : non tested

* : X2 = 15.3; p < 0.001

2 - 3 : Third question : does metal sensitization appear after all metal-metal arthroplasties ?

Table III shows that only aseptic loosening is significantly associated with metal sensitization. When the orthopedic results are good, the percentage of positivity remains approximately the same as in a control population (7.4 % versus 5.3 %). Moreover, in patients with aseptic loosening (table IV), a very high percentage of sensitivity to cobalt is observed : 30 % versus about 3 % in a control population or in patients with a good orthopedic result. Positive patch tests for nickel cobalt and chromium together, indicate a loosening in patients with metal-metal hip arthroplasties. Conversely even if the positive patch tests are clearly related to the prosthese, one finds no skin lesions in these patients.

In summary, only aseptic loosening metal-metal arthroplasties are significantly associated with metal sensitivity.

TABLE III : METAL SENSITIVITY (Ni, Cr, and/or Co) AMONG PATIENTS WITH
METAL ARTHROPLASTY ACCORDING TO THEIR ORTHOPEDIC RESULTS
(SEPSIS EXCLUDED)

AUTHOR	ORTHOPEDIC RESULT	
	GOOD	ASEPTIC LOOSENING
BENSON et al.	11.6 % (43)	66.6 % (15)
BROWN et al.	NT	0 % (20)
DEUTMANN et al.	0 % (6)	14 % (14)
DRY et al.	9 % (33)	53.8 % (13)
ELVES et al.	15 % (27)	73.7 % (19)
EVANS et al.	0 % (24)	64.3 % (14)
JONES et al.	0 % (30)	86 % (7)
MUNRO_ASHMANN et al.	NT	46 % (35)
TOTAL	7.4 % (163) *	46 % (137)

NT : non tested

* : X2 = 59.3; p < 0.001

2 - 4 : Fourth question : what is the relationship between aseptic
loosening and metal sensitization ?

In the first papers addressing this problem (2, 8, 9), in
1974, it was asserted that sensitization probably induces loosening.
By now, the reverse is generally accepted as true. Langlais et al. have
addressed this problem (14, 15, 17, 25, 26) : in a follow-up of more
than 1500 total hip prostheses made of stellite they have found atypical
loosening in 30 cases, all of them requiring further surgery. The
loosening was aseptic and non-mechanical but between the cement and
the bone layed a thick weak "pannus" made up of macrophages containing
metallic particles (25).

Pathological examination of this pannus showed a granuloma
made of mononuclear macrophages with engulfed metal particles. This aspect
does not relate to pathological lesions of eczema (Gell and Coombs type IV).

Bone resorption appears to be induced by the granuloma. The
amount of periprosthetic metallic debris is 100 times more important
in aseptic loosening of metal-metal arthroplasties than in metal-plastic
and a least tenfold more abundant than in septic loosening of metal-
metal prostheses (25).

TABLE IV : SENSITIVITY TO COBALT AMONG PATIENTS WITH ASEPTIC LOOSENING
OF METAL TO METAL HIP ARTHROPLASTY

AUTHOR	% POSITIVE	NUMBER OF PATIENTS
BROWN et al.	0	20
DEUTMANN et al.	7	14
DRY et al.	23	13
EVANS et al.	57	14
JONES et al.	86	7
MUNRO-ASHMANN et al.	37	35
TOTAL	30	103

In metal-metal arthroplasties, the occurence of loosening may be summarized as follows :

- release of metallic particles which are ingested by macrophages. The particles size is an important factor in this macrophagia.
- release by these cells of toxic products such as proteolytic enzymes of lysosomal origin with bone resorption.
- sensitization to metals is then induced : some lymphocytes and even lymphoid formations are observed at the periphery of the granuloma.

However, this hypothesis is probably too simple. Some recent works on bone resorption favour an indirect role of the histiocytic granuloma, the major element being the osteoclasts (10).

CONCLUSION

Metal sensitization is frequent, in a general population, mostly for nickel but allergic reactions to metal remain exceptional in orthopedic. surgery. Arthroplasties cannot induce sensitization, except with metal-metal protheses which are loosening by an aseptic mechanism. In these latter cases sensitization appears to be the consequence and not the cause of this loosening probably due to some "toxic" effect to metal particles on bone cells and/or on macrophages.

These problems due to metal sensitization are less of concern since metal-metal hip arthroplasties are now seldom used. Metal-plastic prostheses are a prefered choice. However some aseptic loosening are now described in total knee arthroplasties with the same mechanisms.

REFERENCES

1   BAER R.L., RAMSEY D.L., BIONDI E. - The most common contact allergies 1968 - 1970, JAMA 1973, 108 : 74 - 76.

2   BENSON M.K.D., GOODWIN P.G., BROSTOFF J. : Metal sensitivity in patients with joint replacement arthroplasties. Br. Med. J. 1974, 4 : 374 - 375.

3   BROWN G.C., LOCKSHIN M.D., SALVATI E.A., BULLOUGH P.G. - Sensitivity to metal as a possible cause of sterile loosening after cobalt chromium total hip-replacement arthroplasty. - J. Bone Joint Surg. 1977, 59 A : 164 - 168.

4   CARLSSON A.S., MAGNUSSON B., MOLLIER H. - Metal sensitivity in patients with metal-to-plastic total hip arthroplasties. Acta orthop. Scand. 1980, 51 : 57 - 62.

5   DEUTMANN R., MULDER T.J., BRIAN R., NATER J.P. : Metal sensitivity before and after total hip arthrosplasty. J. Bone Joint Surg. 1977, 59 A, : 862 - 865.

6   DRY J., LEYNADIER F., BAUX S., KINDERMANS A. : Réaction immuno-allergique au matériel métallique de prothèse : influence du descellement aseptique. - Sem. Hôp. Paris 1977, 53 : 2449 - 2450.

7   DRY J., LEYNADIER F., BERNANI A., PIQUET P., SALAT J. : Intra-uterine copper contraceptive devices and allergy to copper and nickel. Ann. Allergy., 1978, 41 : 194.

8   ELVES M.W., WILSON J.N., SCALES J.T., KEMP H.B.S. : Incidence of metal sensitivity in patients with total joint replacements. Br. Med. J., 1975, 4 : 376 - 378.

9   EVANS E.M., FREEMAN M.U.R., MILLER A.J., VERNON-ROBERTS B. : Metal sensitivity as a cause of bone necrosis and loosening of the prothesis in total joint replacement. J. Bone. Joint. Surg. 1974, 56 B : 620 - 642.

10 FOREST M., COURPIED J.P., LEFLOCH P., CARLIOZ A., ABELANET R., POSTEL M. : La hanche opérée : réactions tissulaires locales. Ann. Pathol., 1985, 5 : 3 - 18.

11 HAMMERSHOY O. : Standard patch tests results in 3225 consecutive Danish patients from 1973 to 1977. Contact Dermatitis, 1980, 6 : 263 - 268.

12 JONES D.A., LUCAS H.K., O'DRISCOLL M., PRICE C.H.G., WIBERLEY B. Cobalt toxicity after Mc Kee hip arthroplasty. J. Bone Joint Surg. 1975, 117 : 289 - 296.

13 KUBBA R., TAYLOR J.S., MARKS K.E. : Cutaneous complications of orthopedic implants. A two year prospective study. Arch. Dermatol. 1981, 117 : 554 - 560.

14 LANGLAIS F., POSTEL M., LE CHARPENTIER Y., LE MAIGRE G. : - Hip protheses and stellite allergy. Ann. Orthop. Ouest, 1973, 10 : 59-65.

15 LANGLAIS F., POSTEL M., BERRY J.P., LE CHARPENTIER Y., WELL B.J.: L'intolérance aux débris d'usure des prothèses. - Int. Orthopedics, 1980, 4 : 145 - 153.

16 LAUGIER R., FOUSSEREAU J. : Les dermites allergiques à distance provoquées par le matériel d'ostéosynthèse - Gaz. Méd. Fr. 1966, 73 : 3409 - 3418.

17 LE CHARPENTIER Y., CARLIOZ A., LE MAIGRE G., ABELANET R., LANGLAIS F. POSTEL M., BERRY J.P., GALLE P. : Déterminisme possible du descellement aseptique des prothèses totales de hanche métal-métal - Nouv. Presse Med., 1978, 7 : 4144 - 4145.

18 MAGNUSSON B., MOLLER H. : Contact allergy without skin disease. Acta. Dermatovener, 1979, 59 (suppl. 85) : 113 - 115.

19 MALTEN K.E., NATER J.P., VAN KETEL W.G. : Patch testing guildelines. Nijmegem, The Netherlands, Dekker and Van De Vegt 1976.

20 Mc KENZIE A.W., AITKEN C.V.E., RIDSDILLSMITH R. : Urticaria after insertion of Smith - Peterson vitallium nail.Br. Med. J. 1967, 4 : 36.

21 MENNE T. : Relationship between cobalt and nickel sensitization in females. Contact Dermatitis, 1980, 6 : 337 - 340.

22 MERRIT K., BROWN S.A. : Metal sensitivity reactions to orthopedic implants. Int. J. Dermatol. 1981, 20 : 89 - 94.

23 MUNRO - ASHMAN D., MILLER A.J. : Rejection of metal to metal prothesis and skin sensitivity to cobalt. - Contact Dermatitis, 1976, 2 : 65 - 67.

24 PELTONEN L. : Nickel sensitivity in the general population. - Contact Dermatitis, 1979, 5 : 27 - 32.

25 POSTEL M., LANGLAIS F. : L'usure des prothèses totales de hanche en stellite. - Mesures "in vivo" et conséquences cliniques. - Rev. Chir. Orthop., 1977, 63 (suppl. II) : 84 - 94.

26 POSTEL M., LANGLAIS F. : Problèmes diagnostiques et thérapeutiques des descellements de prothèses de hanche métal sur métal. -rev. Chir. Orthop., 1978, 64 (suppl. II) : 22 - 25.

27 ROOKER G.D., WILKINSON J.D. : Metal sensitivity in patients undergoing hip replacement. A prospective study. J. Bone Joint Surg., 1980, 62 B : 502 - 505.

28 TISLEY D.A., ROTSTEIN H. : Sensitivity caused by internal exposure to nickel, chrome and cobalt. Contact Dermatitis, 1980, 6 : 175 - 178.

NICKEL, CHROMIUM, COBALT DENTAL ALLOYS AND ALLERGIC REACTIONS

AN OVERVIEW

Hartmut F. Hildebrand, Christian Veron, and Pierre Martin

Institut de Medecine du Travail, Faculte de Medecine
1, Place de Verdun, F-59045, Lille-Cedex, France

INTRODUCTION

Dental alloys containing nickel, cobalt and/or chromium are widely used for dental restorations. Ni-Cr alloys contain generally 69-81% nickel, whereas Co-Cr alloys contain mostly less than 1% of nickel. The main components of these alloys are Co (60-65%), Cr (27-30%) and Mo (5-6%). Another group of non-precious dental alloys are stainless steels with ca. 18%Cr and 8% Ni.

Although considered as inoxidizable, these alloys may corrode in biological milieu. Several authors have described the corrosion of orthopaedic implants (17, 37) or of dental restorations (6, 11, 30, 32, 42). It is possible that Ni, Co and Cr are released from prostheses and dental restorations, and thus induce unwanted reactions of toxic or allergic origin (19, 28).

The cytotoxicity of nickel salts has been demonstrated in the last few years by several investigations on renal epithelial cells of the african Green Monkey (Cercopithecus aethiops) (13), on human gingival cells (22) and on human pulmonary epithelial cells (7). Chromates have an even higher toxic effect, and soluble cobalt and nickel salts have similar cytotoxicity (13). A certain cytotoxicity has also been demonstrated for pure metals and ferromagnetic Pd-Co dental alloys which also contain some Ni and Cr (23).

Concerning sensibilization by these metals, nickel takes the first place for all metal-induced contact dermatites in humans and it induces more allergic reactions than all other metals together (35). Chromium takes the second place. Cobalt, the third metallic allergen causes mostly associated sensibilization with chromium and especially with nickel.

The present paper deals with possible correlations between allergic reactions and the bearing of a dental prosthesis or restoration made from one or more of the above cited metals. These reactions may appear either locally as stomatitis or distantly in form of general or local contact dermatitis.

CONTACT DERMATITIS

Von Pirquet defined in 1906 allergy as a pronounced reaction of an individuum to a substance when the latter is reintroduced into the organism. This definition has been modified with the progress in immunological research. The sensitizing substance is called antigen. The antigen is a molecule or a cell which, once introduced into the organism, induces the formation of antibodies or specific defence cells. These antigens interact with the antibodies or defence cells. Coombs and Gell (8) defined four different classes of allergy types.

The allergic reactions induced by stainless steel or Ni, Cr and/or Co containing alloys are called contact dermatitis and belong to the type IV of the above cited classification. For this form of allergy the allergen or hapten is a substance with a low allergenic power (9). It is at first strongly bound to certain endogenous proteins in order to form a strong antigenic macromolecule. After a here very simplified mechanism the new-formed antigen is captured by macrophages and memorized by certain T-lymphocytes (3).

After a new contact the formerly sensitized lymphocytes produce different substances, e.g. lymphokine, which they liberate into the organism and such provoke certain tissue reactions. The hypersensible effect in contact dermatitis appears generally on the skin in form of eczema. Mucosal reactions and especially stomatitis are possible.

Allergies of type IV thus appear after celluler mediation and without production of antibodies. Different criteria have been established in order to prove contact dermatitis (14, 34). For our bibliographic and personal studies we consider especially the three following criteria:
1- A clinical feature in form of eczema, redness, ulceration, etc.
2- Healing after removal of the allergen.
3- Positive skin reaction after epicutaneous test.

TEST METHODS

The most usual test is the epicutaneous test or "patch-test" or even better expressed: the "Finn Chamber Test". This test consist in the application of the allergen on the skin. The allergen is diluted in water or petrolatum if the allergen is a salt or a powder. Pure metals, alloys, polymers or mixtures of different substances, e.g. tooth paste, can be applicated directly. The allergen is put into small aluminium-chambers or "Finn Chambers" which are directly applicated onto the skin. This method allows to keep the allergen in a direct and constant contact with the skin (33). The interpretation of the test is practised after 48 and 72 hours. Some very sensible persons may have a first reaction already within some hours.

Laboratories for dermatologic and allergic examinations use standard test-batteries of the most current substances. In all these batteries are contained the different metal salts. In table I, we present the salt concentrations of Ni, Co and Cr recommended by different research groups. The table shows that the concentrations of the three metal salts vary very little from one group to another. Indeed, the difference of concentration of nickelsulfate (2.5 or 5%) has no influence on the frequency of revealable allergy. In some laboratories nickelsulfate is replaced by nickelchloride. Also this change does not influence the reliability of the tests.

TABLE I. Allergen concentrations (%) of standardized Test-
Batteries (In Petrolatum)

BATTERY	$NiSO_4$	$K_2Cr_2O_7$	$Cr(SO_4)_3$*	$CoCl_2$	$CoSO_4$*
ERG	5	0.5	2	1-2	2
NACDG	2.5	0.5	-	2	-
ICDRG	5	O.5	-	1	-

ERG: European Research Group

NACDG: North America Contact Dermatitis Group

ICDRG: International Contact Dermatitis Group

*non included in standardized batteries

TABLE II. Statistical assessment of sensitization to Ni, Cr,
and Co in the consultant population (%)

ALLERGEN	REPORTED CASES	♂%	♀%	TOTAL %
Nickel	37,849	3.1	12.9	9.6
Chromium	36,914	12.7	7.1	9.3
Cobalt	31,330	4.7	5.3	6.0

For both, epicutaneous and epimucosal tests, the same
rules have to be respected. The application must stay on the
oral mucosa in continuous contact during 24 hours.

SENSITIZATION TO NICKEL, COBALT AND CHROMIUM

A large number of statistical and epidemiological investigations about contact dermatitis were performed in order to
establish the frequency of allergies. In tables II and III we
summarize the datae of a previous german paper (20), where we
reviewed 21 statistical studies about allergy in the consultant
population (table II) and 5 statistical studies of the general
population (table III).

The consultant population is constituted by patients with
a dermatological problem. The general population is constituted of volunteers without obvious allergic reactions who under-
went epicutaneous tests.

The primary cause of contact dermatitis to these metals
is the occupational exposure in electrogalvanic, metallurgic
and chemical industries, masonry, textile and paper manufac-
tories, agriculture, hospital employment, etc.

The second cause of sensitization is environmental non
occupational exposure: paste jewellery, kitchen utensils,
chromate tanned leather or metallic vestimentary requisites,
domestic cleaners and detergents and a lot of other possibilities. In addition to the external source, we take up a considerable part of metal ions by our alimentation. Foodstuffs
like beans, peas, strawberries, rice, bread, meat and fish contain sometimes more than 1mg of nickel per kg of ware (5).

ALLERGY TO DENTAL RESTORATIONS

Allergic reactions relative to dental interventions may
have different origins. Generally three groups of allergens
have to be considered: medicaments, synthetic materials and
alloys.

Medicamentous allergies may arrise during or after dental
treaetment. The most current substances to induce a medicinal
rush are local anaesthetics, sulfonamides, certain antibiotics
especially penicillin derivatives, and finally analgesics (15).

A large number of sensitizations to *synthetic dental
restorations* have been descibed. The most frequent allergic
reactions are due to sulfonesters and catalysers contained in
matrix materials and provisional restorations (15, 16). Base
materials used for final prostheses also can induce sensitizations: acrylderivatives (44, 46), hercolite, vulcanite, stabi-
lizers such as hydrochinone (15), accelerators such as benzoyl-
peroxide (44), and colour pigments.

*1. Dental base alloys of Ni-Cr and Co-Cr:
   Corrosion and dissolution*

Several observations about the behaviour of prostheses of
non-precious alloys in biological milieu have been described
(39, 40, 47). Bergman et al. (4) implanted subcutaneously four
different alloys with high Ni-content from 64 to 74%. Analyses
of tissues and organs after 5 months of implantation revealed
a high Ni-concentration in kidneys and lung.

TABLE III. Statistical assessment of sensitization to Ni
and Cr in the general population (%).

ALLERGEN	TESTED PERSONS	♂ %	♀ %	TOTAL %
Nickel	3,207	1.5	8.9	4.2
Chromium	822	2.0	1.5	1.7

TABLE IV. Reported cases of allergy to Ni, Cr and/or Co in
relation to dental prostheses.

Number of cases	149
Sex-ratio ♀/♂	2/1
Local symptoms	111
General symptoms	32
Allergic background	28
Healing after withdrawal of prostheses	127

Similar investigations have been performed on humans in order to assess the release of Co and Cr in the oral cavity (11, 42). Already some days after the placement of the the prosthetic material, the authors measured a significant increase and accumulation of Co and Cr in saliva and tongue tissue.

Another not less important investigation was performed by Newman et al. (30) about corrosion of Biobond C&B and Unibond with Ni/Cr contents of 79%/13% and 64%/22% respectively. Although both alloys have an excellent resistance to corrosion in artificial saliva, a significant Ni-leakage was evidenced in autoclaved human saliva.

Temperature and pH have a great influence on the dissolution of stainless steel (31). The release of Cr-ions is increased in an acid medium at elevated temperature with reference to water. This fact is important to be mentioned since it concerns directly the relation between dental restorations and nutritional habitudes.

All the above cited studies demonstrate clearly the existence of corrosion of so-called inoxidizable alloys in a biological milieu. Ni, Cr and Co are released into the oral cavity and are recovered in more or less important concentrations in proximate tissues or in various organs (6, 18, 27, 43).

2. Clinical cases

The metal ions released into the organism are now available to induce allergic reactions. Already many years ago, Foussereau and Langier (12) have described allergic eczemas from metallic foreign bodies of different origin: dental prostheses, orthopaedic implants, war particles. These materials contain usually Ni, Cr and Co. Nickel is omnipresent in all non-precious dental alloys, and even "nickel-free" Co-Cr alloys contain enough Ni to induce allergic reactions in Ni-sensitized persons (38). These Ni-amounts may be so low that even an examination with Dimethylglyoxim (DMGO) gives a negative answer. Most authors now agree on this point and reject since several years the criterium of the positive DMGO-test in order to know whether an alloy releases Ni and thus should be considered as sensitizing or not sensitizing.

In a previous extensive bibliographic study (20) we reviewed the clinical cases from 1934 up today. These cases are summarized in table IV. In total, we have recorded 149 cases responding to the three above defined criteria. All patients have shown a positive test reaction to at least one of the three metals or the applicated alloy. 127 patients healed completely and rapidly after withdrawal of prostheses or restorations. Clinical local symptoms in the oral cavity or peribuccal region were observed in 111 patients. Distant localized or generalized symptoms appeared in 32 patients, and 17 patients showed symptoms in the oral cacity and anywhere else on the body.

It should be underlined, that antecedent signs of allergy to the three metals were observed only in 28 patients, i.e. 20% of all cases. Indeed, the statement of antecedents was no longer possible after appearance of clinical symptoms in the presence of a dental restoration. 80% of negative or uncertain answers on this question remains an important rate. Thus we must consent that a High number of the reported cases were sensitized by the dental restoration. This supposition is enhanced by the high number of patients (111 cases) showing clinical symptoms in the mouth.

The latter remark is contradictory to the general concept that allergic reactions on oral mucosa are extremely seldom (21). Indeed, these reactions may be of mechanical, electro-galvanic or microbial origins, so that epicutaneous and/or epimucosal tests and clinical observations by allergologists must be performed to confirm any allergic manifestation.

3. Healing probability of metal induced contact dermatitis

As shown in table IV, 127 patients healed after removal of their dental restoration. This healing is very delusive. The elimination of the allergic agent is in most cases sufficient to heal the clinical symptoms. The sensitization, however, may persist over many years. Several research groups (11, 17, 45) have followed their patients up to 4 to 7 years by one or more control tests. All have stated that allergy persists at a 76 to 93% rate. In some patients allergy persisted even more than 17 years (10). The authors conclude that the cessation of nickel or chromium exposure has no evident effect on the duration of sensitization.

CONCLUSIONAL REMARKS

The present review deals with incompatibility to dental restorations only from the allergic point of view. Other pheno-mena may induce diseases of similar severity: toxicity, elec-trogalvanism, mechanical and microbial irritation, etc.These aspects are volontarily dropped since they are not in direct relation with allergy. Some remarks shall be made, however, on this subjects.

One of the essential causes of oral diseases is electro-galvanism. Since MacDonald (25) and Moutier (29) so many investigations about this subject were published that it is impossible to cite them in the present work. We should not forget, however, that metal ions released by electrogalvanic effects may be responsable for sensitization and allergic reactions.

Another important reason of dizeases is mechanical irri-tation which disappears - like allergic reactions - after removal of the dental restoration. Since such irritations are clinically quite similar to allergic stomatitis, epicu-taneous or epimucosal tests are indispensable for the dia-gnosis of sensitization.

The physiological origin and the frequency of Ni, Co, Cr contact dermatitis is well known in general and consultant population. In contrast, no statistical or epidemiological investigations were yet carried out to elucidate existing correlation between non-precious dental alloys and allergic reactions. The present review of described clinical cases just may help to understand and to seize the problem but cannot have any further statistical value.

The lack of any statistical study is surprising since most authors recognize the problem and discuss it in their publications. They generally indicate several cases observed in their office without describing them precisely (1, 12, 15, 24, 28, 32, 37). Occur sensitization to dental restorations so frequently, that publication and precise description of clinical cases is no longer needed? Thus Seymour (38) would say right by suggesting that about 35% of all Ni-allergic persons should search the origin of their diseases in dental restorations of non-precious alloys.

Some authors reject the existence of allergic reactions to dental alloys (2, 41), but the problem remains the same. A reaction must not appear immediately but only some months or some years after the first contact with the allergen (26).

Furthermore, we should not forget that dental restoration is made for a long duration. It may be replaced easily by another and better one. This is not so easy for implant material used in stomatology, maxillofacial surgery and orthopaedic surgery.

REFERENCES

1- T.Axell, B.Bjorkner, S.Fregert & B.Niklasson
Standard patch test series for screening of contact allergy to dental materials.
Contact Dermatitis 1983, 9, 82-84

2- H.Barriere, H.L.Boiteau, C.Geraut & C.Metayer
Allergie aux détergents et allergie au nickel
Ann. Dermatol. Vénérol. 1979, 106, 33-37

3- C.Bénézra & M.F.Leroy
Fundamental aspects of contact sensitivity to nickel.
In this issue

4- M.Bergman, B.Bergman & R.Söremark
Tissue accumulation of nickel released due to electrochemical corrosion of non-precious dental casting alloys.
J. oral Rehabil. 1980, 7, 325-330

5- R.Brun
Nickel dans les aliments et eczéma de contact.
Dermatologica 1979, 159, 365-370

6- D.Brune, A.Kiaerheim, A.Hensten-Pettersen & L.Marion
Corrosion of dental alloys studied by implantation and nuclear tracer technique.
Acta Odontol. Scand. 1983, 41, 129-134

7- M.Collyn d'Hooghe, J.P.Kerckaert, M.C.Herlant-Peers & H.F.Hildebrand
Biological effects of nickel-compounds on human embryonic pulmonary cells in culture
In: "4th Internat. Symposium on Trace Elements" M.Anke, W.Baumann, H.Bräunlich & C.Brückner edts. Jena 1983,45-52

8- R.R.Coombs & P.G.H.Gell
Classification of allergic reactions responsible for clinical hypersensitivity and disease.
In: "Clinical aspects of immunology" P.G.H.Gell, R.R.Coombs & P.J.Lachmann edts. Blackwell Scientific Publications, Oxford 1975, p.761

9- P.David, C.Ginocchio & P.Klein
Immunologie bucco-dentaire.
Procodif. Edt-Sèvres, 1983

10-A.A.Fischer & A.Shapiro
Allergic eczematous contact dermatitis due to metallic nickel.
J.Am.Med.Assoc. 1956, 161, 717-721

11-J.Fontes de Melo, N.R.Gierdet & E.S.Erichsen
Metal release from cobalt-chromium partial dentures in the mouth.
Acta Odontol. Scand. 1983,,41, 71-74

12-J.Foussereau & P.Laugier
Allergic eczemas from metallic foreign bodies.
Transactions St John's Hosp. Dermatol. Soc. 1966,52,220-225

13-M.E.Frazier & T.K.Andrews
   In vitro clonal growth assay for evaluating toxicity of
   metal salts.
   In: "Trace Metals in Health and Disease", N.Karasch edt.
   Raven Press, New York, 1979, p 71-81
14-S.Fregert
   Manual of Contact Dermatitis.
   Munksgaard edt. Copenhaguen, 1974
15-F.Gasser
   Allergische Patientenreaktionen auf zahnärztliche Behand-
   lungen und Materialien.
   Die Quintessenz 1983, 5, 1035-1044
16-W.E.Glendenning
   Allergy to cobalt in a metal denture as cause of hand
   dermatitis.
   Contact Dermatitis Newsletter 1971, 10, 225-226
17-H.F.Hildebrand, M.C.Herlant-Peers, M.Stoeppler, U.Bagschik,
   & J.Roggatz
   Les effets lointains d'un enclouage centro-medullaire en
   acier chrome-nickel.
   In: Actes des XVIIe Journées Nat. de Méd. du Travail, edts:
   J.M.Haguenoer & D.Furon, Technique et Documentation, Paris,
   1982, p. 124-126
18-H.F.Hildebrand, C.Veron, J.P.Fernandez, M.C.Herlant-Peers,
   J.P.Berry & P.Martin
   Biological effects of metals used in restorative dentistry
   with special reference to Ni, Cr and Co.
   In: "4th Inten. Symp. on Trace Elements" edts: M.Anke,
   W.Baumann, H.Bräunlich & C.Brückner, Jena 1983, p. 81-88
19-H.F.Hildebrand, C.Veron, M.C.Herlant-Peers, J.P.Fernandez,
   & J.P.Kerckaert
   Les conséquences biologiques de la présence d'ions métal-
   liques dans la cavité buccale.
   Rev. d'Odontol. Stomatol. 1984, 13, 41-56
20-H.F.Hildebrand & C.Veron
   Zahnersatz aus Nichtedelmetall-Legierungen und Allergien.
   Edt: Fachvereinigung Edelmetalle e.V., Dûsseldorf 1985
21-F.Jakobs
   Über Epikutan- und Schleimhauttestungen.
   Derm. Wochenschrift 1953, 127, 446-450
22-N.Jacobsen
   Epithelial-like cells in culture derived from human gingiva
   Response to nickel.
   Scand. J. dent. Res. 1977, 85, 567-574
23-Y.Kawata, M.Shiota, H.Tsutsui, Y.Yoshida, H.Sasaki &
   Y.Kinouchi
   Cytotoxicity of Pd-Co dental casting ferromagnetic alloys.
   J; Dent. Res. 1981, 60, 1403-1409
24-E.Lipman Cohen
   Aspects of Allergy.
   Brit. dent. J. 1955, 99, 185-191
25-W.J.MacDonald
   Chemical and electrogalvanic burns of tongue.
   New England J. Med. 1934, 211, 585-589
26-F.N.Marzulli & H.I.Maibach
   Contact allergy: predictive testing in man.
   Contact Dermatitis 1976, 2, 1-17
27-D.C.Mears
   Electron-probe microanalysis of tissue and cells from
   implant areas.
   J. Bone Joint Surgery 1966, 48B, 567-576

28-J.Moffa & Council on Dental Materials, Instruments and
   Equipment
   Biological effects of nickel containing dental alloys
   J. Am. Dent. Assoc. 1982, 104, 501-506
29-F.Moutier
   Polymétallisme dentaire et dysesthésies linguales
   (Stomatite électro-galvanique)
   Soc. Gastro-Enterol. Paris, 1944, Séance du 8 mai,133-135
30-S.Newman, R.T.Chamberlain & L.J.Nunez.
   Nickel solubility from nickel-chromium dental casting alloys.
   J. biomed. Mater. Res. 1981, 15, 615-617
31-E.G.Offenbacher & F.X.Pi-Sunyer
   Temperature and pH efffects on the release of chromium from
   stainless steel into water and fruit juice.
   J. Agricult. Food Chem. 1983, 31, 89-92
32-H.Y.Park & T.R.Shearer
   In vitro release of nickel and chromium from simulated
   orthodontic appliances.
   Am. J. Orthod. 1983, 84, 156-159
33-V.Pirilä
   Chamber test versus patch test for epicutaneous testing.
   Contact Dermatitis 1975, 1, 48-52
34-N.H.Rickles
   Allergy in surface lesions of the oral mucosa.
   Oral Surg. 1972, 33, 744-754
35-E.J.Rudner, W.C.Clendenning, E.Epstein, A.A.Fisher, O.F.
   Jilson, W.P.Jordan, N.Kanof, W.Larsen, H.Maibach, J.C.
   Mitchell, S.E.O'Quinn, W.F.Schorr &M.B.Sulzberger.
   Epidemiology of contact dermatitis in North America.
   Arch. Dermatol. 1973, 108, 537-540
36-M.H.Samitz & S.A.Katz
   Study of chemical reactions between chromium and skin.
   J. Invest. Derm. 1964, 43, 35-43
37-M.H.Samitz & S.A.Katz
   Nickel dermatitis hazards from prostheses. In vivo and in
   vitro solubilization studies.
   Brit. J. Dermatol. 1975, 92, 287-290
38-M.S.Seymour
   A warning on nickel alloys.
   Brit. dent. J. 1981, 150, 84
39-R.Söremark
   Some biological effects caused by prosthetic materials.
   Swed. dent. J. 1979, 3, 1-7
40-R.Söremark, M.Diab & K.Arvidson.
   Autoradiographic study of distribution pattern of materials
   which occur as corrosion products from dental restorations.
   Scand. J. dent. Res. 1979, 87, 450-458
41-E.Spiechowicz, P.O.Glantz, T.Axell & Ch.Mielewski
   Oral exposure to a nickel-containing dental alloy of persons
   with hypersensitive skin reactions to nickel.
   Contact Dermatitis 1984, 10, 206-211
42-T.Stenberg
   Release of cobalt from cobalt chromium alloy constructions
   in the oral cavity of man.
   Scand J. dent.Res. 1982, 90, 472-479
43-T.Stenberg & B.Bergman
   Release and uptake of cobalt from cobalt chromium alloy
   implants
   Acta Odontol. Scand. 1983, 41, 149-154

44-A.J.W.Turrell
   Allergy to denture-base materials. Fallacy or reality.
   Brit. dent. J. 1966, 129, 415-422
45-T. van Jost & J.J.E. van Everdingen
   Sensitization to cobalt associated with nickel allergy:
   Clinical and statistical studies.
   Acta Dermatovener. 1982, 62, 525-529
46- B.G.Wakkers-Garritsen, L.H.Timmer & J.P.Nater
   Etiological factors in the denture sore mouth syndrome:
   An investigation of 24 patients.
   Contact Dermatitis 1975, 1, 337-343
47-R.D.Woody E.F.Huget & J.E.Horton
   Apparent cytotoxicity of base metal casting alloys.
   J. dent. Res. 1976, 56, 739-743

ACKNOWLEDGEMENT

   This work was supported by a grant of the French association PROTHESOR.

ALLERGIC CONTACT STOMATITIS FROM Ni-ALLOYS

A HISTOLOGICAL, IMMUNOHISTOLOGICAL AND ELECTROCHEMICAL RELATION

L.A.J. Van Loon[1], P. W. Van Elsas[1], P.P.E. Duysters[1],
J. D. Bos[2], and C. L. Davidson[1]

[1]Departments of Clinical Materials Science and of Masticatory Function,
Academic Centre for Dentistry, University of Amsterdam, The Netherlands
[2]Department of Dermatology, Academic Medical Centre
University of Amsterdam, The Netherlands

INTRODUCTION

Various metals, in diverse alloys and combinations are used in dentistry. From several of these metals, the capacity to cause allergic contact dermatitis or stomatitis is well documented. It has to be stressed that the diagnosis of an allergic contact reaction in the mouth is usually more difficult to establish than an allergic skin reaction.

Apprehension of the dental alloy's biocompatible character is of considerable importance in view of the many inconveniences (including financial ones) involved in possibly inevitable removals of cemented dental casts. Anticipation to three major questions is required, for a proper diagnosis whether the patient's complaints can apparently be attributed to a true metal contact allergy. Firstly : which metal-ions-in-particular can evoke an allergic contact reaction in that particular patient ? Secondly : are these ions present in the alloys used ? Thirdly : is there any reason in a particular patient for excessive corrosion by which enough harmful ions can be released ? Subsequently, the desired but suspect dental alloys will have to be tested in the mouth for possible allergenic effects.

Allergic contact stomatitis mostly belongs to the delayed type of (or cellular) hypersensitivity in which sensitized lymphocytes, the so called Thymus dependent cells (T-cells), play an important role. Characteristic features of allergic contact reactions are spongiosis, this is a sponge-like swelling of the epidermis of the skin or mucosa. Moreover, at the place

213

of the allergic reaction, the skin or mucosa shows vasodilitation, edema and perivascular lymphocytes infiltration. This lymphocyte infiltration penetrates in the epidermis. Through proliferation of these T-lymphocytes and the released lymphokines, a chain reaction is created in already sensitized persons, because of which a cell infiltration arises in the dermis and epidermis of, in particular, T-lymphocytes.

Besides these lymphocytes, the quantity of which reaches its maximum mostly after 18-48 hrs., the Langerhans cells play an important role as antigen-presenting cells. The study of the elicitation-phase of the allergic contact reactions may be done by means of an immunopheno-characterization of inflammatory cells in situ with the aid of monoclonal antibodies. The diagnosis of an allergic contact reaction with a patient is determined by the clinically visible symptoms as well as by the above described histological pattern. The clinically visible symptoms, such as redness ans slight swelling, are not very noticeable in the oral cavity. Therefore the histological pattern is also important for a proper diagnosis. As the complaints show much resemblance, this histological pattern can also assist in distinguishing whether the subjective complaints are caused by contact allergy or galvanism.

In order to protect the patient from dental-metal-restorations that might cause allergic contact stomatitis, composition and corrosive properties of the alloy of choice have to be known as well as the patient's sensitiveness towards the possibly relaesed ions.

There is a great variety of commercially available cast alloys. These are seldom accompanied by a proper description regarding composition, corrosivity and biocompatibility. Whatever the kind or brand, such an alloy consists of some combinations taken out of a series of about 20 different metals. The series include Au, Pt, Pd, Ru, Ir, Ag, Hg, Cr, Co, Ni, Fe, Cu, Zn, Mo, Sn, Ga, In and Si. In a pilot study, 12 different ions were selected in a first attempt to establish a routine standard metal battery for reliable patch testing (table I). This experimental test battery for dental materials has been tested epicutaneously on a number of patients. In connection to this, foils of some relevant pure metals or alloys were fixed for several days on teeth of patients, volunteering in this project, to investigate clinically and histologically their capacity to elicit allergic contact stomatitis.

Parallel to this, in a laboratory study, the corrosion rates of several dental cast alloys were investigated at various environmental circumstances.

In this way, the actual stomatological reactions can be studied in relation to the corrosion rate on a number of ions from table I.

Table I

Experimental dental test battery for patch testing on allergic sensitivity

Testmaterial	Concentration (% aq)
gold chloride (Au)	1.0
palladium chloride (Pd)	2.5
zinc chloride (Zn)	2.0
ammonium heptamolybdate (Mo)	1.0
stannium chloride (Sn)	0.5
gallium chloride (Ga)	10.0
indium chloride (In)	10.0
cobalt chloride (Co)	2.0
chromium chloride (Cr)	5.0
chromium trioxide (Cr)	0.5
potassium bichromate (Cr)	0.5
chromium sulphate (Cr)	5.0
nickel sulfate (Ni)	5.0
ferro chloride (Fe)	5.0
ferri chloride (Fe)	2.0
silicum tetrachloride (Si)	2.0

I IN SITU TESTING OF Ni-SENSITATION

Patients were selected on the basis of their history of allergic contact reactions to metals. They were patch tested (I.C.D.R.G. procedure) with diverse chemicals (see table I). Out of the group with a positive history of metal allergy, patients were selected with a positive skin reaction to metal ions. Patients without any allergic reaction to metals were used as a control group.

To test the effect of metallic Ni and Ni containing dental alloys on the mucosa, 3 x 5 mm foils were cut (thickness 0.5 mm), bonded to premolars and tested in the mouth. Biopsies (diameter : 3 mm, depth :

1 mm) were taken from the mucosa at the contra-lateral side. On this contra-lateral side, always pure Pd foil was applied as a control to exclude any reactions of an either chemical or mechanical irritant nature.

Stomatological reactions to pure Ni

A. Clinical observations

In a first experiment, 17 patients with a positive history to an allergic contact dermatitis were involved.

They were devided into three groups :

group 1 : 5 patients with a positive patch test to Ni-ions. In the oral cavity, Ni-foil was fixed on a lower premolar.

group 2 : 3 patients with a positive patch test to one or more of the metal ions Ni, Co, Cr and Pd. In the oral cavity Pd-foil was fixed on a lower premolar.

group 3 : 9 patients without any allergic skin reaction to metal ions, except 1 patient with a positive patch test to Cr. In the oral cavity, Pd-foil was fixed at 3 patients out of this group and Ni-foil at the remaining 6 patients.

group 2/3 : These 12 patients were also used as a control group.

In this study, all 5 subjects allergic to Ni who had a Ni test foil in their mouths, developed a local allergic stomatitis in the contact area. The oral reaction was less obvious in one subject, but the histological pattern showed a clear allergic reaction. The mucosa on the contra-lateral side was normal in all cases. Not one of the subjects who developed a positive skin reaction to Pd, developed any reaction on the oral mucosa. In the control group (group 3) of 9 subjects, the oral mucosa was normal at both sides. The signs of a local allergic contact stomatitis to Ni for the 5 subjects can be described as follows. Bimanual palpation showed the mucosa adjacent to the Ni-foil as well as the gingiva at the test site to be somewhat swollen and red in comparison with the control side. One subject developed a small papule on the face around the mouth during the period of the test. Within a week after removal of the foil, the signs vanished. One subject had a relapsing vesicle on the mucosa in contact with the plate, but not until after the foil had been removed. This effect disappeared after 3 weeks.

The subjects had no signs within 72 hrs. after application of the foils, but always later. This contrasts with skin symptoms. Because of the surrounding colour, the demarcation of the reaction area was less clearly perceptible than with the skin tests. The reaction area on the oral mucosa was wider than the actual contact area.

B. Histological observations

In all 5 subjects positive to Ni and tested orally, a spongiotic stomatitis was seen in the mucosa adjacent to the Ni plate. There was increased lymphocytic infiltration in the epidermis and extra-cellular edema; the papillae were disintegrated and there was an increased perivascular cellular infiltration. The cells were mainly lymphocytes with some granulocytes.

The histological pattern of the cheek mucosa on the contralateral side (control side) in the 5 Ni-positive subjects was normal. The mucosa of the other 12 subjects (group 2 and 3) was also normal, but at both sides (the test side and the contralateral side). This means that the Pd-test foil caused no stomatitis. No subject showed a stomatitis reaction to Ni if also when the patch test to Ni was negative.

II STOMATOLOGICAL REACTIONS TO Ni-CONTAINING ALLOYS
   CLINICAL AND HISTOLOGICAL OBSERVATIONS

In the next experiment, 12 patients were involved with a positive history to allergic metal contact reactions and a positive patch test to Ni.

With 5 patients a foil of the Ni-alloy, Ceramalloy was tested in situ with another 5 patients a foil of the Ni-alloy NP2 was tested and with the remaining 2 patients a foil of pure Ni was used.

Three biopsies were taken of each of the patients. One biopsy was taken at a Ni-alloy site, one at the Pd-site and one at a control site, which is located in the upper yaw region at the Pd-side.

In contrast to pure Ni, Caramalloy as well as NP2 elicit no allergic contact stomatitis at all with these patients. The allergic reaction to the pure Ni is accompanied by an enormous cell infiltration in the dermis and the epidermis. This infiltration mainly consists of lymphocytes. Also the structure of the basic membrane is desintegrated.

The number of lymphocytes in this allergic reaction to pure Ni is approx. 10 to 15 times higher than in a healthy oral mucosa. However, the control biopsies show clearly that more lymphocytes are present in a normal healthy skin.

In spite of the fact that thses Ni-alloys do not cause an allergic contact stomatitis, statistical calculations are necessary to determine if these two alloys may influence the amount of different types of lymphocytes and the Langerhans cells in the epidermis and dermis of the oral mucosa.

III IMMUNOHISTOCHEMICAL EXPERIMENTS

Next to the usual histological techniques with normal hematoxylin and eosin colouring, immuno histochemical techniques were used for characterization of the inflammatory cells, in particular the immuno competent cells in the oral mucosa adjacent to the test foils. In this region eventually an increase was measured of different types of lymphocytes (T-suppressor/cytotoxic cells and T-helper/inducer cells) as a result of Ni-ions from the slowly corroding Ni-containing alloy.

The biopsies were frozen in liquid nitrogen and sections of 6 micrometer thickness were cut and fixed on glass slides, for microscopical study. Monoclonal antibodies were chosen to characterize the lymphocyte subclasses and the Langerhans cells (see table II).

Positively stained cells were quantified by counting all cells in the epidermis and dermis in some hystological slides.

Table II

Monoclonal antibodies applied in the study of
in situ distribution of immunocompetent cells.

Trade designation	Reaction with	Source
Leu 217 (T8)	T-suppressor/cytotoxic cells	Becton
Leu 3a-3b (T4)	T-helper/inducer cells	and
Leu 4 (pan-T)	all peripheral T-cells	Dickinson
Okt. 6 (T6)	Langerhans cells	Orthoclone

IV THE DETERMINATION OF THE CORROSION RATES OF SOME Ni-CONTAINING DENTAL ALLOYS

To establish whether the used Ni-containing alloys release a substantial amount of ions, a study was carried out on the rate of corrosion for a series of differently formulated alloys. As bulk metal does not elicit an allergic stomatitis, knowledge about the rate of ion release is of great importance. Apprehension of the dental alloy's biocompatible character can thus be obtained by establishing the corrosion rate. As corrosion of alloys is greatly dependant of the environmental circumstances, also extreme conditions which may occur in the oral cavity should be incorporated in a characterization of the dental corrosion. Especially the base-metal-alloys may loose their passifying oxyde layer at high acidity or low oxygen concentration. The former may be caused by bacteria and lead to pH values as low as 4, whilst the later will occur at sites which cannot be reached by the oxygen from the air.

Table III

The alloys used in the corrosion study
and their composition (wt%)

Alloys	Total	Ni	Cr	Mo	Co	Ga	Fe	Other
1. Exp. 100[c]	100	100	--	--	--	--	--	--
2. Exp. 85[c]	100	85	15	--	--	--	--	--
3. Exp. 78[c]	100	78	22	--	--	--	--	--
4. Ultratek[a]	99	80	11	2	--	--	2	2Al,2Be
5. Ceramalloy[b]	100	72	19	4	--	--	--	1B,4Si
6. Xerxes-220[c]	97	65	22	8	--	--	--	2Nb,Al
7. Resistal-P[d]	98	64	21	9	--	--	--	4Nb
8. NP2-Microbond[e]	100	60	13	8	--	7	7	5Mn
9. Vitallium[e]	100	--	33	7	60	--	--	--

[a] Metals for Modern Industry, U.S.A
[b] Johnson & Johnson, U.S.A
[c] Elephant BV, Netherlands
[d] degussa, F.R.G
[e] Austenal, U.S.A

Table III compiles the materials investigated together with their respective compositions. For better understanding of the role of Cr as a corrosion protective factor, an experimental Cr-free and two Cr-rich alloys are included.

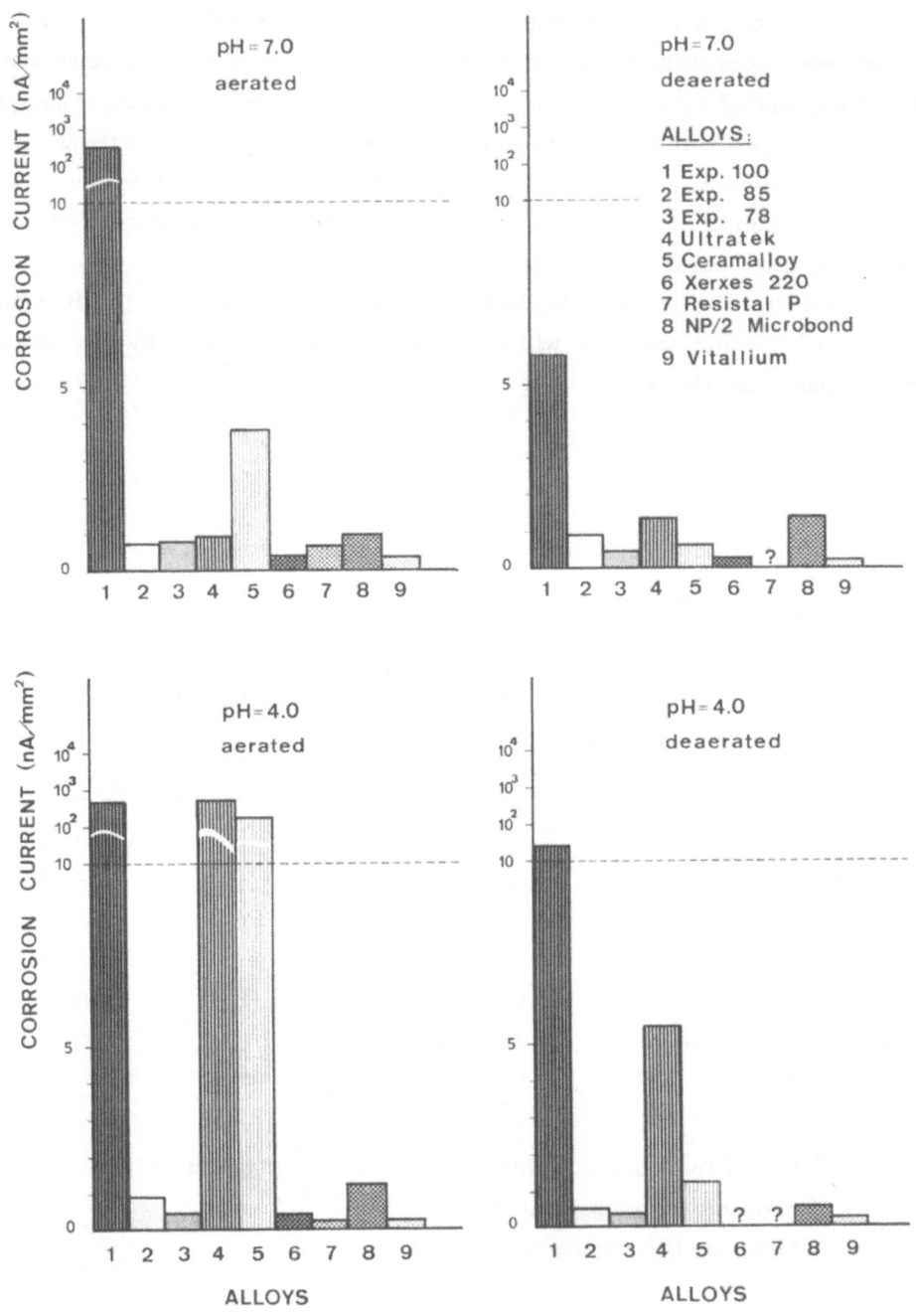

Fig. 1
The corrosion current of various base metal-cast
alloys under four environmental conditions
in an experimental set up.

The results from the laboratory corrosion tests are compiled in fig.1. The corrosivity is expressed in the corrosion current, being proportional to the amount of ions released from 1 cm^2 surface, under standarized conditions. these figures can be helpful to compare the various products with respect to corrosion. (Note that, in fig.1, the vertical scale above 10 nA/mm^2 is a logarithmic one). Under aerated circumstances at pH = 7, the corrosion of pure Ni is retarded by a factor 2500 (!) by adding only 15% Cr. However under more severe conditions, e.g. at pH = 4, Cr-containing alloys like Ceramalloy and Ultratec loose their passivation and corrode as much as pure Ni. The alloys Ultratec and Ceramalloy show a questionable corrosion resistance under the four different environmental conditions. Obviously even the, relatively high, Cr-content in Ceramalloy could not guarantee a proper passivation.

Under the described electro-chemical conditions, also pure Pd was investigated. The corrosion rate under all circumstances proved to be neglectable (less than 0.01 nA/mm^2). On this basis Pd-foil was selected as a non-irritant reference in our clinical trials.

CONCLUSIONS

Based on these studies one could eventually select alloys if extreme caution is taken with nickel sensitive patients. Allergic reactions on some Ni-alloys have been reported in literature and in this study we could only demonstrate clinically as well as histological contact stomatitis on pure Ni with nickel sensitive patients. In contrast to this, under the experimental conditions of this clinical study, no significant activity could be shown with patients using passified Ni-alloys, such as Ceramalloy and NP2. It cannot be excluded that, e.g. Ceramalloy, might cause allergic reactions under more severe electrochemical circumstances. One might conclude that corrosivity is a parameter for tissue reaction if the biocompatibility of the alloy is questioned on the basis of its composition. One would expect that oxygen passivied base metal alloys, would suffer from absence of environmental oxygen. This could not be demonstrated in our relatively short term study. Some alloys passify spontaneously under all conditions. In long term corrosion studies some alloys showed degeneration of the passifying layer, indicating the occurence of pits or cracks in the protective oxyde surface. Therefore yet, a general rule on the influence of the environmental conditions on the corrosion behaviour cannot be given. Some base metal alloys investigated in this study were stable under all conditions and therefore also very likely will function without any problem in vivo. A full proof criterion on the acceptable corrosion in relation to allergic stomatitis can only be given when dermatology establishes MAC (Maximum Acceptable Concentration) values for the various ions, as there exists for toxicology. So far the different commercially available products should be accompanied by information on the corrosivity based on independently established standards.

LITERATURE

BOHNENSTENGEL G., 1965, Kaliumbichromat-Testungen in Verschiedenen pH-Bereichen, Dermatologishe Wochenschrift, 29 : 797.

BOS J.D. and EMSBROEK J.A., 1985, Immunocompetent cell interactions in allergic contact dermatitis, J. for Drugtherapy and Research, 10, 6 : 770-773.

BOS J.D., HULSBOSCH H.J., KRIEG S.R., BAKKER P.M. and CORMANE R.H., 1983, Immunocompetent cells in psoriasis in situ immunophenotyping by monoclonal antibodies, Arch. Dermatol. Res., 275 : 181-189.

CRONIN E. , 1980, Contact Dermatitis, Edingburg, London and New York : Churchill Livingstone.

DAVIDSON C.L., 1983, Das Kaleidoskopartige Bild des internationalen Marktes für zahnärtzliche Geisslegerungen, in : Medizinische und technologische Aspekte dentaler Alternativlegerungen, R. Herbert, ed., Berlin : Quintessence Verlag, 15-21.

DAVIDSON C.L., 1985, Corrosie in de mond, in : Het Tandheelkundig Jaar 1985, W.A.M. van der Kwast et al., eds. : Bohn, Scheltema & Holkema, Utrecht, 14-26.

FISCHER T., RYSTEDT I., 1985, False-positive follicular and irritant patch test reactions to metal salts, Contact dermatitis, 12 : 93-98.

FRANZ G., 1983, The frequency of allergy to dental materials, Journal of the Dental Association of South Africa, 37 : 805-810.

KANERVA L., RANKI A. and LAUHARANTA J., 1984, Lymphocytes and Langerhans cells in patch tests, An immunohistochemical and electron microscopic study, Contact Dermatitis, 11 : 150-155.

KETEL W.G. VAN, NIEBOER C., 1981, Allergy to palladium in dental alloys, Contact Dermatitis, 7 : 331-357.

LOON L.A.J. VAN, ELSAS P.W., JOOST Th. VAN, DAVIDSON C.L., 1984, Contact Stomatitis and Dermatitis to nickel and palladium, 11 : 294-297.

LOON L.A.J. VAN, ELSAS P.W. VAN, JOOST Th. VAN and DAVIDSON C.L., 1986, Test battery for metal allergy in dentistry, Contact Drerlatititz, 14 : 158-161.

MAGNUSSON B., BERGMAN M., BERGMAN B., SOREMARK R., 1982, Nickel allergy and nickel-containing dental alloys, Scand. J. Dent. Res., 90 : 163-169.

MOBACKEN H., HERSLE K., SOLBERG K., THILANDER H., 1984, Oral Lichen planus : hypersensitivity to dental restoration material, Contact Dermatitis, 10 : 11-15.

NAKAYAMA H., 1982, Hypersensitivity to palladium is linked to oral lichen planus, Dermatology News, February.

PERNIS B., VOGEL H.J., 1980, Regulatory T-Lymphocytes, Academic Press, New York.

RAPSON W.S., 1985, Skin contact with gold and gold alloys, Contact Dermatitis, 13 : 56-65.

RANKI A., KANERVA L., FORSTROM L., KOUTTINEN Y. and MUSTAKALLIO K., 1983, T and B Lymphocytes, Macrophages and Langerhans cells during the course of Contact Allergic and Irritant Skin reactions in man, Acta Derm. Venereal (Stockolm), 63 : 376-383.

SILVERNOINEN-KASSINEN S., NIINIMATI A., 1983, Gold sensitivity blast transformation, Contact Dermatitis, 11 : 156-158.

TURK J.L., 1975, Delayed hypersensitivity, North Holland Publishing Co., Amsterdam/Oxford, 2nd ed., vol. 4.

WATERHOUSE J., 1982, Langerhans Cells in Oral Mucosa and Skin, in : Oral immunogenetics and tissue Transplantation, G.R. Riviere and W.H. Hildemann, ed., Elsevier North Holland Inc.

IMMUNOLOGICAL ASPECTS OF CONTACT HYPERSENSITIVITY

TO NICKEL AND CHROMIUM IN DENTAL ALLOYS

Karel.J.J. Vreeburg, Klaas de Groot, Cees K.H. v.d. Burg
and Rik.J. Scheper

Dept. of Biomaterial Science and Dept of Pathology, Free
University, P.O. Box 7161, 1007 MC Amsterdam, the Netherlands

INTRODUCTION

Sensitization to nickel and to chromium is very common. Approximately
10% of the female population and a smaller, but increasing number of the
male population is hypersensitive to nickel (1.). Chromium is known as the
most important cause of occupational contact dermatitis in man, the frequency of which (11% in Europe) is still increasing with the increasing use
of chromium containing products in our daily life (2,3.). Despite the high
prevalence of nickel and chromium hypersensitivity in the population, cases
due to dental treatment with nickel and chromium containing alloys are rare.
Until now only a few cases of nickel and chromium hypersensitivity have been
reported, but also for gold, palladium, copper, mercury and other metals
(4-7). To better understand for pathogenetic and predictive reasons, the
mechanism behind metal allergy, there is a need for various studies to explore the immunological aspects of this disease.

IMMUNOLOGICAL REVIEW

Classification. According to the classification of Coombs and Gell (8.),
there are four types of hypersensitivity reactions. The first three types,
type I,II and III are antibody-mediated or immediate-type allergic reactions
which are transferable from one animal to another by serum components. The
fourth type, T cell-mediated or delayed-type hypersensitivity reaction, can
only be transfered from one animal to another by means of sensitized T lymphocytes. Metal contact hypersensitivity, as discussed in this paper is a
form of cell-mediated or delayed-type hypersensitivity.

Contact hypersensitivity. In contact hypersensitivity two phases of the
immune response to a contact sensitizer may be observed. The **inductive phase**,
which is the primary response after antigenic stimulation and the so called
eliciting phase which is the secondary response. From many studies in the
field of experimental and clinical contact dermatitis it is known that several different cells play a role in the development, elicitation or suppression of a contact sensitivity. A specialized skin macrophage, the Langerhans
cell is the antigen presenting cell in both the inductive and the eliciting
phase in contact hypersensitivity (9, 10). Many contact allergens, like
metals, do not react as an antigen themselves, but are first as a hapten

conjugated to HLA-D region dependent molecules which are present on the surface membrane of Langerhans cells in the dermis. After primary antigenic stimulation (sensitization) Langerhans cells travel via the afferent lymphatics (as "veiled cells") to the regional lymphnodes. Within the paracortical area of these lymphnodes, those T cells that specifically bind to the antigen presenting cell complex will be activated to release several mediators (interleukines) and to proliferate. Depending on the sensitization potential of the antigen, two different cell pools may appear within a few days after sensitization. T effector cells which are responsible for the primary response and T memory cells which spread over the entire body and wait for a second antigenic stimulation.After a second antigenic challenge, T memory cells are transformed into T effector cells or proliferate to form new T effector cells and a new pool of T memory cells. Other cells which are of importance in contact hypersensitivity are the T suppressor cells. T suppressor cells may have a suppressive effect on the sensitization and on the elicitation of T effector cells and they may have a non-antigen specific regulatory function after an immunological response. For mechanism and routing of antigen presentation, see Fig 1. and Tables 1. and 2.

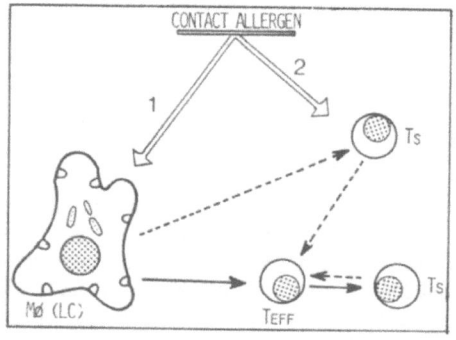

Mø	Specialized macrophage e.g. epidermal Langerhans cells.
T eff	T effector cells
T s	T suppressor cells
1.	Epidermal route
2.	Oral/Intravenous route

Fig. 1. Mechanism of type IV antigen presentation.

Mediators in contact sensitivity. Cellular interaction and expression in contact dermatitis is mediated by specific membrane receptors on Langerhans cells and on T cells and by soluble lymphokines. Allergens are first associated with HLA-D region dependent molecules (from the Major Histocompatability Complex) which are present on Langerhans cells. T cells bear T cell specific receptors on their surface. Activation of T effector cells starts after interaction of these receptors with the MHC/antigen receptor on Langerhans cells. After this antigen specific modulation, the T effector cell may proliferate or start producing several lymphokines. The many different lymphokines that are released by T cells can be characterized by their function. The most important functions of lymphokines in contact sensitivity are regulation and modulation of other cells. Some key functions such as regulation of specific T helper cells or specific T suppressor cells, are antigen specific. The major part of the functions however is non-antigen specific. Lymphokines may have a stimulatory effect on other lymphocytes, on mononuclear phagocytes and on mast cells. The release of the lymphokine Migration Inhibition Factor is used in an in vitro test as a prove of T cell reactivity to a specific antigen. In this test the migration of macrophages is inhibited in the presense of antigen and sensitized lymphocytes.

Primary response. As Langerhans cells are present at almost every site of the skin in man and animals, primary topical application of an allergen in non-presensitized individuals may result in the development of a contact hypersensitivity. If however the antigen bypasses the Langerhans cell system in non-presensitized individuals the T suppressor cell system may be activated which may result in the development of an antigen specific immunological tolerance. From animal studies it is known that oral and intravenous administration of certain allergens or topical application on a Langerhans cell depleted skin (e.g. mouse tail) may have such a tolerogenic effect. Subsequent attempts to sensitize such pretreated animals with the same allergen will fail, because activated T suppressor cells suppress the ability of T effector cells to become sensitized (10,11.). (See Table 1 and Fig 1.)

Table 1. Different immunological effects depending on the route of antigen administration as a _primary_ response.

Route	Mechanism	Effect
Normal skin	T effector cell activation by Langerhans cells	Sensitization
Orally, i.v. LC depleted skin.	T suppressor cell activation	Immunological Tolerance

Secondary response. If an individual or an animal has already been sensitized, skin challenge with that certain antigen will result in an elicitation of a type IV delayed type hypersensitivity reaction at the antigen application site. Oral or intravenous challenge in hypersensitive individuals may result in so called "flare reactions" of old reaction sites or even "rash reactions" of several parts of the skin. Macroscopically a contact sensitivity reaction is characterized by erythema and swelling at 24-72 hours after challenging. Microscopically the skin reaction can be classified as a mononuclear cell infiltration consisting of T lymphocytes and macrophages. 10-20% of the T lymphocytes in an infiltrate may be antigen-specific and may persist in the skin for several months. This might explain why old reaction sites show erythema and swelling after retest reactions in other parts of the skin or after oral challenge with the antigen. Rash reactions are probably the result of circulating antigen that reaches T effector cells in the skin. This might happen after oral ingestion or after intensive skin contact with a large amount of allergen. (See Table 2. and Fig 1.)

Table 2. Secondary responses after challenge on normal skin or by oral or intravenous route.

Route	Mechanism	Effect
Normal skin orally, i.v.	Activation of T memory and T effector cells.	Type IV reaction[a]

[a]Type IV reactivity in this Table includes the so called "Flare reactions" and " Rash reactions".

EXPERIMENTS

In an attempt to assess if dental alloys have the same sensitizing or tolerogenic properties as the normally used experimental antigens like picryl-chloride and DNCB (dinitrochlorbenzene), we have been working on several studies in the field of experimental contact dermatitis to nickel and chromium. In our first experiments we have been working on orally induced immunological tolerance to nickel and chromium in a guinea pig model (12.). Other experiments in guinea pigs were maximization of nickel sensitization procedures and flare reaction tests. As however the currently available guinea pig models in especially nickel allergy suffer from reproducibility and a lack of immunochemical reagents in this species we have been working on a mouse model for experimental metal allergy (13.) In an epidemiological study in a group of apprentice nurses we looked for the influence of a dental treatment on the development of nickel allergy (14.).

MATERIALS AND METHODS

Guinea pig experiments. In these experiments guinea pigs received an oral dental applience (occlusal splint) or were fed with nickel and chromium containg food pellets for a period of at least 30 days. Negative control groups received normal food. All pretreated guinea pigs as wel as untreated control groups, were then tested for the possible presence of hypersensitivity to nickel and chromium. Following these treatments, the animals were sensitized with a strongly immunizing emulsion of nickel and chromium in Freund's Complete Adjuvant, and were skintested again two weeks later. The time tables and food procedure of these experiments are shown in Fig 2.

Mice experiments. An attempt was made to sensitize mice to nickel and to chromium. Two different mouse strains were used, CBA (H-2k) mice, assumed to perform well in contact sensitivity studies, and BALB/c (H-2d) mice because of their wide availability in the Netherlands. For sensitization epicutaneous techniques and techniques involving the use of Freund's Complete Adjuvant were used. For epicuteneous sensitization nickel or chromium salt in petrolatum was applied to the shaved abdominal skin and covered with a bandage of moistened Paris plaster which was wrapped around the abdomen and lower region of the thorax. All casts were removed after 9 days, mice were challenged with $K_2Cr_2O_7$ or $NiSO_4$ 4 days later. Some animals received pretreatment with Cyclophosphamide in order to enhance the development of hypersensitivity to nickel or chromium . The animals were skin tested on the ears. The intensity of the reaction was assessed by measuring the increase of earthickness with an engineers micrometer.

Epidemiological study. Groups of apprentice nurses and hairdressers were tested for the presence of nickel allergy at the beginning of their apprententiceship. In both groups data were collected of previous contacts with nickel from wearing dental prostheses in orthodontic treatment during childhood and on ear piercing. The incidence of nickel hypersensitivity was compared to the individual history of orthodontic treatment. In both groups the influence of the moment of ear-piercing on the development of nickel hypersensitivity was studied in relation to the moment of orthodontic treatment.

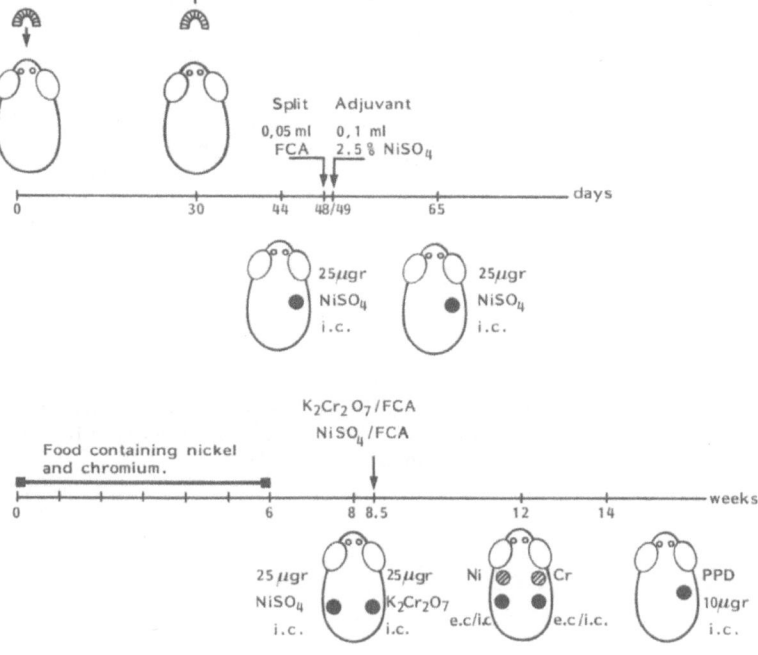

Fig. 2. Time schedule and experimental set-up of the guinea
pig experiments. Exp. 1. Orally exposure to a Ni-Cr
dental device. Exp. 2. Three groups received daily
1 or 5 mg Ni-Cr powder or 5 mg Ni-Cr salt in their
food during 6 weeks. Positive and negative control
groups received normal food only.

RESULTS

Guinea pig experiments. In the guinea pig experiments the animals
were for a period of at least 30 days orally exposed to both nickel and
chromium. In the first experiment by means of an oral dental appliance, in
the second experiment by feeding different amounts of nickel and chromium.
Challenge with nickel or chromium showed that none of the animals had been
sensitized by this pretreatment. In both experiments next an attempt was
made to sensitize pretreated as well as non-pretreated animals with nickel
and chromium. A second epicutaneous challenge after sensitization showed
that most of the pretreated animals had **not** become hypersensitive to nickel
or chromium. In contrast, virtually all the sensitized animals that had
not been pretreated were clearly hypersensitive to one or both metals.
An additional skintest with PPD (purified protein derivative) showed no
differences between pretreated and non-pretreated groups. All animals that
had received the mycobacterium-containing Freund's Complete Adjuvant exhibited similarly strong reactivity to the corresponding antigen. Results of
the guinea pig experiments are shown in Fig. 3. and 4. In further experiments with the sensitized guinea pigs, repeated oral administration of
nickel and chromium containing food did not elicit a so called flare reaction. New challenges with both metals however showed that the animals remained hypersensitive to both metals for a long time (Unpublished data).

Mice experiments. Induction of contact hypersensitivity to nickel
or chromium in mice has never been accomplished. Recent attempts of Möller

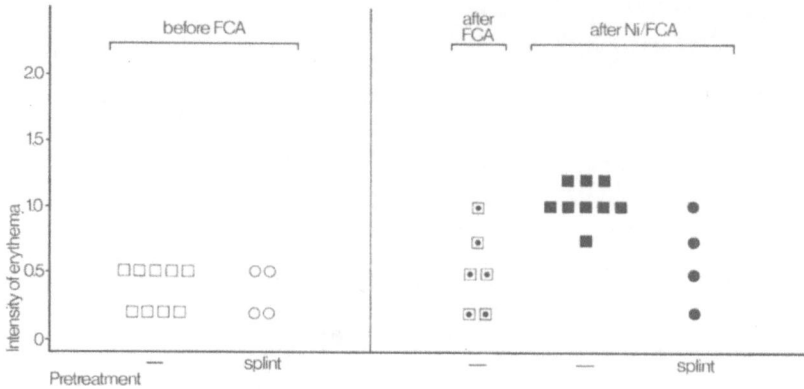

Fig. 3, 24 hours skin reactivity to nickel as measured by
intradermal testing before and after immunization with
FCA either or not containing $NiSO_4$.

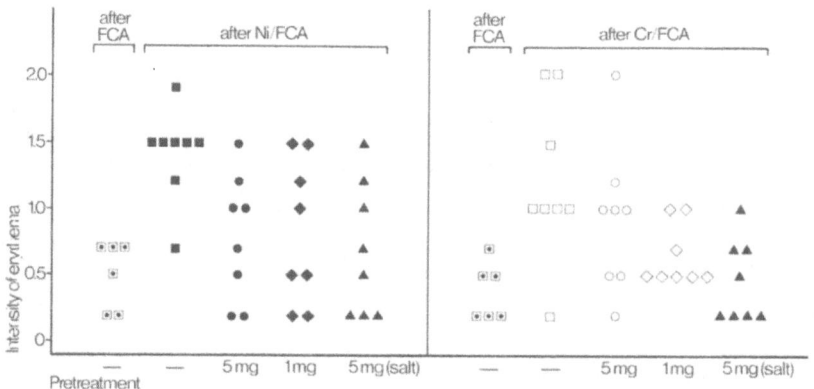

Fig. 4. 24 hours skin reactivity to nickel and to chromium as
measured by epicutaneous testing <u>after</u> immunization
with nickel or chromium.

with NMRI mice demonstrated a failure of intramuscular methods using FCA,
whereas marginal nickel allergy could be induced by repeated epicutaneous
application of a strong nickel solution (15.). Tigelaar and coworkers de-
monstated earlier that oral ingestion of epicutaneous applied hapten could
"unwittingly downregulate the development of contact hypersensitivity"
(16.). We therefore explored epicutaneous sensitization in mice in whom
oral grooming of the sensitization site was prevented by wrapping a plaster
cast around the abdomen and lower thorax. The result of this experiment
was that CBA mice could be sensitized to nickel using the FCA-method and to
a lesser degree using the epicutaneous casting method. Chromium hypersensitivity could not be induced in this strain. In contrast both nickel and
chromium contact hypersensitivity could be readily induced in the BALB/c
strain. Epicutaneous sensitization resulted in strong delayed hypersensitivity in all animals tested. The ear reactions in nickel and chromium
hypersensitive BALB/c mice not only showed strong swelling, but also a
clear erythematous response. All reactions showed a protracted time course,
with strong swelling detectable for at least 72 hours after challenging.
The results of epicutaneous sensitization in BALB/c mice are shown in Fig 5.

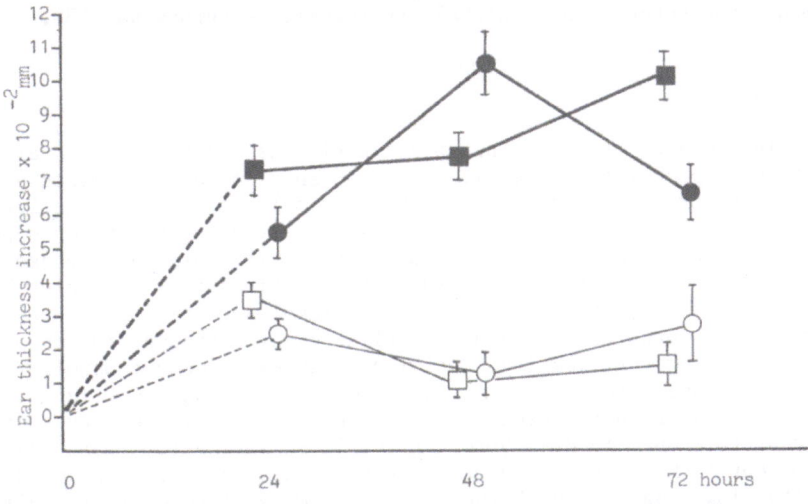

Fig 5. Time course of the ear swelling response ± SEM to nickel
and chromium in BALB/c mice sensitized epicutaneously by
casting. Closed symbols: Mice sensitized and challenged
with nickel (■—■) or chromium (●—●), open symbols: control, non-sensitized mice, challenged with nickel (□—□)
or chromium (○—○).

Epidemiological study. Potential relationship between the development of nickel allergy and previous ear piercing or orthodontic treatment 111
with nickel-containing dental appliances in groups of nurses and hairdressers at the beginning of their apprenticeship were evaluated. In the group
of junior nurses 86% of the students had their ears pierced, in the group
of hairdressers the percentage of students with pierced ears was 95%.
Data obtained on the history of previous orthodontic treatment revealed a
very low incidence of nickel sensitivity in the junior nurses who had orthodontic treatment during childhood (see Table 3.). No effect of orthodontic treatment was observed on the high frequency (28%) of nickel sensitivity in the junior hairdressers. In both students groups, orthodontic
treatment took place around the age of 12. In the group of nurses, 74 nurses (40%) had a positive history of a previous orthodontic treatment with
a Ni-Cr appliance. In this group the percentage of nickel positive skin

Table 3. Correlation between previous orthodontic treatment and present nickel sensitivity in female
aspirant nurses.

Group	History of orthodontic treatment	number	Nurses with nickel positive skin tests number	percentage
1	positive	74	4	5%[*]
2	negative	114	21	18%

[*]Frequency decreased (x^2 test p   0.05 as compared to
group 2).

tests was 5%. In the group with a negative history of previous orthodontic treatment, the percentage óf nickel hypersensitive nurses was 18%.

DISCUSSION

The results of our guinea pig experiments indicate that oral administration of nickel or chromium, either as an alloy or as a mixture of salts, may suppress the capacity of these animals to develop an allergic response to these metals. This suppression proved to be antigen specific as the guinea pigs were still able to mount a cell mediated response to mycobacterial antigens (PPD). In the retrospective study in junior hairdressers and nurses, the low incidence of nickel sensitivity in the junior nurses who had worn an orthodontic appliance may represent the same phenomenon. Assuming that wearing dental devices can induce immunological tolerance, and that ear-piercing is the major sensitizing event in nickel allergy, sensitizing in the group of junior nurses may have been prevented by a state of tolerance induced before possible sensitization could take place. The apparent lack of such prevention in the junior hairdressers may be related to the lower age at which they had their ears pierced. It is well known that oral contact with an allergen in already sensitized individuals induces transient desensitization and/or flare up reactions rather than immunological tolerance. To better understand the underlying mechanism of metal allergy, much more further animal studies are needed. As currently available guinea pig models in metal allergy suffer from reproducibility and a lack of immunological reagents in this species, we made an attempt to induce contact hypersensitivity to nickel and chromium in mice. The present BALB/c casting model, which was very succesful, may prove to be of great value in further studies in the field of metal allergy. This is particularly true considering that metal allergy in humans is induced in an analogous way, that is, via persistant skin-contact. Obviously, further mice studies must be carried out to better define the immunological basis of the presently described metal hypersensitivity model in BALB/c mice. The lack of an appreciable role of humoral antibodies in nickel and chromium allergy in man and guinea pig, as well as the delayed time course of the presently observed hypersensitivity reactions support a major role of T cell mediated immunity in this mouse model.

REFERENCES

1. Menné, T. Borgan, O. and Green, A. (1982); Nickel allergy and hand dermatitis in a stratified sample of the Danish population. Acta Derm. (Stockh) 62: 35
2. Fregert, T. (1975); Occupational dermatitis in a 10 year material. Contact Dermatitis. 1:96
3. Cronin, E. (1971); Contact dermatitis, XV Chromate dermatitis in men. Brit. J. Dermatology. 85: 95.
4. Fenton, A.H. Jeffrey, J.D. and Sound,P. (1978); Allergy to a partial denture casting. Dent. J. 44: 466.
5. Fregert, T. Kollander, M. and Poulsen, J.(1979); Allergic contact dermatitis from gold dentures. Contact Dermatitis 5: 63.
6. Duxburry, A.J. Ead, R.D. McMurrough, S. and Watts,D.C. (1982); Allergy to mercury in dental amalgam. Br. Dent. J. 152:47.
7. Loon, L.A.J.van. Nieboer, C. and Ketel, W.G. van. (1982); A case of local and general complaints caused by palladium hypersensitivity. Ned. Tijdschr. Tandheelk. 89: 50.
8. Coombs, R.R.A. and Gell, P.G.H. (1975) Classification of allergic reactions responsible for clinical hypersensitivity and disease. In Gell, Coombs and Lachman, Clinical aspects of immunology. 3rd ed, pp761.

232

9. Roit, I. Brostoff, J and Male, D. (1985); Immunology. Gowers Medical Publishers. London. New York. ISBN 0-906923-35-2.

10. Polak,L. (1980) Immunological aspects of contact sensitivity. Monographs in Allergy, Vol 15. ISBN. 3-8055-3050-1.

11. Asherson, G.L. Zembala, M. Perera, M.A.C.C. Mayhew, B and Thomas, W.R. (1977); Production of immunity and unresponsiveness in the mouse by feeding contact sensitizing agents and the role of suppressor cells in the Peyer's patches, mesenteric lymphnodes and other lymphoid tissues. Cell. Immunol. 33:145

12. Vreeburg,K.J.J. Groot, K.de. Blomberg,M von. and Scheper, R.J. (1984) Induction of immunological tolerance by oral administration of nickel and chromium. J. Dent. Res. 63: 124.

13. Vreeburg, K.J.J. Groot, K de and Scheper, R.J. (1986); Successful induction of allergic contact dermatitis to nickel and chromium in mice. (submitted for publication)

14. Burg, C.K.H. v.d. Bruynzeel, D.P. Vreeburg,K.J.J. Blomberg, B.M.E.von and Scheper, R.J. (1986); Hand eczema in hairdressers and nurses; a prospective study. I. Evaluation of atopy and nickel-hypersensitivity at the start of apprenticeship. Contact Dermatitis (Accepted for publication).

15. Möller, H. (1984); Attempts to induce contact allergy to nickel in the mouse. Contact Dermatitis. 10:65

16. Tigelaar; E.R. Bergstresser, P.R. Lonsberry, L.M. Elmets, C. Wood, P.J. and Streilein, J.W. (1982); Oral ingestion of epicutaneously applied hapten in mice may unwittingly induce down-regulation of contact hypersensitivity. J. Immunol. 129:1898.

CHAIRMEN SUMMARY: IMMUNOPATHOLOGY

C. Benezra and K. Merritt

The fundamental aspects of allergic contact dermatitis (ACD) to
nickel were discussed placing Ni-allergy in the general allergy context
(C. Benezra and M.F. Leroy). In particular what is the skin sensitizer :
the metal itself (which is also able to complex with proteins) or Ni^{++}
salts ? Some speculations are cited about the possible role fo Ni in the
skin such as replacing other naturally occuring metals on proteins (like
Fe, Zn, Cu) and dramatic charge in proteins transforming them into antigens
because the preferred geometry of Ni is octahedral and of the other metals
is tetrahedral. Oxidation potentials (low for Ni, high for Pt or Au),
solvation etc... also explains the strong sensitization power of nickel.

Hypersensitivity reaction to nickel, cobalt and chromium was presented by K. Merritt. Sensitivity was assessed using the *in vitro* test of
inhibition of leucocyte migration (LIF). 25% of the population will have
a sensitivity to nickel implants. The correlation between the allergy and
the *in vitro* test was assessed using rabbits (LIF test compared to skin
test). Lymphocyte transformation test was also used in rabbits. Chemotactic
properties of Ni and Co were also assessed. Corrosion of implants probably
explain sensitization to implants.

Metal sensitivity in patients with total hip arthroplasties was
presented by F. Leynadier and F. Langlais. A brief review on Ni and Co
sensitivity was made, as for Ni sensitivity in the general population.
In a gynaecology clinic, 6% of the women were sensitive to Ni. The percent
of Ni and Co sensitive people in patients with total hip arthroplasties
can be high. The induction of ACD is probably not due to hip arthroplasty.

ACD can be due to loosening in metal-to-metal arthroplasty, and metal ions released in tissues which are ingested by macrophages. Metal-plastic prostheses are now more used so that the problem is not so important.

Dr Shapcott commented on chromium species : Cr(III) is immediatly hexahydrated at physiological pH.Cr(III) is unstable and gives a complex which is <u>inert</u>. It has high affinity for nitrogen. The form of chromium is important. Cr(VI) is liposoluble, dissolves in the membrane, and is immediatly reduced to Cr(III). From a chemical point of view it is no use to say that corrosion of implants would give Cr(VI) or Cr(III) (comment on Dr Merritt's paper).

A survey pf ACD to Nickel, Chromium and Cobalt was presented by Drs H.F. Hildebrand, C. Veron and P. Martin. The patch tests were discribed. The incidence of such ACD was mentioned. Electron microscopy of lymphocyte cultures were used to assess differences in sensitized and unsensitized persons. After incubation with Ni-salts, lymphocytes of sensitized patients incorporated these salts under a metabolized form. These metabolites were rarely observed in lymphocytes of unsensitized persons, which formed uropodes instead.

Dr van Loon described his experiments on stomatitis in patients sensitive to nickel. He placed a band of nickel foil on a tooth. A band of palladium on the opposite side was used as a control. Lesions appeared only in tissue contacting the nickel band.

He also described the results of *in vitro* corrosion studies on nickel containing alloys. The amount of released nickel was not in proportion to the amount in the alloy. He showed some alloys with very great release of nickel.

A discussion of corrosion rates followed the presentation. The question of how to assess corrosion from dental implants was raised. This needs to be studied.

Dr Vreeburg discussed the oral ingestion of allergens. The conclusion was that if potential allergens are contacted by a route which does not bring it in contact with macrophages, tolerance results. If sensitivity already exists, these routes will elicit a response. A discussion of induction and elicitation of allergic responses by the oral route followed.

It seemed generally agreed that elicitation occured. No agreement was reached on induction of hypersensitivity by the oral route. Further studies are needed.

CLINICAL ASPECTS

ANALYSIS OF TOTAL JOINT REPLACEMENTS (HIP AND KNEE)

Bernard Bloch

Consultant Orthopaedic Surgeon
Sydney Hospital Australia

Results from 22 Sydney metropolitan hospitals years 1969-79 were outlined. Over 7,000 histories were examined. A follow-up by questionnaire or telephone contact or from up-dated medical records was obtained in 63% of cases.

BACKGROUND

Over the last three decades, the practice of orthopaedic surgery has increasingly been directed to the disabling effects of chronic disorders such as arthritis. Here work capacity, physical independence and self care are in jeopardy. Inflammatory conditions such as rheumatoid arthritis or the degenerative changes of osteoarthritis are the disease processes commonly involved. Empirical "biological" surgery such as soft tissue releases of muscles or joint capsule or the corrective osteotomy of a mal-aligned bone also have limited application. A more drastic mechanical solution is required where gross irreversible disease processes have affected the articular surfaces.

In earlier methods of joint reconstruction, only one or part of an articular component was replaced. The hip joint was the first to be totally replaced. The techniques were pioneered in the main in the United Kingdom (McKee-Farrar of Norwich, Charnley at Wrightington) and in Europe, (Muller at Berne). A major development was the use of polymethylmethacrylate bone cement to anchor the components to the skeleton. As experience increased, the techniques were increasingly adopted. The 1960's saw limited application but by early 1970's the operation was freely practised in many countries. Techniques for replacement of the knee joint were well developed by this time. In the USA, the Federal Drug Authority only released the bone cement for use in 1973. By the late 1970's that country had become the major exponent of joint replacements.

The numbers of total joints replaced are difficult to obtain with any accuracy. In 1978, international authorities estimated a world wide figure of between 500,000 to one million replacements, the lower level being more favoured, for that year.

MEDICAL RECORD CODING

Until the 1960's codes used for the procedures of total joint replacements (TJR's) were open to individual interpretation by the records officers and varied between institutions. In the 1960's health authorities reproduced a code based on the General Register Office of England and Wales. The code number 810 in the 1968 edition would have included many cases of prosthetic replacement for fractures of femoral neck as well as TJR's!

The World Health Organization harmonized the classifications of diseases and procedures. A 9th revision of International Classification of Diseases and Procedures has been in general use since 1980. The International Code does clarify to a greater and clearer extent the TJR procedures and complications. In this new Coding the numbers have been altered. Thus number 810 in ICD8 becomes 5-815 in ICD9. The NSW Health Commission annual morbidity statistics only commenced in 1979.

IN THE WRITERS EXPERIENCE THE SAME CONFUSIONS AND INADEQUACIES PERTAIN AT NATIONAL AND INTERNATIONAL LEVEL

TECHNIQUES OF TJR

There are many variations in design in both hip and knee replacements. The surgeon in choosing a specific type considers in the main, his own experiences in a particular method. No doubt degree of complications, results and further up-dated information play a role.

Both in total hip replacement (THR) and total knee replacement (TKR), there have been hundreds of different models available. This initial exuberance has now been curtailed by commercial as well as surgical rationalization. These many variations in shape, size or contour of a socket or articulation component reflect an exponent's belief in a specific improvement either in the fixation or function of the device.

CURRENT MATERIALS

The socket of a total hip replacement (THR) is made from ultra high molecular weight polyethylene (UHMWPE). The femoral component is of either a cobalt chrome or

stainless steel alloy, less frequently of titanium. Similar
materials are used in total knee replacement (TKR), again one
articulation being of the UHMWPE. The relative success of
these procedures relates to the anchorage of components to the
skeleton by the polymethylmethacrylate cement (PMMA).

Note too, that the UHMWPE components are gamma
irradiated (2 rads), that the monomer of the PMMA cement can
only be Seitz filtered, the powder being irradiated. Metal
components are sterilized in the routine manner.

The PMMA cement is inserted in a dough-like consistency,
after mixing powder and monomer. There is a high exothermic
reaction during final polymerization in situ, temperatures of
72C have been recorded. Free monomer enters the circulation.

In a THR, the following is a guide as to the bulk of
foreign material introduced:-

UHMWPE 17g. PMMA cement some 80g plus 40 ccs of monomer.
Femoral component (cobalt chrome) 193g.

Certain TKR's may have a combined UHWMPE on metal
component of 151g on one side of the joint and a metal
component of 313g on the other, plus 80g cement.

A number of patients have several or more joints
replaced.

Latterly, ceramic component(s) are increasingly used and
PMMA cement is avoided where components have porous surfaces
to allow bony ingrowth and anchorage. Surface area of foreign
materials is much increased.

A considerable amount of foreign materials thus remains
in the body tissues. Same are subject to forces of function,
degradation, corrosion, effects of host tissues and reactions.
These "permanent" prostheses are intended to perform
satisfactorily during the life span of the host.

THE BIOLOGICAL EFFECTS OF Ni Cr, Co AND OTHER METAL ION
UPTAKE PRESENTED AT THIS CONFERENCE IS HIGHLY SIGNIFICANT

There is no such thing as a fully compatible synthetic
biomaterial. All implanted materials degrade, corrode. The
cyclical forces of function impose further stresses. At best
a permanent implant will exist in a state of benign tolerance
with host tissues, performing its intended purpose for the
life span of the patient.

There can be no mechanical solution to a biological
problem particularly in the absence of full knowledge of the
biomechanical physiological and pathological factors involved.
Yet decades of orthopaedic experience suggest that total joint
replacement (particularly THR's) is an efficient method of
treatment for many painful and disabling forms of arthritis.

DATA PROCESSING METHOD

The detailed information obtained from each case record was entered into a data sheet. The latter contained 19 fields covering required information. This included personal details, previous procedures, diagnosis, type of operation performed, immediate and late complications and an anlysis of costs involved.

By written questionnaire or telephone contact details were obtained of any subsequent problems and more importantly of the quality of life enjoyed by the patients.

All records have been retained in the hope of continuing the study as a register or cohort now exists for epidemiological study.

EXTRAPOLATION OF RESULTS

7073 total joint replacement of hip or knee were performed at 22 Sydney metropolitan hospitals between the years of 1969-79 inclusive. 5561 patients were involved.

AGE

64% of patients were under the age of 70 years, the vast majority being in their sixth decade. 36% of patients were over the age of 70. The youngest patient was aged 19, the oldest 92 years.

SEX

56.5% of cases were females, 43.5% males. These proportions are in keeping with the population imbalance between the sexes with increasing age.

PREVIOUS TJR'S

24.4% of cases had one or more total joints replaced or a failed one corrected. While osteoarthritis is usually monarticular both hips may need replacing. In rheumatoid patients multiple replacements can occur. This explains why there were more replacements than patients.

AETIOLOGY

74% of the series had osteoarthritis, in 12% there had been previous major trauma. 14% replacements were for rheumatoid arthritis. In 7% a previous prosthesis had to be replaced. In under 3% there were disabling congenital or developmental causes.

MORTALITY

11% of patients were known to have died from hospital

records. Immediate post-operative mortality was 0.7%. Later patient follow-ups updated the overall mortality to 17%.

DAYS IN HOSPITALS

The average for the whole series was 34 days but this included cases severely ill with rheumatoid arthritis and other reasons for lengthy hospitalizations.

SUMMARY OF COMPLICATIONS

% INCIDENCE COMPLICATIONS IN 7073 TJR's AT 22 HOSPITALS
1969-79

	TOTAL AVERAGE % EARLY/LATE CASES	% VARIATIONS 22 HOSPITALS
INFECTION	6.2%	2.3 - 12.2%
URINARY TRACT INFECTION	6.0%	0.9 - 21.8%
DEEP VEIN THROMBOSIS	7.0%	3.9 - 14.6%
CARDIOVASCULAR	10.8%	4.4 - 19.8%
OTHER SPECIFIED	15.1%	3.8 - 34.4%
DISLOCATION	4.6% (all incidents)	0 - 10.8%
LOOSENING	6.2% (all incidents)	2.3 - 13.5%
NEUROLOGICAL	1.4%	0 - 6.2%
OSSEOUS	2.2%	0.7 - 4.1%
TROCH. OSTEOT. COMPLICATIONS	6.5%	0 - 90.0%
HETEROTOPIC CALCIFICATION	1.3%	0 - 2.5%
REMOVAL TJR	7.2%	2.8 - 14.6%
BROKEN COMPONENTS	0.7%	0 - 1.5%

COSTS

The average cost for the in-patient period for the series was $2,856 per procedure, or nearly $21 million for the series. The costs of expensive medication, transfusions, physiotherapy, rehabilitation, transport are not itemized here and would be additional. The vast majority of cases were not active in the workforce.

Actual contact by phone or mailed questionnaire occurred in 63% of the cases. If finances had allowed, the incidence of contact would have been far higher.

We were interested in the QUALITY OF LIFE experienced by the patients.

Up-dated figures from the replies show that on average:-

26% were totally free from pain, fully ambulant, active, independent, not using any walking aids. = Category A

36% had some restriction of a relatively minor degree, were still active and independent, managed public transport
= Category B

16% required the use of a stick, were not fully active, had some pain and difficulty with transport, were still independent in life style. = Category B-

5% of cases were still very or even more handicapped, requiring constant help and care, had limited movement in a walking frame or were confined to bed. = Category C

Thus 62% (Categories A and B) of previously handicapped patients were more than satisfied with the outcome in a stated period of time.

Survivorship analyses are in progress to determine the average time to failure of a TJR. Removal of such replacement is the ultimate criterion of failure. A removal hip replacement leaves a shortened limb and limp, but a reasonable range of function, if further replacement is impossible. The salvaged knee is a more difficult problem requiring arthrodesis. If this or replacement is impractical disability is markedly increased.

TYPES AND MATERIALS IN TJR PROSTHESES

These have been listed for each case and it is beyond the scope of this paper to itemize such details. Althought manufacturers in some cases claim that the materials comply or conform with public standards IN NONE WAS THERE A KITE MARK OR SPECIFIC LICENCE WITH A SAID STANDARDS BUREAU CERTIFYING THAT THIS WAS THE CASE. The implications are obvious to those dealing with retrieval analysis of failed implants or who investigate implant/host reactions.

IN NONE OF THESE CASES OF REMOVAL WAS RETRIEVAL ANALYSIS PERFORMED. This lack of investigation is world-wide and makes a mockery, in my opinion, of implant studies. Knowledge of longterm effects of metal ions is dependant on such analyses in the large number of "human experiments" being performed. Our knowledge of biocompatibility would increase. There is the ever present fear of longterm carcinogenesis.

CONCLUSION

Health regulatory authorities show increasing interest in benefit analysis of surgical as well as other procedures. While centres of expertise continue to publish results, same are not necessarily effected at national or international levels.

Of concern, is the quality of life that ensues as well as survivorship of the prosthesis and the role that the recipient is able to play in daily or communal living. Few of the elderly patients return to the workforce.

Many of these statistics obtained in this survey such as use of manpower, future projections of requirements, numbers of cases---these and many others are on record. A registry of such cases is in my opinion essential for prospective study particularly in regard to longterm major problems including those of immuno-pathology and possible carcinogenesis.

Participation by clinicians, scientists, administrators and manufacturers in national standards bureau activities and those at international level dealing with surgical implants should be strongly encouraged.

ASPECTS OF THE FAILURE OF THE IMPLANTED SYSTEMS
IN ORTHOPEDICS AND TRAUMATOLOGY

Y. Andrianne, F. Burny, J. Quintin, and M. Donkerwolke

Department of Orthopedics and Traumatology
Cliniques Universitaires de Bruxelles
Hôpital Erasme 808, Route de Lennick, 1070 Brussels, Belgium

INTRODUCTION

A surgical treatment in Orthopedics and Traumatology is often defined in terms of device : plate, nail, prosthesis.

The word "osteosynthesis" (suture osseuse), first used by Albin Lambotte in 1907, is defined in the french dictionary "Larousse" as a "surgical procedure to mechanically stabilize the bone fragments of a fracture with a metallic device to allow the bone union with callus formation". We define the osteosynthesis as "the fixation of a fracture by direct anchorage of a device in the bone". The definition includes external fixation. In 1917, Delbet defines the endoprosthesis as "a definitive, non degradable implant, for a mechanical purpose.

To our best knowledge, no attempt was ever made to define implants on a general basis. In 1983, we introduced (Burny et al.) the concept of the "implanted system" as a set of biological and artificial components linked by interfaces and submitted to biological and mechanical enviroment.

It would be possible to apply to the implanted system the "set theory" as developed by the german mathematician Cantor at the end of the 19th century in order to generalize the concept. Our aim is to construct a logical model of an implanted system, in the field of Orthopedics and Traumatology that possibly could be extended to an implanted system in Biology and Medicine.

The application of the symbolic and formal logic to the biomedical field was pioneered by Woodger (1937, 1939) : any data or conceptual information which can be expressed in ordinary language can be translated into symbolic logic. Statements can be quantitative, descriptive with logical conclusions. Operators are used to qualify a statement or to show the relationship of one statement to another one.

The next step of our research will be to demonstrate the applicability of our approach to actual data.

A living organism is an open system. The flow of energy, mostly mechanical in orthopedics, passes continuously through. If we consider the implanted system of an osteosynthesis, an organizational homeostasis must always remain intact despite the reciprocal aggression host-implant.

The implanted system is separated into morphologically distinct compartments, where different functions are carried out. The individual pieces represent an assembly line in order to achieve the final product : the healing of the fracture, for instance.

As a result of an injury or of a degenerative disease, the implanted system must adapt to biological and mechanical conditions. It should be desirable to evaluate functional deficit on the basis of observed or measured information. It is necessary to describe accurately the possible failures or alterations of an implanted system in order to be able to avaluate the risk.

THE IMPLANTED SYSTEM IN ORTHOPEDICS AND IN TRAUMATOLOGY

A genaral implanted system in Orthopedics or Traumatology is presented in figure 1 ; the different elements are described in table 1.

Table 1 : Elements of a genaral implanted system
in Orthopedics and Traumatology

B (1--->n)	Bone fragments
C (1--->n)	Connecting implants
P (1--->n)	Principal implants
IBB (1--->n)	Interface bone-bone (callus, time related)
ICB (1--->n)	Interface bone-connecting implants
IPB (1--->n)	Interface bone-principal implants

(continued)

IPP (1--->n)   Interface between principal implants

ICP (1--->n)   Interface connecting-principal implants

ICB (1--->n)   Interface bone-connecting implants

ICS (1--->n)   interface soft tissues-connecting implants

IPS (1--->n)   Interface soft tissues-principal implants

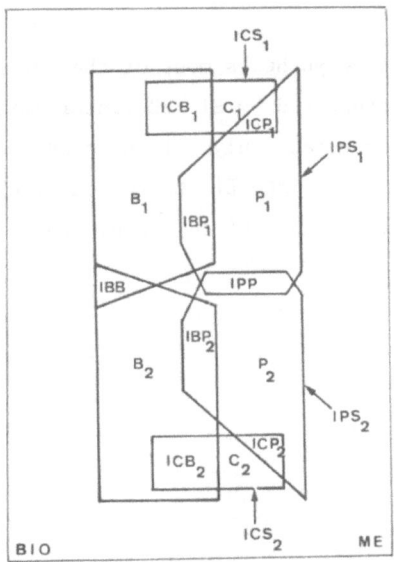

Figure 1

Standard and simplified representation of a general
implanted system in Orthopedics and Traumatology

The Bio-Mechanical environment is represented by the main frame of
fig. 1 and each element is noted in this preliminary paper by BIO or ME
such as MEC(2) meaning the mechanical effects on the connecting implant
number 2, BIOICB(2) the influence of biological factors at the interface
Bone-Connecting implant number 2. As biological and mechanical influences
affect all the system, both will be represented as the main element BIO-ME
of the implanted system.

This paper illustrates only some mechanical aspects of the failure
of an implanted system in Orthopedics and Traumatology.

Osteosynthesis by external fixation

Figure 2 represents an Hoffmann external fixation osteosynthesis.
The connecting implants C are represented most of the time by 6 independant
elements (the pins) with corresponding interfaces :

```
    C    1----->6
ICB    1----->6    (E1  E2)
ICS    1----->§
```

The contact "intermediary implant-principal implant" is represented
by :

```
ICP 1----->6
```

In an external osteosynthesis most of the interfaces between the
components could be represented by E1 E2 (intersection of elements 1 and
2), if we consider an interface only. If the conditions of implantation are
ideal, E1 E2 will transform into E1 E2 (union of elements 1 and 2) and
again change into E1 E2 in case of loosening from biological or mechanical
etiology.

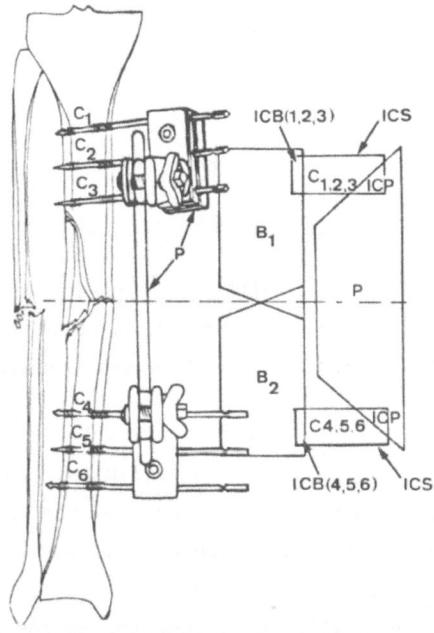

Figure 2

Hoffmann external fixation osteosynthesis

The principal implant itself may be represented by a succession of
elements and interfaces, the external fixation being one of the most complex
system (figure 3). In such a complex example, the content of the table
should be extended.

The pin-bone and pin-soft tissue interfaces are the most critical
and may lead significant complications (figure 4).

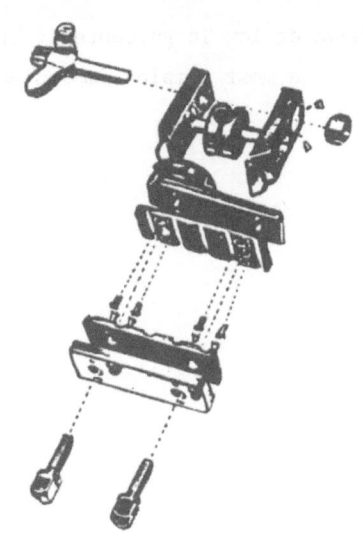

Figure 3

Diagram of a Hoffmann external fixation ball joint

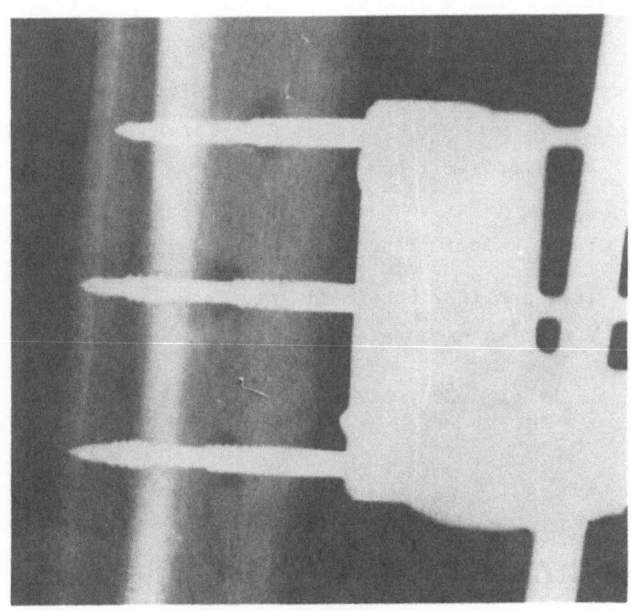

Figure 4

Osteolysis arround a pin

The reactions are however low in percentage. This type of external fixation represents one of the most complex implanted system because of the percutaneous aspect.

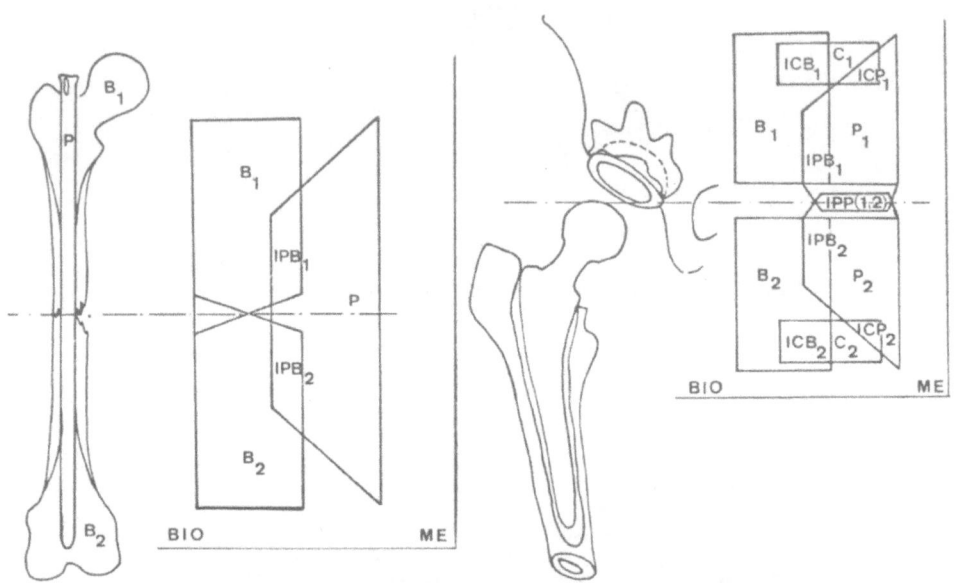

Figure 5   Figure 6
The centromedullary nail   Total hip prosthesis

B (1) proximal bone fragment
B (2) distal bone fragment
P   main implant (centromedullary nail)

IPB1 and IPB2 bone/nail interfaces
(friction anchorage)

Centromedullary nailing

It represents a different aspect of the implanted system : in usual centromedullary nailing, the intermediate implant does not exist, the anchorage depends on the interfaces IPB 1 and IPB 2 (figure 5).

Joint Replacement

The same representation is applied to partial or total joint replacement (figure 6). The different elements are defined in table 1.

FAILURE OF AN IMPLANTED SYSTEM

Three modes of identification seem to be required :

- based on the nature of teh failure (mechanical, chemical, biological, ...)
- based on the level (components and/or interfaces) of the failure
- related to the clinical repercussions of the failure (retrieval or not, replacement or not of the implant,...)

Each component or interface of an implanted system can be responsible for a failure of the bone/implant/soft tissue system. Several elements may contribute to the failure. The concept of the "Implanted System" is useful for the detailed analysis of the failures.

1. The interface IBB and the main implant P

At the time of surgery, the stability of a fracture depends on the quality of the reduction and of the interfragmentary contact as already noticed by Lambotte (1913) : "J'estime qu'il faut ériger en loi la nécessité de la réduction mathématique". After correct reduction, the fracture gap is often small or non existent. In that situation, added rigidity results due to the apposition and rugosity of the surfaces. In the "tension band" osteosynthesis, Pauwels (1965) takes advantage of the interfragmentary contact : the bone is submitted to compression, the implant to tension.

An example of the importance of the interfragmentary contact is given by Halleux (Burny, 1980) for a subtrochanteric fracture treated by a McLaughlin nail plate. For an axial load of 100 kgf, without interfragmentary contact, the calculated stress exceeds the ultimate strenght of the material.

The bone fragments contribute to the strenght of the synthesis by decreasing the bending moment applied to the implant. An increased bending moment may result in the breakage of the main implant by overloading or fatigue failure (figure 7).

Because of callus formation, time-related biochemical changes occur at the interface IBB. Two main patterns of bone healing are described.

The primary bone union (called "callus per primam", healing without callus) was described by Lane and obtained by Danis (1949). The healing is the result of bone remodelling, developing microscopic connection (osteons) between the fragments. We explain this type of union by the lack of mechanical periosteal stimulation (Burny, 1977).

If the bone-implant complex is not rigid, different mechanical conditions exist at the interface IBB producing an elastic mounting. Perioteal callus formation occurs (figure 8). The callus is induced by mechanical stimulation. For many authors, the amount of periosteal callus is closely related to the interfragmentary mobility.

Figure 7
Fatigue fracture of a nail plate

In both cases, the vascularization of B1, B2, ... Bn is supposed to be intact (or restaured) to allow bone union.

Rigid fixation slows down the healing process of a fracture leading to a conflict between the need of long term fixation (to protect the bone union) and the risk of fatigue fracture of the implant due to cyclic loading.

The risk of refracture at IBB also exists after removal of the implant.

In the case of elastic fixation, the problem is to know how much interfragmentary mobility that can be accepted before taking a risk of developing a hypertrophic pseudoarthrosis.

The stresses on the implants decrease quickly when bone healing progresses and callus stiffness increases (figure 9) (Burny et al., 1972).

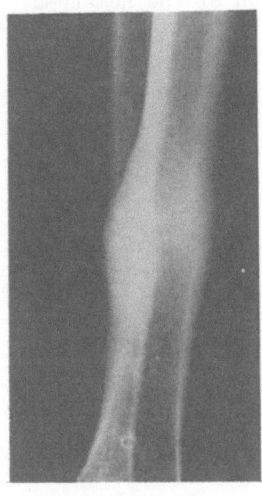

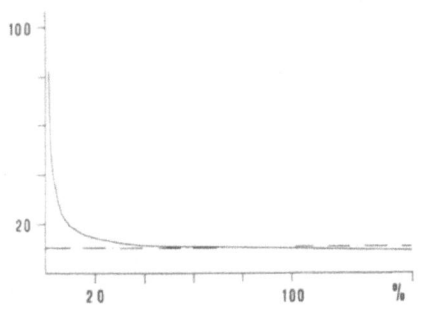

Figure 8

Periosteal callus

Figure 9

Relationship between % of healing (abscissa) and deformation of a plate (ordinate)

At the early beginning of the fracture healing of an unstable fracture all the stress flows through the osteosynthesis. Callus formation decreases the load on the implant.

Bone healing of long bones fractures treated by elastic fixation is often characterized by the formation of a periosteal callus with an exponential profile. We established a modified mathematical model for the callus formation (Burny, 1977). The basic assumption is the development of a callus of equal strenght, obtained by variation of form and dimension.

The interfragmentary motion was measured in patients treated by elastic external fixation, using a strain gauge bonded to the fixation rod (Burny, 1968, 1977, 1979 ; Bourgois, 1984). The deformation plotted against time represents the evolution of the mechanical situation of the fracture (figure 10).

An increase of the interfragmentary mobility (the maximum happening at 3 weeks) was observed in 43% of the fractures and could be explained as the result of :

- micromotion at the fracture site due to the elasticity of the mounting, inducing a grinding of the fracture surfaces and decreasing the interfragmentary contact.

- solubility of hydroxyapatite related to the pH of the hematoma.

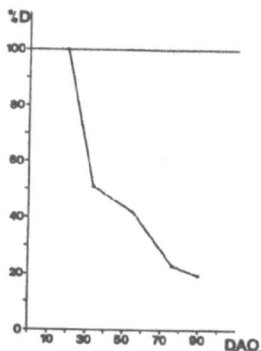

Figure 10
Typical curve of normal healing
Abscissa : day after operation
Ordinate : maximum deformation of the fixation rod

The increase of mobility does not appear in the case of an anatomical reduction or of very comminuted fractures (the mobility is already maximum). The decrease of the mobility as a result of the building process (periosteal callus construction) overcoming the early destruction phase. The slope of the decrease is regular if measurements are repeated every 14 days. The pattern of the curve is very different if daily measurements are performed : large variations of mobility are observed related to actual changes in the stability of the fracture. The dispersion of the values of the deformation calculated on a two weeks period significantly decreases from the begining to the end of a normal evolution.

Because of mathematical considerations, the changes of mechanical characteristics durong the remodelling phase after bone union are not detected with strain gauge measurements technique : the curve reaches a plateau.

The general improvement of the stability is the result of sequential phases of stability and of instability demonstrated by daily measurements of the mechanical properties of the callus. The refracture of the callus seems to be an exageration of this physiological process.

2. The bone fragments B(1...n) and the principal implant P

All implants modifie the biomechanical environment of the bone. The implant should respect the blood supply of the bone and should allow mechanical stimulations in physiologic limits. Blood supply is essential for the normal remodelling of the bone tissue but also for periosteal callus formation.

Most of the internal fixation devices are interfering with the normal blood supply of the bone. Bone necrosis can occur after extended lifting or destruction of the periosteum by a trauma.

After centromedullary nailing or bone cement introduction, the blood supply is disturbed. Experimental plating of a intact long bone produces non perfused areas of the cortex with slow recovery within ten weeks (Gunst, 1980).

In fractures treated by plate, the complex bone/plate must provide a good stability during the revascularization of the cortex. Bone necrosis can be responsible for failure of the implanted system.

Implants also alter the bone remodelling processes significantly. This situation, recognized by Schenk et al. (1964), was attributed by Uhtoff et al. (1971) to the "protection" of bone by the implant. The more rigid the implant, the more disturbed bone physiology (Tonino, 1980). Rigid plates decrease the torsional strenght of the rabbit tibia down to 50% of the initial value after 12 weeks. Recoverying after removal of the plate requires 6 weeks (Strömberg et al., 1980).

At the extremities of an internal principal implant, stress gradient effect produces bone formation, fatigue fracture or common fracture, according to the nature of the machanical stimulus.

The main implant represents a potential pitfall for the survival of the implanted system and for the mechanical characteristics of the bone after implant retrieval.

3. The principle of symmetry

From a mechanical point of vew, an implanted system is a succession of elements, the strenght of the mole whole being that of the weakest element (Burny et al., 1980).

The different elements of the strenght of the osteosynthesis are summetrically distributed at each part of the principal implant P and of the fracture site.

In table 1, B represents a bone fragment but also the mechanical characteristics of the bone fragments ; C represents the connecting implant but also the mechanical characteristic of the connecting implant,...

If S1 represents the set of mechanical characteristics related to the proximal fragment or group of fragments, S2 that of the distal group, idealy :

$$S1 = S2$$

This principle of symmetry also apply to total joint replacement (figure 6). It also applies to partial implants such as Moore prosthesis or dental implants.

A total joint or an osteosynthesis represents a succession of elements symmetrically distributed proximally and distally to a plane of symmetry represented by the fracture line or by the prosthetic joint. This "symmetry" is not a geometrical but a symmetry of the mechanical properties of the implanted system. If the principle of symmetry is not respected, failure occurs at the weakest element.

4. The interfaces ICB and IPB

To support the load applied to the skeleton, the implant is to be firmly connected to the bone. Sudden failures of the implanted system are sometimes related to the loosening of the achorage system at the interfaces ICB or IPB. The anchorage is obtained by the association of the connecting

implants (C, interface ICB), by the main implant (P, interface IPB) or by
both. The anchorage is by friction (nails), by thread cutting (screws and
pins), by bone ingrowth or by a combination of different fixation types.

The anchorage of an osteosynthesis is usually obtained by threated
implants. There is a relationship between the holding power of a threated
implant (measured by the shearing torque) and the thickness of the cortical
bone (Burny et al., 1983 ; Wagenknecht et al., 1984). The bone thread can
be destroyed, not only by overloading, but also by inadequate technique or
insertion of the threated connecting implant. this problem is complex for
self drilling and self tapping pins of external fixation. Osteolysis at
the interface ICB can lead to a total loss of anchorage of the synthesis.

In external fixation, the problem also exists, especially in the
osteoporotic bone. Bone cement prevents this type of failure (figure 11).

The quality of the anchorage is untill now a question of clinical
experience. The ideal number and the types of screws or pins, the optimal
diameter of a centromedullary nail, the best type of femoral tip of a total
hip prosthesis are unsolved problems. There is no procedure to measure the
bone quality at the time of insertion (Burny et al., 1983).

5. Interface ICP

When present, it may by responsible of numerous failures in internal
fixation as well as in external devices.

Typical examples are given by Burny et al. (1972, 1980). The failure
of the bolt of the Mc. Laughlin nail plate is well documented (figure 12).

Excessive shearing stresses at the screw heads in the case of plate
fixation lead to fracture of the screws.

CONCLUSIONS

The relation between the bone and the implant is a difficult problem
that should be considered in the framework of "the implanted system",
including the biological and the mechanical environment.

The concept of the implanted system seems to be useful for the
description of failures of orthopedic implants.

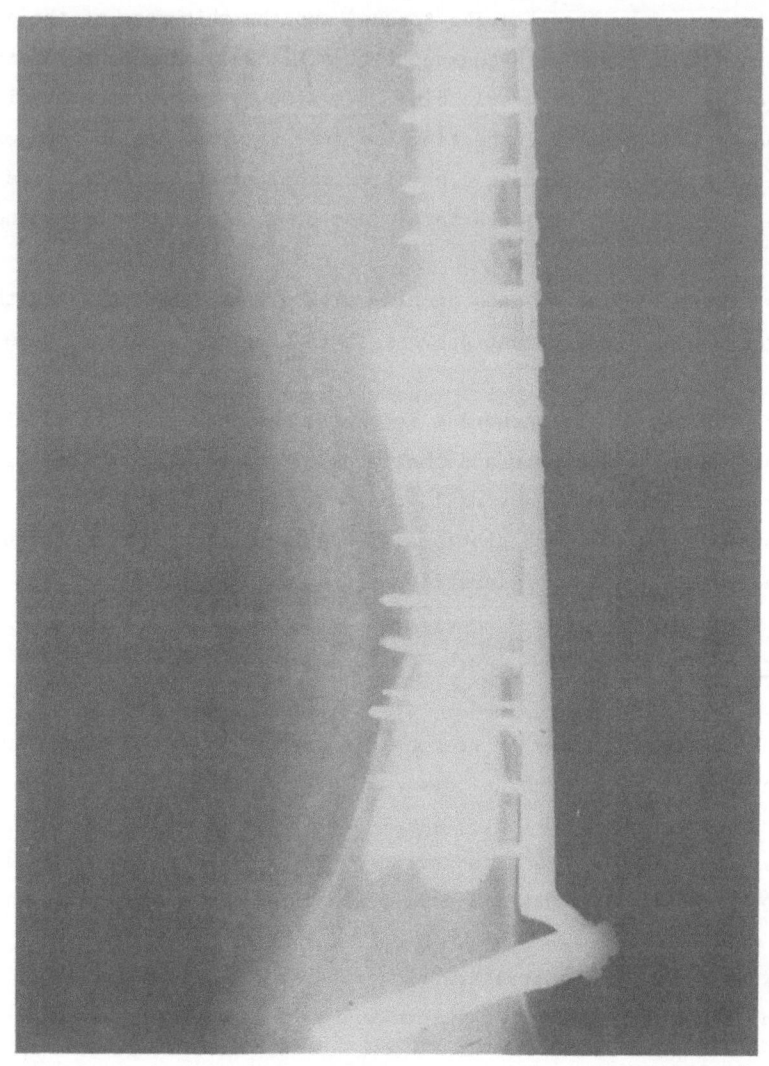

Figure 11

Bone cement used to prevent the failure of the screw
anchorage in osteoporotic bone

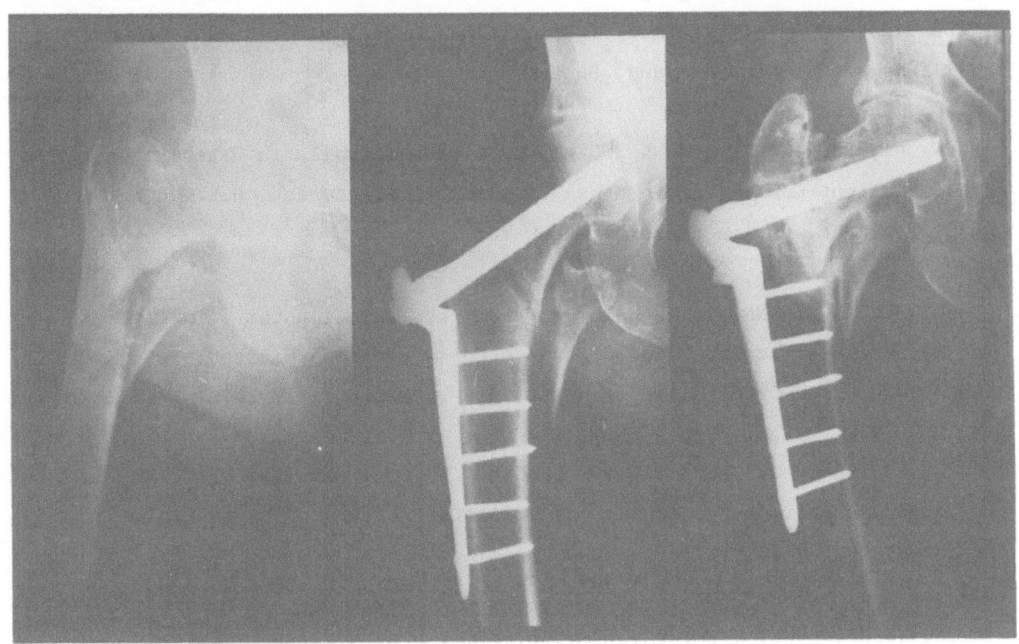

Figure 12

Failure at the interface ICP of a Mc. Laughlin mail-plate

We just present here examples of mechanical failures, some including the effect of the biological environment.

The proposed presentation of the possible failures of an implantes system in Orthopedic and Traumatology should be the basis of a classification of the failures.

BIBLIOGRAPHIE

BURNY F., 1968, Etude par jauges de déformation de la consolidation des fractures en clinique, Acta Orthop. Belg., 34 : 917.

BURNY F., BOURGOIS R., AUBRIOT J.H., 1972, etude théorique et expérimentale des défaillances du matériel d'ostéosynthèse. Corrélation avec la réalité clinique et conclusions pratiques, Proceedings of the 12th Congress of the SICOT, Tel Aviv, International Congress Series n° 291 (ISBN 90219 0213 3) : 167.

BURNY F., 1977, Mesure de la rigidité du cal au moyen de jauges extensométriques, Rev. Chir. Orthop., 63, Suppl. 11 : 816.

BURNY F., 1977, L'organisation mécanique du cal, Rev. Chir. Orthop., 63, Suppl. 11 : 178.

BURNY F., 1979, Strain gauges measurements of fracture healing. A study
of 350 cases, in : "External fixation; The current state of the art", The
Williams and Wilkins Company, Baltimore.

BURNY F., BOURGOIS R., DONKERWOLCKE M., 1980, Pitfalls in Osteosynthesis,
in : "Current Concept of internal Fixation of Fractures", H.K. Uhthoff,
ed., Springer, Berlin Heidelberg New York.

BURNY F., ANDRIANNE Y., QUINTIN J., 1983, Les défaillances d'implants :
essai de classification, communication to : "Riunione Nazionale di
Biomeccanica in Ortopedia e traumatologia", Pesaro, Italy.

BURNY F., SARIC O., BOURGOIS R., DENKERWOLCKE M., ANDRIANNE Y, 1983, In
vivo measurement of the mechanical properties of bone tissue. Preliminary
results, Eng. Med., 12 : 43.

BOURGOIS R., BURNY F., 1984, Feasibility of bone healing measurement with
external fixation : experimental study, Orthopedics, 7 : 673.

DANIS R., 1949, Théorie et pratique de l'ostéosynthèse, Masson, Paris.

DELBET P., 1919, Endoprothèse en caoutchouc armé pour perte de substance
du squelette. Mém. Acad. Méd. (Paris).

GUNST M.A., 1980, Interference with Bone Blood Supply through Plating of
Intact Bone, in : "Current concept or Internam Fixation of Fractures",
H.K. Uhthoff, ed., Springer, Berlin Heidelberg New York.

LAMBOTTE A., 1913, Chirurgie opératoire des fractures, Lamertin, Bruxelles.

PAUWELS F., 1965, Gesammelte Abhandlungen zur funktionellen Anatomie des
Bewegungsapparates. Springer, Berlin Heidelberg New York.

RHINELANDER F.W., 1980, Vascular proliferation and Blood Supply during
Fracture Healing, in : "Current Concept of Internal Fixation of Fractures",
H.K. Uhthoff, ed., Springer, Berlin Heidelberg New York.

SCHENK R. and WILLENEGGER H., 1964, Zur Histologie der primäre Knochenheilung,
Langenbecks Arch. Klin. Chir., 308 : 440.

STROMBERG L., DALEN N., LAFTMAN P., SIGURDSSON F., 1980, Atrophy of Cortical

Bone caused by Rigid Plates and its Recovery, in : "Current Concept of Internal Fixation of Fractures", H.K. Uhthoff, ed., Springer, Berlin Heidelberg New York.

TONINO A.J., KLOPPER P.J., The Use of Plastic Plates in the Treatment of Fractures, 1980, in : "Current Concept of Internal Fixation of Fractures", H.K. Uhthoff, ed., Springer, Berlin Heidelberg New York.

UHTHOFF H.K. and DUBUC F., 1971, Bone Structure changes in the dog under rigid fixation, Clin. Orthop., 81 : 165.

WAGNEKNECHT M., ANDRIANNE Y., BURNY F., DONKERWOLCKE M., 1984, Study of the mechanical characteristics of external fixation pin anchorage : preliminary results, Orthopedics, 7 : 629.

WOODGER J.H., 1937, The Axiomatic Method in Biology, The University Press, Cambridge, Mass.

WOODGER J.H., 1939, The technique of theory construction, Encyclopedia of Unified Science, vol. 2, n° 5, University of Chicago press, Chicago.

CLINICAL INTOLERANCE TO PROSTHESIS MATERIAL

INCLUDING METALLIC IMPLANTS

M. Donazzan, M. Chanavaz, L. Duret, C. Véron
and J.P. Fernandez

Service de Chirurgie maxillo-faciale, C.H.U.
Hopital B   59037 Lille Cedex, France

INTRODUCTION

Excavations of Egyptian or Chinese tombs have brought to
light every now and then the existence of metallic fragments
in the human body apparently used as therapeutic devices.
Apart from this, the rational use of metals in traumatic
pathology dates back to the last century when about 1880
LISTER used a silver wire to splint a fractured knee-cap,
which was followed some six years later, in 1886, when
HANSMANN placed screwed plates for the surgical treatment
of long bone fractures.

The earliest studies on the biocompatibility of metal
were carried out by LAMBOTTE in 1909. As from 1930, ERDLE and
PRANGE finalized the formula of Vitallium which was the first
chromium - cobalt alloy used in 1934, for the making of fixed
cast prostheses. The same alloy was used in 1936 for the first
time in osteosynthesis by San Antonio surgeons VENABLE and
STUCK. As for nickel-chromium alloys, their presence in odonto
stomatology goes back only about thirty years.

ADVANTAGES

Owing to their remarkable mechanical properties, these
alloys containing different proportions of chromium, cobalt
and nickel have been extensively used either in cast dental
prostheses or in various devices employed in orthodontics, or
in plates, screws and other fixtures used in maxillo-facial
surgery. Some current periosteal and more rarely endosseous
dental implants are made of these alloys. The diversity of
their application in the field of odonto-stomatology has
aroused increasing interest in these alloys, especially as
they are also much less costly than precious metal alloys.

DISADVANTAGES

The frequent use of Co, Cr and Ni alloys has made it
possible to evaluate a number of disadvantages closely related
to different clinical intolerances observed. These clinical
manifestations, which are usually very discreet but occasionally

acute and persistant, may induce the surgeon to suggest the removal of the appliance in question.

The therapist must always have in mind that these materials may
1. cause or undergo mechanical stresses within the neighbouring soft or hard tissues either physiologically or pathologically ;
2. suffer corrosions of variable degree in contact with biological environment (saliva, bone and soft tissues). These corrosions, liberating Ni, Cr or Co may induce reactions of intolerance because of their toxic or allergic origine.
3. in the oral cavity in particular, the presence of different metallic dental restorations may stimulate the appearance of electro-galvanic currents.

In the paper, we shall report successively :
- the clinical symptomatology frequently non-specific,
- the different aspects of etiological diagnosis,
- and, finally, the preventive measures prescribed by the clinician.

CLINICAL SYMPTOMS OF INTOLERANCE

These symptoms are often insiduous, progressive or on the contrary sudden and intense, requiring careful study in order not confuse them with other symptomatologies. They are not specific. They could thus be seen in the oral cavity or at a distance from it. We shall endeavour in this paper to make a composition study of all the symptomatology as extracted from the literature.

1. Intra-Oral Symptoms

As from 1951, Edwin SIDI and F. CASALI declared that "the intolerance appearing in different parts of the mucosa of the buccal cavity shows a great degree of polymorphism".

a. Subjective Manifestations

In 1934, Mac DONALD asserted "The subjective signs are many and varied. The commonest is bad breath. I wonder how many people possessing bad breath have questioned the efficiency of their entire gastro-intestinal tracts when the blame could be found at the very doorway. A metallic taste is quite common, together with a sense of fullness in the mouth... Individuals who smoke complain of burning tongue. Those who never smoke also state that their tongue is easily irritated by simple foods. The tongue appears to be hypersensitive".

In addition to this metallic taste in the mouth, the patient may complain of the sensation of his gums retracting whilst his mouth and pharynx feel completely dry. Some patients suffer from a simple metallic contact with their mouth. The mere contact of a fork, needle or aluminium wrapping paper used for example in chocolate packing with their mouth triggers or maintains the impression of metallic taste.

More often, the patients are alarmed with the installation of itching or pricking feeling in their rear throat, which brings them to consult an ENT specialist whose examination turns out to be negative. These abnormal sensations such as stinging, tickling or existence of a foreign body in the mouth may be accompanied with the exaggeration of sensitivity to salty or spicy food (D. LYON, Thesis from Strasburg, 1978). There may even exist sudden sharp pain in contact with a fork or spoon, or provoked by simple placing of an amalgam in a prepared dental cavity.

We have had the opportunity of observing a patient, about sixty years old, who complained of such a severe sensation of pricking in his pharynx that he virtually stopped feeding himself and led a very perturbed life. This oral symptomatology which accompanied a general pruritus was found to be related to the existence of dental restorative poly-metallism, and especially a stainless steel screw-post in one of the upper lateral incisors. The patient recovered a perfectly normal life immediatly after the screw-post was removed.

Severe glossodynia may also occur. These are often in form of burning sensation without any notable mucosa alteration. When associated with buccal dryness they may be confused with Sjogren syndrom, with the risk of starting not only a very costly examination procedure, but also a needless one.

Smoking may aggravate the burning sensation of the tongue.

According to MOUTIER in 1944, "it is not exceptional in the case of the patients with all these subjective, strange and tyrannical sensations to be taken for cranks and abandoned to their "obsessions". Some patients can literally become "possessed" by the abnormal sensations and end up as real psychopaths. This is unfortunately the frequent suspected etiology when examining a newly retired man or a recently menopaused woman suffering from these symptoms.

This subjective symptomatology should always bring in mind the possibility of existence of endobuccal polymetallism, but also the general application of alloys in such as screw-posts etc... in odonto-stomatology.

b. Objective Manifestations

Soft tissue : this consists essentially of inflammations of variable intensity involving the mucosa of oral cavity, to which sometimes the lesions of dyskeratosis are associated.

The most visible of these manifestations is cheilitis, which could sometimes assume a pemphigoïd appearance (FISCHER).

The erythro-stomatitis is recognized by the existence of painful and red mucosa around and under dental prostheses. This stomatitis is particularly frequent in the palate and is usually very well-defined by the contour of the metallic plate compressing the palatal soft-tissue ; the patient complains invariably of having very inflamed gums, which are graduelly relieved as he progressively abandons wearing his prostheses.

But whilst carrying the prosthesis, its metal surface

might cause extensive area of soft tissue irritation within hours after insertion, consisting of small clear vesicles with an erythematous border, which will eventually break, leek, become an open ulcer and is finally covered by agrey yellow slough (FENTON).

With regard to the tongue, glossitis is often associated with some oedema and sometimes even aphtoïd ulcerations, or depapilated zones, especially around the edge of the tongue.

Some authors such BANOCZY and SOREMARK have described bilateral superficial white lesions corresponding to zones of dyskeratosis on the inner wall of the cheeks, the lingual edges and the gums.

Tooth : these manifestations may involve either :

    1. tooth roots bearing artificial posts, in which case
        the pain is of desmodontitis type. The clinical examination might reveal a colour changing of the tooth and sometimes the neighbouring gingival mucosa which tend to become darkish. The retro-alveolar X-ray data show either :
- irregular contours or
- limited loss of substance or
- total resorption of the intracanal pivot and sometimes actual root fractures.

    2. the studies of SOREMARK have shown the penetration of
        metallic ions in the already decalcified dental enamel in contact with metallic denture anchorages.

Bone : all local stress or agression to a bone is met by bone
        resorption or osteolysis :

    1. metallic osteosynthesis, either using wires or plates,
        usually cause none or very little negative side effect in bone consolidation, in very rare cases when there may appear a slight sign of intolerance, the simple removal of metallic appliance is enough to allow the bone healing to be completed.

    2. as much as modern dental implantology offers greater
        predictibility, involving very improved surgical procedures, the failure in stability and longitivity of these implants are more dramatical in consequences. The literature is relatively rich in explaining the causes of implant failure.

Implants : they can also be the sources of bone resorption
        when they transmit to the adjacent bone the inadapted occlusal stresses caused by the loading prosthesis.

Some recent reports have also brought to the light the possibility of metallic induced bone resorption. Most of these cases have a direct relationship with the mechanical over-stresses applied to the implants for the purpose of shaping them before insertion. These stresses are considered to cause metal fatigue which in due course triggers metal corrosion. Some immediate corrosion of iron containing implants (ramus blades) especially when the implants have been scratched whilst bending them without immediate passivation are most demonstrative. This has probably been the main motivation for the manufacturers who have recently replaced these alloys by Titanium containing implants.

As a matter of fact the Titanium alloys (with exception of ceramic, sapphires and hydroxyl apatite implants) are almost the only alloys used in dental implantology.

Salivary : are very rare and extremely difficult to prove.

2. Extra-Oral Symptoms

a. Regional Manifestations

BORK 1978 reported cases of chronic rhinitis, connected with an allergy to Ni in dental prosthesis, which disappeared once the appliance was removed.
GRIMALT and ROMAGURA, 1980 reported a case of palpebral oedema.

b. General Manifestations

Itching : either localised or diffused, without an associated skin lesion, which might induce the clinician to a diagnosis error.
Eczema : which might occur either locally or at a distance from oral cavity.

In most of these cases the simple removal of intra oral metallic material is sufficient to allow the disappearance of itching and eczema. When the healing process is slow, it might be valuable to carry out Patch Tests in order to determine specific sensitivities.

3. Infra-Clinical Symptoms

These symptoms ara related to tissular microscopic alterations under the prostheses. HILDEBRAND et al. reported ultrastructural studies of the mucosa under Co-Cr removable prostheses. The irritated zones displayed important alterations of the epithelium and connective tissue. In the epithelium, frequent ruptures of the basal lamina, widening of lymphocytes, Langherans-like cells and other non-epithelial cells were observed. In the connective tissue, the cell density was increased and lymphocytes and plasmocytes were frequent. In all biopsies, electron dense particles (diameter 50 nm) were present in epithelial cells as well as in cells of the connective tissue.
The particles were located in the cytoplasm of all cells and in their nuclei. A relationship could be made between the number of particles and the irregularity of the nucleus shape. The electron probe microanalysis revealed that chromium was the main component of these asteroïd particles. It is noteworthy to emphasize that the particles had penetrated not only into irritated mucosa but also into mucosa without any clinical symptom of intolerance.

CONCLUSION

The clinician must always have in mind and preferably recognize various intolerances to metal alloys. He should then endeavour to confirm his diagnosis by the appropriate tests.

And then adopt the only course of treatment which is either
the simple removal of the particular metallic appliance or
its replacement with a different biomaterial.

REFERENCES

1. B. Angmar-Mansson. Root fractures due to corrosion. Odental
        Revy, 1969, vol. 20.

2. B.J. Banoczy, B. Roed-Petersen, J. Pindborg and J. Inovay.
        Clinical and histologic studies on electrogalvani-
        cally induced oral white lesions. Oral Surgery,
        1979, 48,

3. M. Bergman, B. Bergman and R. Soremark. Tissue accumulation
        of Nickel released due to electrochemical corrosion
        of non-precious dental casting alloys. J. of Oral
        Rehabilitation, 1980, 7 : 325-330.

4. D.L. Brendlinger and J.J. Tarsitano. Generalized dermatitis
        to sensitivity to a chrome cobalt removable partial
        denture. J. Amer. Dent. Ass., 1970, 81 : 392-394.

5. P.F. Caitucoli et S. Levy. L'électrogalvanisme buccal et
        sa pathologie. Encycl. méd. chir., Paris, 1984,
        Stomatologie II, 23063D10.

6. B. Duperrat. L'allergie au Nickel. Presse Médicale, 1962,
        70, 47 : 2213-2214.

7. A.A. Fisher. Safety of stainless steel in Nickel sensitivity.
        J.A.M.A., 1972, 221, 11.

8. A.A. Fisher. Contact dermatitis, Philadelphia, 1975, Lea
                                                and Febiger.

9. A.A. Fisher. Nickel : omnipresent contact allergen. Cutis,
        1979, 3, 21 : 143-149.

10. S. Fregert. Allergy to chromium, nickel and cobalt. Acta
        Derm.-Vénéréol., 1966, 46.

11. H.F. Hildebrand, C. Véron, M.C. Herlant-Peers, J.P.
        Fernandez et J.P. Kerckaert. Les conséquences
        biologiques de la présence d'ions métalliques
        dans la cavité buccale. Revue d'Odonto-Stomatologie,
        1984, 13, 1 : 41-56.

12. J.E. Lemons. Biomaterial considerations for dental implants.
        J. of Oral Implantology, 1975, 4 : 503-515.

13. M.F. Moutier. Polymétallisme dentaire et dysesthésies
        linguales (stomatite électrogalvanique). Soc.
        Gastro-Entéro., Paris, 1944, 133-135.

14. K.B. Petersen. Longitudinal root fracture due to corrosion of an endodontic post. <u>J. Canada Dent. Ass.</u>, 1971,2.

15. M. Rougier et R. Kuffer. Allergie es Stomatologie. <u>Encycl. méd. chir.</u>, Paris, 1983, Stomatologie, 22050M10.

16. E. Sidi. Les intolérances de la muqueuse buccale. <u>Presse Médicale</u>, 1951, 59, 35.

17. W.R. Schriver, R.H. Shereff, J.M. Domnitz, E.F. Swintak and S. Civjan. Allergic response to stainless steel wire. <u>Oral Surgery</u>, 1976, 42, 5 : 578-581.

18. R. Soremark. Penetration of metallic ions from restorations into teeth. <u>J. Prosth. Dent.</u>, 1968, 20, 6.

CLINICAL ASPECTS ON THE CORROSION OF DENTAL ALLOYS

Rune Söremark

Karolinska Institutet

Stockholm, Sweden

CORROSION IN THE ORAL CAVITY

In the human oral environment the conditions are such that none of the dental alloys used can stay completely resistant against corrosion. In the mouth it can be considerable fluctuations in

- the quantity and quality of saliva
- pH
- the amount of bacterial plaque
- attrition, abrasion, and mechanical stress
- the physical and chemical properties of food and liquid
- the temperature
- the intake of drugs
- the general and local health conditions, etc

Under such circumstances it is not surprising that the alloys corrode. On the contrary, it is surprising that the alloys do not corrode more than they do; especially when considering that most oral cavities have a mixture of various types of dental alloys.

Dental alloys in the humid oral cavity have different galvanic potentials. There are not only different potential levels between various alloy restorations but also between different areas within the same restoration. The possibility of corrosion occurring in the oral cavity cannot be assessed from potentials only. The potential provides only a measure of the tendency for corrosion to occur. Even if there is a tendency for corrosion to occur, the metallic surface can be passivated by biological films and/or by protective layers of corrosion products. Such films and layers affect the tendency for corrosion to occur and can, for all practical purposes, stop it. Thus the most interesting electrochemical parameter when evaluating the possible effects of corrosion in the oral cavity is the current between the two alloy parts in question and the current density. However, such measurements are technically difficult to do in the mouth. It should also be noted that the small intraoral currents which can occur among patients with intolerance problems caused by corrosion products are of the same order of magnitude as those which occur among groups of persons without any symptoms.

The passivating films can be disrupted chemically or mechnically and then an electric current starts, i.e. the alloy starts to corrode. Depassivation is thus the destruction of the inert protective surface that prevents corrosion. The passivated surface will normally reform, however.

In experiments it has been shown that sealing of amalgam specimens (the anode) had little effect on the corrosion rate. However, when the sealing was placed on the gold specimens (the cathode) the corrosion rate was decreased. The sealing effect on the gold specimen decreased the cathode's access to oxygen and thus reduced the corrosion process. An anodic polarization behaviour is often dependent on the chloride ion concentration available and the cathodic process on the oxygen concentration. However, the cathode/anode surface area ratio, the composition of the alloys, and other factors influence to a high degree the corrosion rate.

The mechanism of corrosion is complex and consists of various components: depassivation, mechanical stress or damage to the metal surface, local chemical factors, area relations between the anode and the cathode, the composition of the alloys, mixture of alloys, the properties of the saliva and food. The properties of the saliva - the unstimulated saliva - seem to play a major role.

When a metallic restoration is placed in the mouth of a patient it takes some time for a passivating layer to be formed on the surface of the restoration. Until such a passivating film is formed there is an active corrosion. Immediately after a removable partial denture with a metal framework is put in function scrapings from the tongue show high concentrations of the metals. After some days the scrapings show less metals. In some patients, however, the concentrations of metals remain high in the tongue scrapings for extended periods of time.

Corrosion products released in one solution are often different from those released in a slightly different solution. Saliva and other body fluids contain corrosive substances as well as corrosion inhibiting substances. The interplay of these two groups of substances is dependent on a number of factors. Acidic food and salt rich food can enhance the corrosion. The mixture of various alloys in the same oral cavity may create more corrosion as compared to the situation when only one type of alloy is present. When amalgam is in contact with gold the amalgam filling may corrode fast if the area of the amalgam filling is small compared to that of the gold restoration.

When bacterial plaque is covering a part of an alloy surface there is a decrease in oxygen under the colony. Thus an anodic reaction (metal oxidation) takes place under the bacterial colony. The cathodic reaction (oxygen reduction) takes place on that part of the alloy restoration surface which is free from plaque and in contact with saliva. In the bacterial plaque the growth of anaerobes is facilitated when the oxygen is depleted and the dissolution of metal continues. This results in an excess of positive charged ions in the area. Metal chlorides are formed and their concentration increases. At the same time bacteria produce organic acids which change the pH of the environment and cause the disintegration of the pellicle.

Some anaerobic bacteria present in the oral cavity can reduce sulphates to sulphides. Sulphate reducing bacteria act as depolarizers or oxidizing agents by removing hydrogen from the metal surface and thereby driving the anodic reaction, the dissolution of the metal. The hydrogen is utilized by the microorganisms for the reduction of sulphate to sulphide. There is also a production of hydrogen sulphide in the crevicular fluid and periodontal pockets. H2S and HS- are weak acids while the sulphide ion S2- a strong alkali. In the temperature conditions of the oral cavity sulphide ions can react with metals and form different sulphides. Other sources of sulphur which can influence dental alloys in situ are various types of food, i.e. egg, and in some areas drinking water.

DISTRIBUTION OF CORROSION PRODUCTS

The uptake of corrosion products in teeth varies and can sometimes be considerable. Autoradiographic studies and neutron activation analyses show the corrosion of the radioactive metal in the alloy and the distribution of the metal into enamel, dentin, and pulp. In three weeks the concentration of Cr and Co is between 10 - 50 ppm in the enamel and the dentin of the tooth which has had a Co-Cr-dental clasp on the intact enamel surface. A combination of stainless steel, amalgam cores, and gold crowns may discolour the roots and the gingiva by corrosion products; in such mixed alloy situations there can be pronounced corrosion.

The corrosion products are taken up by the teeth, the saliva and the oral mucosa. Most of the corrosion products are excreted by faeces and urin. However, a fraction may accumulate in specific target organs and tissues, where the element can interfere with the metabolic processes. Corrosion products can, sooner or later, be harmful to the tissues in the oral cavity and/or in other parts of the body.

By means of radioactive metals we have in experimental animals studied autoradiographically the distribution patterns of Ni, Co, Cr, Cd, and some other metals used in dentistry. It is likely there are similar distribution patterns of the various alloys in humans. (Fig.1)

Chromium was found to accumulate in cartilage, yolk sac, skin, liver, renal cortex, myocardium, and mammary glands.

Cobalt accumulates in fetus, yolk sac, bone, myocardium, liver, and central nervous system.

Nickel accumulates in skin, central nervous system, lung tissues, and kidneys.(Fig.2)

Exposure to nickel, chromium, and cobalt may cause allergic eczematous dermatitis. Such eczematous reactions of the contact type in the skin may be caused by small amounts of the metals reaching the skin after corrosion in the oral cavity of the dental alloy and after being resorbed by the blood. This kind of systematically induced eczematous reaction has the same pathogenesis as the usual exogenic form (external contact eczema), with the exception that the metal comes into contact with the skin from the inside through the blood. It should be kept in mind that these elements seldom give any reactions intraorally.

Although urinary excretion is the main route of elimination, the accumulation of chromium, cobalt, and nickel in the kidney, particularly in the renal cortex, indicates a more specific local interaction. The extent of renal excretion of these metals probably plays a major part in determining their accumulation in the body. However, relatively little is known about the extent to which filtration, reabsorption, and secretion determine the rate of excretion. The sensitivity of the kidney

^{115}Cd   2 days

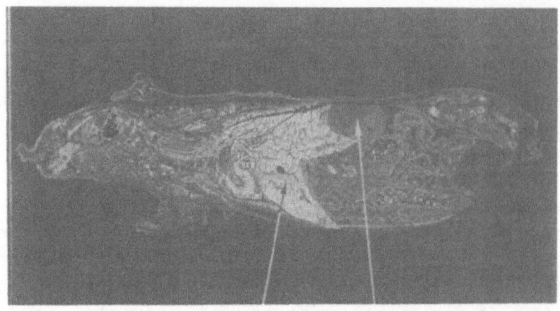

Liver   Kidney

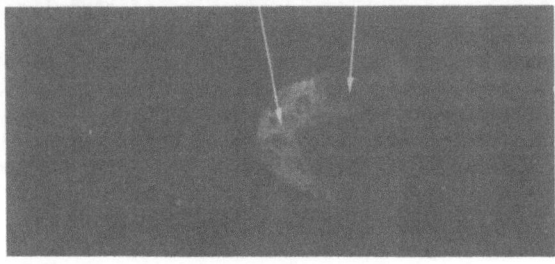

Fig 1. Autoradiogram showing the uptake in a mouse of cadmium in the
liver and in the kidney two days after intra venous injection
of radioactive Cd. The corresponding histological section, a
20 μm thick sagital section, is shown above

Brain   Lung   Spinal cord   Vertebrae

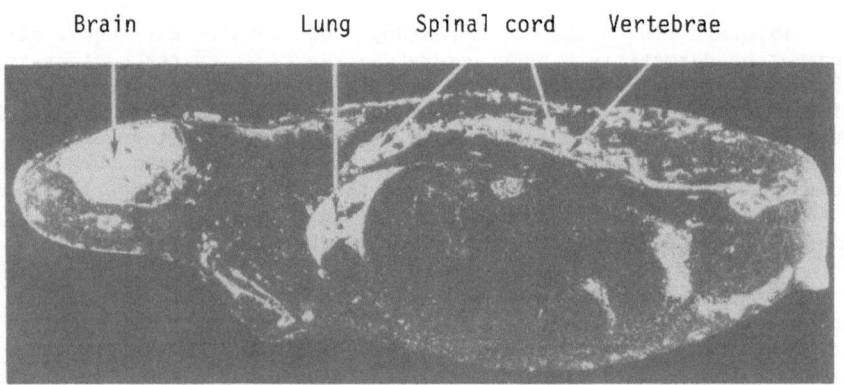

Fig 2. Autoradiogram from a sagital section, 20 μm thick, through a
mouse showing the uptake of radioactive nickel

to toxic effects of heavy metals is well documented, particularly with respect to the effect on kidney tubules.

High concentrations of chromium, cobalt, nickel, as well as many other substances, have been found in the liver. Accumulation in the liver is most probably due to the portal transport system, whereby the liver functions as a collecting pool.

Chromium, cobalt, nickel, and mercury accumulate in the central nervous system. The penetration of mercury through the blood-brain barrier and its specific affinity for certain neural structures has been correlated with clinical findings. The mercuric ions primarily affect certain specific regions of the central nervous system, such as the choroid plexus, that is devoid of tight junctions between the endothelial cells. The specific sensitivity of small cells in the central nervous system to mercury compounds may explain the pattern of damage found in the cerebral cortex and the cerebellum. It has also been reported that mercuric ions affect synaptic transmissions. Very little is known about the effects and the retention of nickel, chromium, and cobalt in the nervous tissues.

The "normal" values of the amount of Cr, Cu, Co etc in the body of healthy adults reported in the literature vary considerably. The values vary with a factor of 10 or more when the same analytical method is used. Including different analytical methods the values reported for a given element vary even more.

Metals in contact with skin (fingerrings, earings, buttons) corrode and metals in the food, water, and air increase the total body burden of the metals. There are also great fluctuations in the concentrations of trace elements, such as those mentioned, from one person to another and from one time to another in the same individual. A certain increase of an element in the body may not cause any biological problem for one individual but can cause problems for another.

INTOLERANCE REACTIONS

Stainless steel, brass, gold coated devices,and other dental alloys may under certain circumstances be subject to a pronounced corrosion and may create intolerance reactions in some patients. Players of brassinstruments sometimes have oral problems when they have dental alloy restorations and particularly if they often drink fluids with low pH. When a temporary crown made of aluminium is used many patients suffer an intense burning sensation in their mouths.

Reports have been presented showing patients with Ni eczema in the skin but without tissue reactions intraorally after having orally been rehabilitated with prostheses with Ni containing alloys. About 15 per cent of all women and about 6 per cent of all men are allergic to nickel. Much higher figures have recently been published in the U.S.A. There is consequently a large number of people hypersensitized to nickel. They must for the rest of their lives always avoid contact with Ni-containing materials, tools, and food and, of course, it is important that they are never treated with any Ni-containing dental alloy.

Diagnosing the etiological factors associated with an autoimmune disease is hindered by the delayed response at the immune system, the gradual onset of symptoms, and the presence of symptoms distant from the site of sensitization. As was mentioned the dermatitis and urticaria associated with nickel sensitivity are most often found distant from the nickel source. Hands, neck, and eyelids are the most common areas of dermatitis and urticaria, regardless of the location of the sensitizing nickel alloy.

There is a great variety of factors that can influence the development of hypersensitivity. The most important are mechanical irritation, tissue maceration, individual susceptibility, temperature, and duration of exposure.

It should be mentioned that in some sensitive patients there have been acute and dramatic reactions with vomiting and fever immediately after the dentists have been grinding on restorations made of nickel based alloys, as well as in amalgam fillings. This is probably due to the great exposure of alloy dust to the mucous membrane of the oral cavity.

Clinical studies have shown that many patients with oral lichen planus are sensitive to mercury. The number of oral lichen planus patients who react of mercury testing is about 62%. Removal of restorations containing the irritant (amalgams) cause a regression and disappearance of the lesion. In addition, it has been demonstrated that the majority of patients with oral lichen planus (87%) exhibits low salivary secretion rates of the unstimulated saliva and a number of these patients (42%) low or very low pH of the unstimulated saliva.

Fig 3. These six photographs show a 74 year old man, who had removable partial dentures in the upper and lower jaw. The frameworks in both dentures were made of a cobalt-chromium alloy. He also had two short bridges, one in the upper jaw which was a four unit gold bridge and one in the the lower jaw, a two unit gold bridge. The removable partial dentures were connected by the frameworks to the bridges.

The dentures and the bridges were made in 1970. In 1973 he developed an eczema all over the body and his dermatologist had given him pharmaceuticals for this. This the patient told us when we saw him for the first time in 1982. He came to us to repair the lower denture which was broken.

Three weeks later the patient returned to our clinic and got the repaired denture back. We then asked the patient about the eczema and he said he was hypersensitive to chromium. For many years he had been a bricklayer. We also found out that he was drinking vitamin C (ascorbic acid) several times a day and had done so for years. The cobalt-chromium frameworks of both dentures were corroded and dark.

We asked the patient if we could keep both dentures for two weeks "in order to improve them". The patient agreed. When he came back he didn't show any eczema. We did not make any comments to the patient on this observation. We just gave him back his two dentures. A week later when he came back for a check-up we noticed his eczema again. Then we told him that the framework in his removable partial dentures were made of a cobalt-chromium alloy.

He now has two removable partial dentures with the frameworks made of a dental gold alloy. He has not any eczema since the new gold dentures were put in. The patient and his dermatologist were surprised that chromium was used in the oral cavity and had never given it a thought.

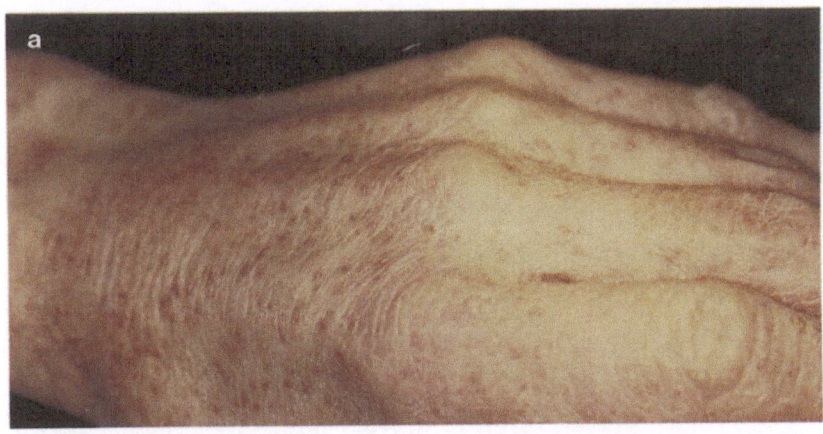

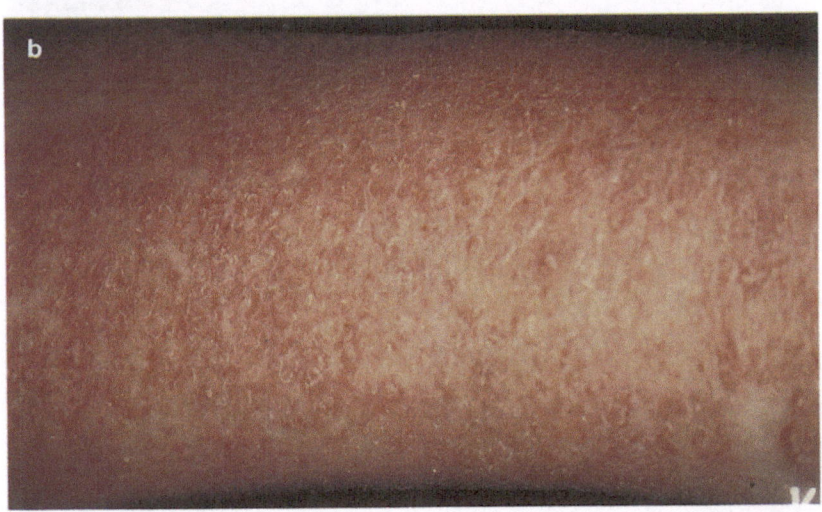

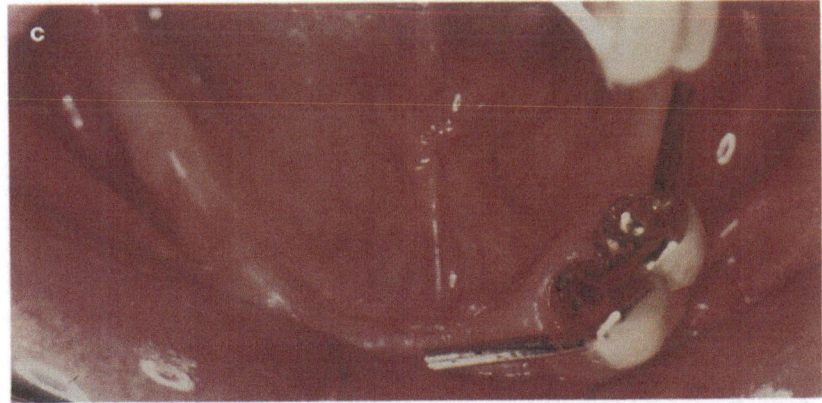

Figure 3. (a,b) Excema on hand and leg. The patient was allergic to
chromium. (c,d) The patient had removable partial dentures
in both the lower and the upper jaws. The dentures were
made of a Co-Cr-Ni alloy. (e,f) When the dentures were not
used, the excema healed.

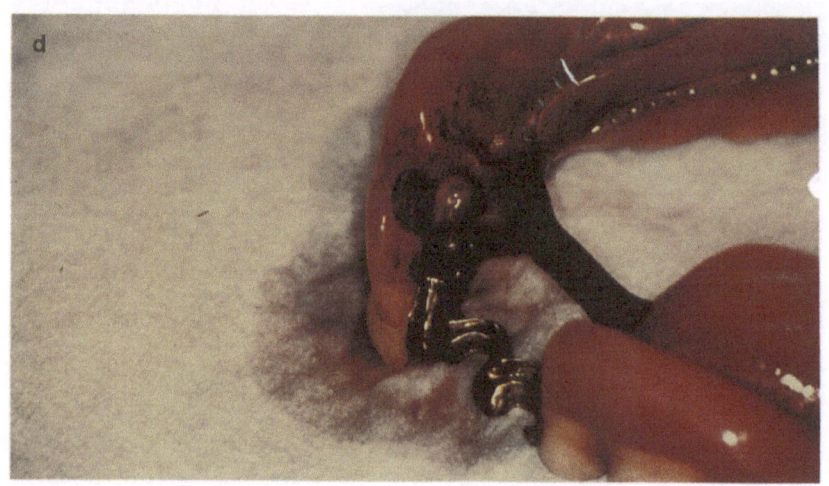

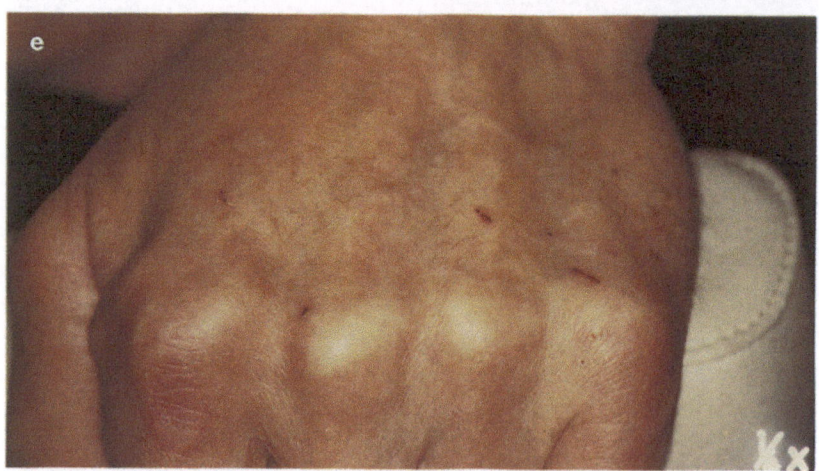

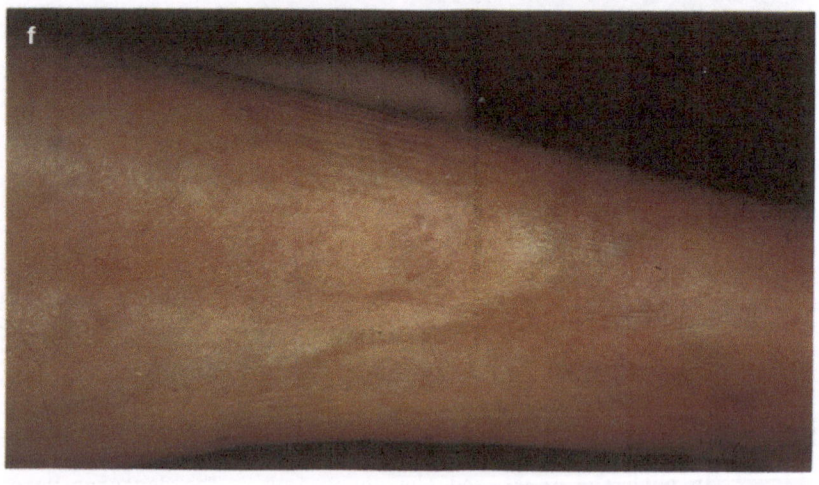

Figure 3. (Continued)

The biological significance of metal pollution caused by dental alloys is sometimes difficult to differentiate from biological effects caused by other sources of pollution, as for instance ordinary tap water, which now contains more and more cadmium, aluminium, copper, mercury etc due to the decreasing pH in our lakes. In addition, effects are frequently complicated by chelating agents, synergetic, and other unforeseeable effects occuring in the body.

Dental alloys may create problems for the patients when electro-corrosive processes are enhanced. Among the various common subjective symptoms are metallic taste, decreasd salivation, mucous irritations, burning sensations, and other more undefined intolerance reactions. It is often difficult to give a correct diagnosis in relation to the patients' complaint. Furthermore, they are not always adequately considered. Therefore there is probably a number of patients suffering from corrosion products who are not being helped . Some alloys give rise to objective symptoms in the soft tissues of the oral cavity or in the skin; these lesions are clearly visible, at least when one knows what to look for.

Patients with symptoms mentioned have in common that they often have one or more of the following restorations

- amalgam in contact with cast gold crowns
- ceramic fused-to-metal restorations
- Ni-based alloys
- a mixture of various alloys
- contaminated cast alloys
    combined with
- a low salivary secretion rate (unstimulated saliva)
- a low buffering capacity of the saliva (unstimulated saliva).

Some specific properties of the saliva may thus be involved in triggering mechanisms for corrosion processes and/or for symptoms to appear.

THE SALIVA

One important question is: How come that so many with for instance nickel-based alloys or cobalt-chromium-based alloys or amalgams don't seem to suffer from or show any side.effects from these alloys? Of all the many persons having such alloys in their mouths there are indeed very few with problems, at least as far as we can identify with the present knowledge. We don't know the complete answer to that question yet. Some factors however, can, be mentioned: In general people are healthy with sufficient power of resistance to pollution. The surfaces of the dental alloys are passivated due to an adequate composition of the saliva and there is a relatively good oral hygiene.

Of all the patients we have seen with subjective and/or objective symptoms from corrosion of dental alloys the following characteristics can be mentioned: Most of these patients are not in good physical and/or psychological health. Most of them grind their teeth. The buffering capacity of the saliva is reduced. There is a mixture of different alloys present in their mouths. The salivary flow rate is low. There are more women than men suffering from the corrosion phenomena. The women are mostly 45 or older, while the men cover all ages above 25.

Phosphate buffer exerts its protective effect by the formation of a film. A positive synergism has been observed between phosphate and biological macromolecules absorbed on alloy surfaces; this is also a corrosion inhibiting mechanism. It has been shown that phosphate buffer exercises a corrosion inhibiting long lasting effect on dental alloys at pH 7 - 4. This is probably due to the protective film formed. This film is apparently stable enough to resist the aggression of sodium chloride for many days. Similar experiments conducted with organic inhibitors did not reveal the same effect. The protective effect of phosphate buffer is concentration-dependent. This implies that the unstimulated saliva of slow saliva secretors exerts impaired protection on dental alloys against corrosion.

Our results have also shown that the inhibiting effect of phosphate buffer can be reduced by low pH, <4. Studies have shown that when the inhibiting phosphate is present in low concentrations it acts as a corrosive factor.

There are factors in the oral cavity which exhibit both corrosive and inhibiting effects. Which of the properties will be demonstrated in each situation depends on the existing conditions. Examples of corrosive factors which may act as inhibitors in the presence of some organic compounds are thiol, chloride, and fluoride ions.

The corrosion of dental alloys in the oral cavity of individual patients is dependent on the interplay between corrosive and corrosion inhibiting factors. Consequently there will be an intensification of the corrosion process when there is an increase of the corrosive factors or when the inhibiting system is impaired.

In the oral cavity there are

| Corrosive | and | corrosion-inhibiting |

factors

Acids	Phosphates
. plaque	Calcium
. food	Carbonic acid -
. stomach	bicarbonate system
Sulphates	Organic components
Chlorides	

In evaluating the interplay of the corrosive and the inhibiting systems the following parameters could be diagnosed in the patient

- the oral hygiene status
- the types of food eaten
- the frequency of food intake
- the presence of sulphur in drinking water and food
- the rate of salivary secretion (unstimulated)
- the buffering capacity of the saliva (unstimulated)
- the pH in the immediate environment
- the intake of drugs retarding the salivary secretion
- the mental status reducing salivary secretion

- general and local health conditions
- abrasion, attrition, and mechanical stress
- measurements of the amount of orthophosphate, calcium, and zinc in the saliva

CLINICAL CONSIDERATIONS TO REDUCE CORROSION AND ITS BIOLOGICAL EFFECTS

Based on what has been discussed above and clinical experience the following recommendations can be given.

* To the patients:

- Keep teeth and oral soft tissues clean . Eat three meals a day and each of the three should have about the same amount and be adequately nutritionally composed
- Avoid all kinds of in-between meals, snacks, candies, sugar etc
- Avoid salt (Cl)
- Avoid acidic food and fluids
- Avoid egg (S), which enhance corrosion
- Drink water often, which has a diluting effect on the accumulation of undesired products
- Eat fiber rich food, which binds metals
- Be sure to have enough Ca and P in the food
- Eat food requiring chewing, which stimulates the salivary secretion
- Be sure the drinking water doesn't contain metals
- Don't store food in metal pots or keep the food in contact with metal objects for extended periods of time
- Avoid fingerrings,earrings, buttons, and other metal objects in contact with the skin if the alloys contain Ni, Cr, Cd, Cu
- When using metal jewelry use alloys with high content of gold/or platinum (white gold often contains Ni!)
- Always ask your dentist what the materials consist of being used in the treatment of your mouth
- When hyperactivities in your mouth (grinding teeth, tooth pressing etc) occur consult your dentist
- Consult your dentist regularly, twice a year
- For patients with unstimulated saliva low in secretion rate and low in buffering capability we have in our laboratory developed a mouth rinse solution. When this solution is used once or twice a day the corrosion rate is drastically reduced.

* To the dentists:

- Always ask the manufacturers for a complete declaration of the content of dental materials
- Give the dental laboratory a detailed prescription of which materials should be used in the fixed or removable prosthesis ordered
- When possible avoid mixing different alloys in the same patient
- Don't use gold soldering alloys or other alloys containing Cd

- Avoid dental alloys containing Ni, Al, Be, Cd, and Cu
- Be sure to use the correct handling procedures of the material in question
- Give the patients full information
- In patients with lichen ruber planus in the oral mucosa replace amalgams with gold inlays (Cu-free gold alloy) and composite fillings
- Use an effective evacuation system when working with mercury and amalgam.

* To the dental technicians:

- Follow strictly the instructions given by the dentist and by the manufacturers of the alloys
- Be sure not contaminating the alloys.

RESPONSIBILITY

The medical act makes the doctor legally responsible for the entire treatment, including e.g. instruments and biomaterials used. However, defects in the materials or the devices used are the responsibility of the manufacturers. The doctor, on the other hand, can defend himself by referring to the manufacturer, laboratory etc. The true legal responsibility is often very complicated to identify between manufacturer and laboratory, auxiliary personal, and doctor.

Consequently the dentist is legally, morally, and professionally responsible for the biocompatibility of the materials he/she uses in the dental office and also for the welfare of the dental laboratory technician who uses dental products following the dentist's prescription. The dentist, and only the dentist, knows the history of the patient, with records of previous treatments and requirements. It is therefore logical that the dentist describes in detail to the dental laboratory the materials to be used in the prosthesis for a particular patient. If the dentist is not specific about what alloy he/she wants, the dental laboratory will use the alloy it prefers or happens to have on hand.

To be able to evaluate the various prosthetic materials the manufacturers must give us complete information on their products. There are strict regulations concerning the declaration of ingredients in drugs and foodstuff and even in cookware. Such regulations forcing manufacturers of dental products to declare all the constituents of dental materials do not exist yet, which is surprising. It seems reasonable to regard and classify dental materials as "substances used for medicinal purpose" and let them be considered and tested in the same way as drugs and pharmaceuticals are. It also seems logical that there is the same legal and moral prescription - communication procedure between the dentist and the dental laboratory as it is between the dentist and the pharmacy.

Dental gold alloys stand for the traditional dental gold alloy with the minimum of 50 atomic per cent (or 75 per cent by weight) of gold and platinum and paladium. The alternative alloys are characterized by their low or no content of gold and platinum. The terms "precious", "non- -precious", "semi-precious", "not-so- precious", "noble", "base" are used differently by different manufacturers and have therefore become completely confusing and we cannot use these terms anymore.

Presently we have new dental alloys being introduced on the market in a frequency and at a speed which makes it impossible to test them adequately from a biological point of view before they are used on humans. When problems are created in patients due to an unfavorable composition or property of a dental alloy the patients and the public will lose confidence and faith in our profession.

Seemingly small changes in the composition of an alloy may drastically change the biocompatibility. When two or more different alloys are used in the same oral cavity we may cause reactions and unpredictable effects. No dental alloy is as perfect in all respects as we would like to see or, in other words, meets all our requirements. Whenever we have alternatives we must use the one with the lowest risk of causing side effects.

Usually new dental alloys show improvement in one or a few properties but neglect others. Not seldom are requirements neglected regarding the bioadaptability and tissue compatibility.

Cases showing reactions on dental alloys reported are, or were sometimes considered as, of negligible general significance, as the frequency of tissue reactions reported was 'assumed' to be low, and as a report of this type merely demonstrated that sooner or later we can find a person who is allergic to any dental material used. Opportunism of this kind is dangerous in many respects. It is dangerous for our relations with the public, and does not favour serious research in this field. It is also important to stress that there are many more cases over and above those reported. Furthermore, tissue reactions to dental alloys and prosthetic materials in general, are not always identified. As our knowledge and interest increase, the more we observe.

The most rational and logical way to avoid and reduce iatrogenic effects from dental alloys is to use alloys with properties being theoretically and clinically considered as safe and biocompatible.

THE INFLUENCE OF INTRAOPERATIVE FORMING

PROCEDURES ON CORROSION OF NI-CR ALLOYS

Eberhard Fischer-Brandies and Eckhard Dielert

Klinik und Poliklinik für Kieferchirurgie
der Universität München
D-8000 München 2

INTRODUCTION

For the temporary reconstruction of defects in mandibular continuity various methods are possible. Allentheses which are covered by soft tissue have considerable advantages over methods employing an enoral or trans-cutaneous approach. Plate and screws enable the surgeon to reconstruct a topographically correct position of the resection stumps of the mandible and the dissected muscles. Postoperative dysfunction is thus reduced and following plastic-reconstructive surgery is facilitated.

In secondary osteoplasty, for example, the bone transplant can be attached to the same reconstruction plate which originally served as an alloplastic bridging of the defect.

COMPLICATIONS

Our own experiences with these bridging systems are not always as could be desired. In literature, also, secondary exposition of the metal devices is reported due to dehiscence (Bowerman 1974), soft tissue necrosis (Stellmach 1978), and perforation through skin and mucosa (Austermann et al. 1977). Furthermore, osteolysis in the screw area, fracture of the plate itself (Ewers a. Joos 1977), and therapy-resistant infections of the implant bed (Schmidseder a. Esswein 1979) necessitate untimely removal.

The question arises, to what extent the intraoperative forming procedures contribute to these complications, as the plate is submitted to cold deformation during the process of adapting the plate to the contour of the mandible. Thereby the originally good mechanical and electrochemical properties of the alloy can be negatively influenced.

CLINICS

The analysis of the plate fractures observed by the authors and the inquiry of the surgeons revealed that shortcomings in the management of the devices may have played a role. During the adaptation procedure of the plate shown in fig. 1, e.g., difficulties had occurred, because mandibular corpus, angle as well as ramus had to be reconstructed alloplastically. The originally straight device had to be formed three-dimensionally, whereby it was bent to and fro.

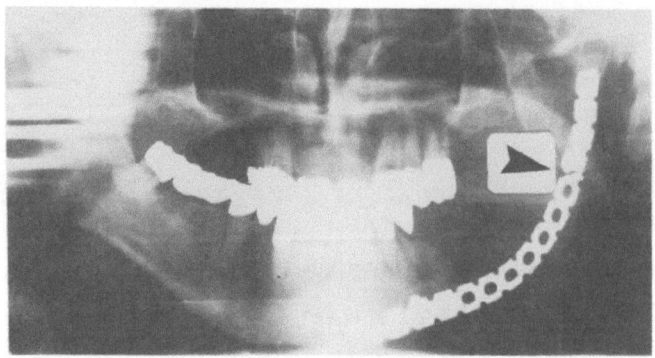

Fig. 1. Situation after resection of the mandible because
of carcinoma. Reconstruction plate with fracture
(--->) which occurred 2 years postoperatively.

Two years after tumor resection and postoperative irradiation plate
fracture occurred at the border to the central mandibular stump.

Because of the short time span elapsed since irradiation, free bone
grafting for the reconstruction of the bone continuity was contraindicated.
Additionally, osteolysis was present around the screws. A renewal of the
alloplasty was not performable because of a lack of bone mass in the short
central fragment.

In this problematic situation, the intricate treatment using an iliac
bone graft with microsurgically artery and vein anastomosis was employed.
The transplant was fixed by a combination of plate osteosynthesis and wire
suturing (fig. 2).

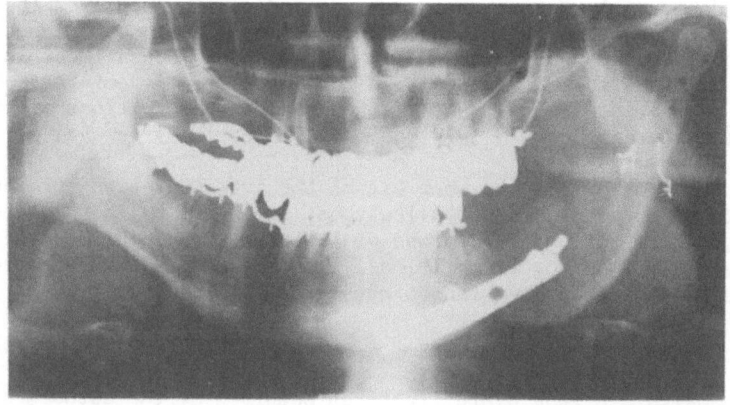

Fig. 2. The fractured plate has been removed. The bone
continuity is reconstructed by inserting a microsurgically anastomozed bone graft from the iliac
crest.

In this case a satisfactory therapeutic result could only be attained by means of microsurgery. Nevertheless, plate fractures should be avoided by correct management of the devices.

METALLURGICAL INVESTIGATIONS

For investigation of the influence of the intraoperative adaptation procedure on the alloy (X5 Cr Ni Mo 18/12, DIN 17006; chemical composition Fe 62, Cr 17.5, Ni 12.0, Mo 2.5, Mn < 3, Si < 3, C 0.03 weight percent) plates were bent to different degrees.

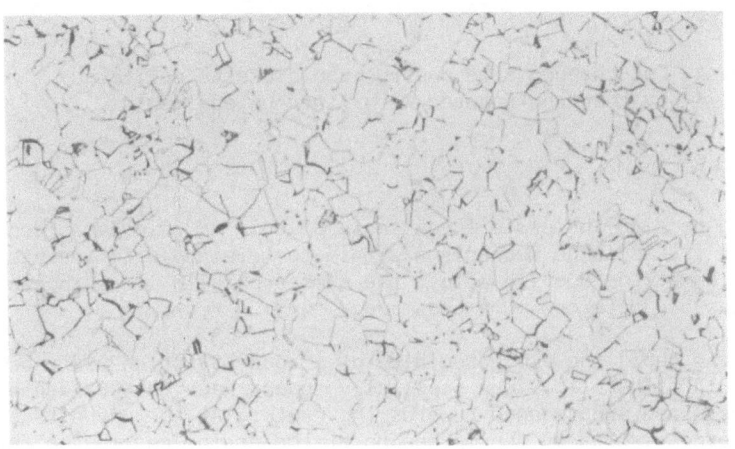

Fig. 3. Structure of the reconstruction plate as delivered
by the manufacturer. Note the parallel lines
representing recrystallization twins.

It has been shown by metallographic studies that the plates delivered by the manufacturer exhibit a cubic surface centered, fine-crystalline homogenous structure (Dielert 1983). Recrystallization twins are results of the previous tempering. At the grain boundaries no separation phenomena (heterogenous formation of a new phase) are detectable (fig. 3). This indicates a precise cooling process. Lined up enclosures result from the rolling and forging of the austhenitic mixed crystal structure of the alloy (fig. 4).

The adaptation procedure leads to further plastic cold-deformation of the three-dimensional crystal lattice. Macroscopically the plate becomes dull in the affected areas.

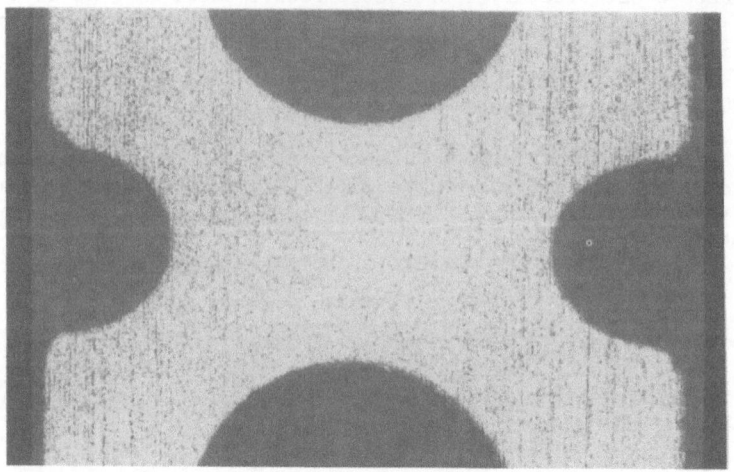

Fig. 4. Cut through a reconstruction plate as delivered
by the manufacturer with lined up enclosures.

When bending the plate in an angle of only 4° slip lines occur
resulting from sliding mechanisms within the grains. With increasing
deformation of the ductile alloy, the number of slip lines increases and
the grains are deformed more and more.

Fig. 5 shows the elongated side of a reconstruction plate bent in an
angle of 18°. The slip lines especially appear near the surface of the
allenthesis and are reduced towards the center. At the opposite side the
acting force results in a compression of the grains and in the formation
of overlapping zones at the surface. Slip lines are also present (fig. 6).

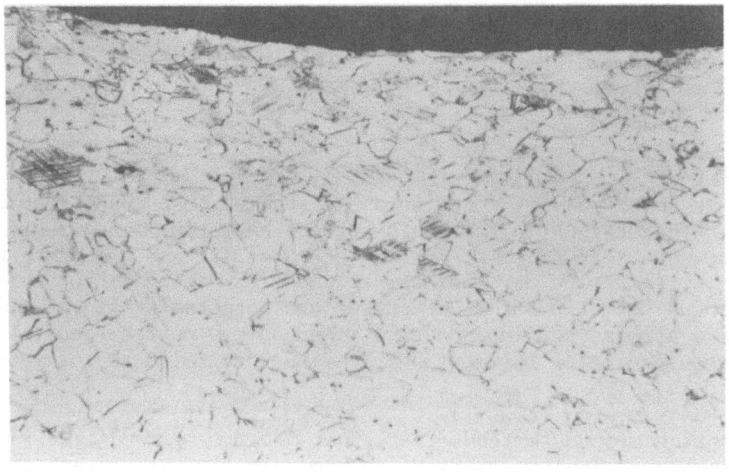

Fig. 5. Elongated side after bending in an angle of 18°
with slip lines.

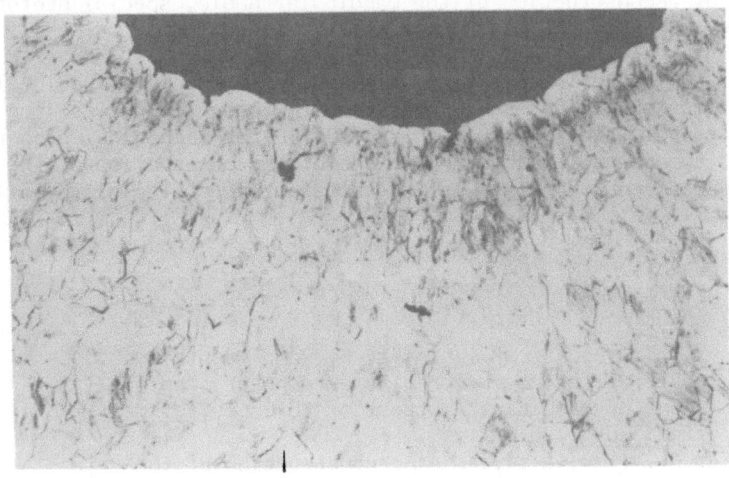

Fig. 6. Compressed side with overlapping zones,
compressed grains and slip lines.

These structural changes result in a reduction of the mechanical
resistance and enhancement of corrosion processes by the body fluids,
especially in the area of slip lines and overlapping zones.

The majority of manufacturers produce a special bending forceps
permitting a bending of plates of up to 20°. According to the authors'
results the freedom of movement of the instrument should be limited
to a lower degree. Even greater accuracy is required when pleiers are
used, as they permit uncontrolled and unlimited bendings of the plate.

CONCLUSIONS

1.) Being implanted in a damaged tissue bed, the therapeutic effects
can be improved by a careful surgical procedure, leading to reduction of
hematoma, seroma and infection. This is of importance, because such complications lead to local lack of oxygen in the tissue favouring implant
corrosion.

2.) For the reduction of local stresses of the allenthesis and bone,
the plate should be fixed with 3 to 5 screws on each side. The risk of
osteolysis in the screw area can thereby be reduced.

3.) To avoid metal transfer during plate bending, only special
instruments recommended by the manufacturer should be used.

4.) Damage to the plate surface, e. g. scratching, by the bending
instrument has to be avoided.

5.) It is not recommendable to bend plates repeatedly in both
directions.

6.) Bending in the region of the screwholes should be avoided.

7.) Deformation over 10° will enhance corrosion in a higher degree.

8.) It is preferable to bend a plate at several points at angles of
10°, if a higher deviation is necessary.

9.) For reconstruction of the mandibular angle, special preformed plates should be used to avoid unnecessary cold-deformation.

Until now there is no allenthesis available which is absolutely resistant against the corrosive attack of the body fluids. As long as no other material is available, which has sufficient mechanical properties, the use of these metal alloys is advisable. The observation of the 9 rules given above can contribute to better therapeutic results of the alloplasty.

REFERENCES

Austermann, K. H., Becker, R., Büning, K., and Machtens, E., 1977, Titanium implants as a temporary replacement of mandible. A report of 30 cases. J. max.-fac. Surg., 5:167.

Bowerman, J. E., 1974, A review of reconstruction of the mandible. Proc. Roy. Soc. Med., 67:610.

Dielert, E., 1983, Risikominderung bei metallischen Unterkiefer-Allenthesen. Dtsch. Z. Mund-Kiefer-Gesichts-Chir., 7:72.

Ewers, R., Joos, U., 1977, Temporäre Defektüberbrückung bei Unterkieferresektionen mit Osteosynthesemethoden. Dtsch. zahnärztl. Z., 32:332.

Schmidseder, R., Esswein, W., 1979, Plastisch-chirurgische Eingriffe zur Rekonstruktion der Kinnregion, in: "Fortschr. Kiefer- u. Gesichtschir. Bd.XXIV," K. Schuchardt, ed., Thieme, Stuttgart.

Stellmach, R., 1978, Die Fixierung des Spans bei der freien Knochentransplantation, in: "Fortschr. Kiefer- u. Gesichtschir. Bd.XXIII," K. Schuchardt, ed., Thieme, Stuttgart.

PILOT STUDY OF ALLERGIC EFFECTS OF

Co-Cr-Ni ALLOYS AMONG DENTAL TECHNICIANS

D. Choudat, F. Vallentin, P. Brochard, T. Berriau, and
J. Proteau

Département de médecine du travail, Faculté de Médecine
Cochin-Port-Royal, 24 rue du faubourg Saint-Jacques
75014 Paris
Laboratoire d'Etudes des Particules Inhalées, 44 rue Charles
Mourreu 75013 Paris
Institut de Médecine du Travail, 15 rue de l'Ecole-de-
Médecine 75006 Paris

In exposed workers, nickel, chromium and cobalt may induce contact
dermatitis and asthma[1,2,3]. The focus of this conference has been on
the risks, allergic or not, to the patients of the use of non precious
alloys. Another group that may be at perhaps even greater risk are the
technicians who prepare the prostheses for insertion. Because dental
technicians are exposed repeatedly to respirable particles of non
precious alloys as well as of silica, we undertook a pilot study* of the
prevalence of allergic manifestations in fifty eight dental technicians.

Dental technicians are responsible for preparing prostheses made of
precious or non precious alloys. The procedure starts with the dentist
making a negative impression of the area to be repaired. From this
negative impression, the dental technician first makes a positive
reproduction of wax or plastic. The wax or plastic pattern is then
placed in a small casting ring. This ring, sometimes asbestos plated,
is filled with a refractory investment material that surrounds the
pattern and produces a mold. The casting ring is then placed in an
oven, and all the wax or plastic is eliminated leaving an empty mold
chamber. The restoration is cast with molden metal forced into the
mold chamber. After the cast is cooled, the mold is broken and
residual material is removed with an abrasive sand blaster or a hand
finishing tool. Additional finishing is done with hand grinding and
polishing.

This work produces great amount of respirable dust and generates
high levels of potentially toxic airborne particles including silica,
asbestos fibers and non precious alloys (Fig. 1).

Supported by a grant from D.R.A.S.S. d'Ile de France and D.A.S.S. de
Paris (crédits régionalisés de prévention).

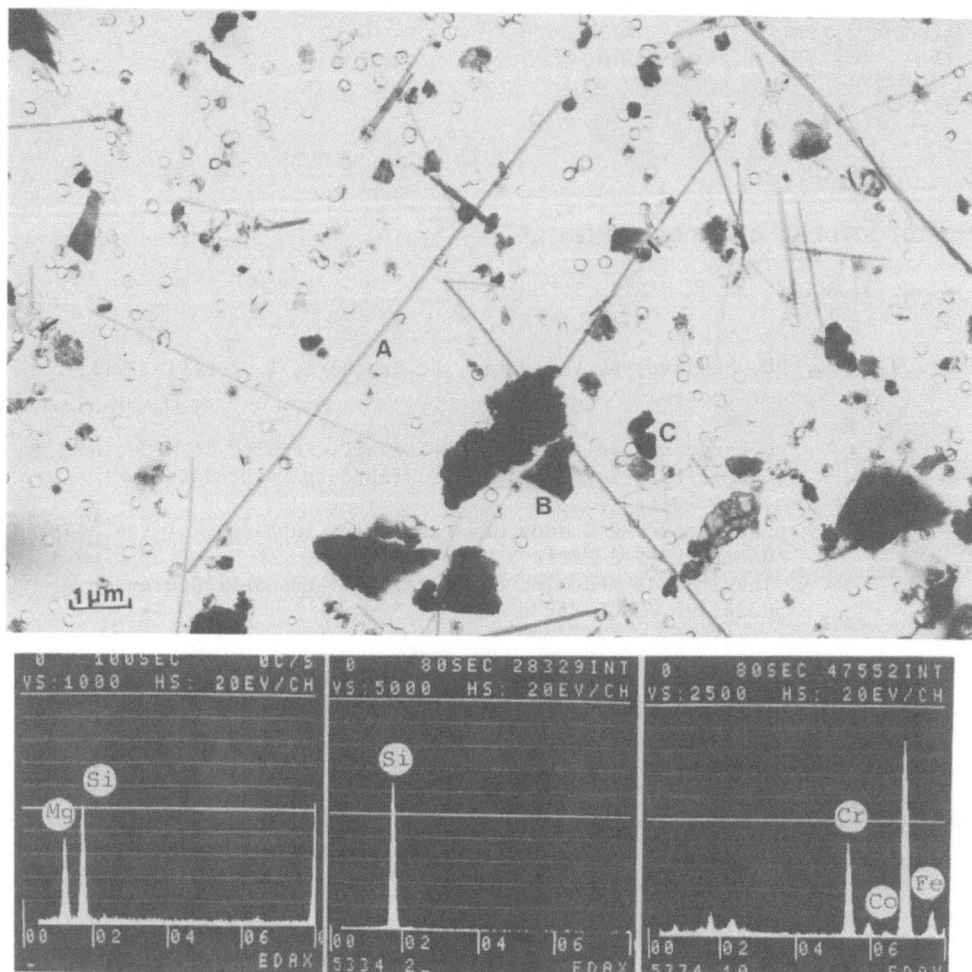

Fig. 1. Airborne particles in a dental laboratory. Electron microscopic
view and analysis of particles by EDAX. A : chrysotile fiber ;
B : silica ; C : particle of Cr-Co-Fe alloy.

Exposure to these dusts is already recognized as causing respiratory
diseases among dental technicians. Silicosis (Fig. 2), asbestosis, and
berylliosis have all been reported.

Potential diseases include hard metal diseases (a form of pulmonary
fibrosis), asthma, hypersensitivy pneumonitis and bronchial carcinoma.
In our study we examined the prevalence of asthmatic symptoms and allergic contact dermatitis to non precious alloys.

The study group of dental technicians included fifty seven men and
one woman. They worked in Paris or its suburbs, and they were volunteers
to participate to the study. Their mean age was 41 years $\pm$ 10 (m$\pm$S.D.)
and the mean duration of exposure was 24 years with a range from 5 to
40 years.

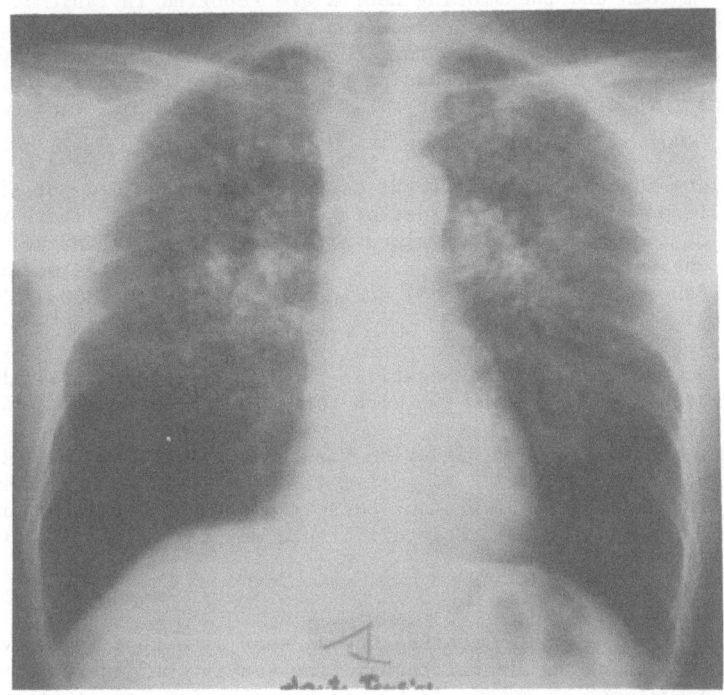

Fig. 2. A postero anterior roentgenogram of this 58 years old dental
technician reveals general involvement of both lungs by small
rounded opacities.

Three types of data were obtained :
1 - on respiratory function, by autoquestionnaire translated and derived
from British Medical Council questionnaire for the prevalence
of respiratory symptoms, by function tests including challenge with spray
of nickel sulfate, by blood gas analysis, and by chest radiography ;
2 - on contact dermatitis by autoquestionnaire and by skin tests with
nickel, chromium and cobalt salts ; 3 - on urinary chromium concentration,
by flameless atomic absorption spectrophotometry.

Table 1. Characteristics of the study group of dental technicians

Characteristics	Number of cases	
Smokers	22	(38%)
Ex Smokers	21	(36%)
Non Smokers	15	(26%)
Cough	14	(25%)
Sputum	9	(16%)
Asthma	9[a]	
Eczema	13[b]	
"Sarcoidosis"	2	
Silicosis	1	

[a]Among these cases, two were observed before exposure

[b]Among these cases, six were observed before exposure

The symptoms and the smoking habits are reported in Table 1. We have found no direct relationship between the occurence of asthmatic symptoms and occupational exposure ; symptoms occured even during the holidays. Two subjects gave a history of having had "sarcoidosis", but among dental technicians this diagnosis may be related to beryllium or organic antigens exposures[4].

The lung function values of the dental technicians did not significantly differ from the reference values of Knudson[5] (Table 2). There was no variation of the lung function values after challenge with nickel sulfate except in one asthmatic subject. Blood gas analysis revealed moderate hypoxemia (mean PaO_2 = 81.4 Torrs $\pm$ 6.8) and normocapnia (mean $PaCO_2$ = 35.6 Torrs $\pm$3.4.).

All the patch-tests were negative. Few prick-tests were slightly positive : three with nickel, two with cobalt, none with chromium.

The mean chromium level in urine was 4.5. $\mu g/l \pm$ 1.4. (m$\pm$S.D.) among the dental technicians[6] and 3.2. $\mu g/l \pm$ 1.2. (m$\pm$ S.D.) among a control group of fifty one patients without occupational exposure to mineral dusts and matched for age (Fig.3). This difference was significant ($p < 0.01$).

DISCUSSION

In this study allergic manifestations such as asthma and eczema were more prevalent among the dental technicians tan in the general population. A similar prevalence of asthma 12.6% was found by Rom and coworkers[7,8] among one hundred and thirteen dental technicians in Salt Lake City. In our study the allergic manifestations did not seem directly work related.

The increase in urinary chromium seems to be moderate in dental technicians compared to other professional groups[9,10]. However, comparisons must take into account the differences between analytical techniques and the different types of chromium exposure. The dental technicians are exposed to metallic chromium for which pulmonary clearance is slow[10]. Thus, measurement of urinary levels of this metal does not allow an estimate of the total amount of chromium retained in the lungs[6].

This study is a pilot study. The dental technicians are not randomly selected and we cannot exclude a selection bias. The population included only fifty eight dental technicians ; this group is perhaps too small to observe a sub group with symptoms work-related. We studied only nickel, chromium, cobalt alloys. The workers are exposed to other potentially toxic or allergenic materials.

Table 2. Lung function values among dental technicians expressed as percentage of the Knudson values

Forced Vital Capacity	104 $\pm$ 12%
FEV$_1$	103 $\pm$ 13%
FEV$_1$/FVC	97 $\pm$ 7%
FEF 25-75	81 $\pm$ 23%
Peak Flow	107 $\pm$ 17%
FEF 75%	98 $\pm$ 27%
FEF 50%	75 $\pm$ 22%
FEF 25%	56 $\pm$ 20%

In conclusion, both asthma and eczema seem slightly more prevalent among dental technicians than in the general population, but the increase in prevalence is modest and allergic symptoms did not appear to be directly related to work exposures. We therefore think that the major respiratory risks are not of allergic disease but of interstitial lung diseases, such as silicosis or granulomatous disorders. Nevertheless, all these potential diseases can be reduced by proper ventilation and good work practices.

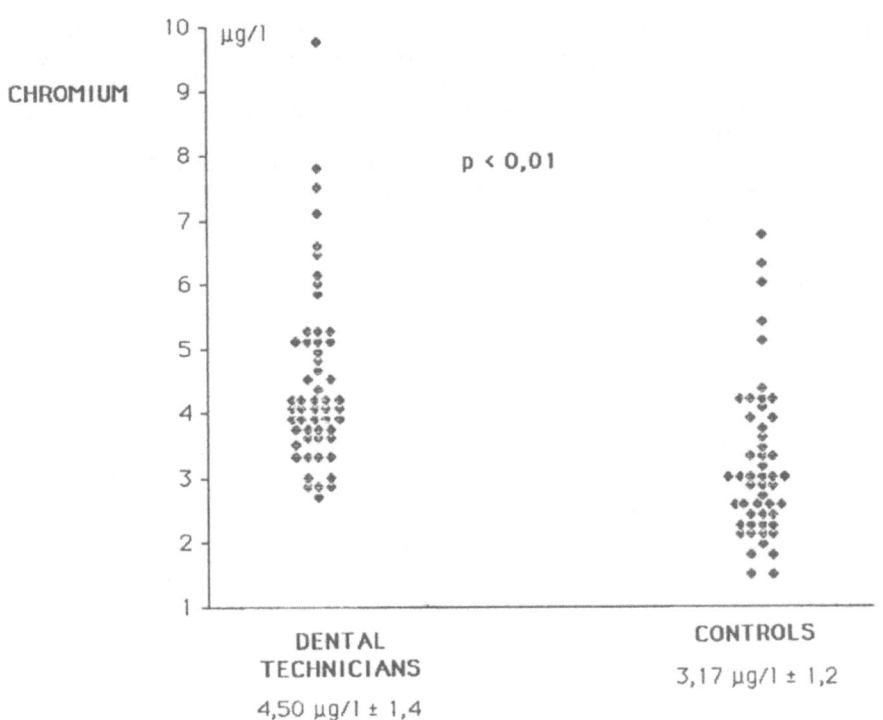

Fig. 3. Urinary chromium concentration among 58 dental technicians and 51 control subjects.

REFERENCES

1 - G.T. Block and M. Yeung, Asthma induced by nickel, JAMA 247 : 1600-1602 (1982).

2 - J.L. Malo, A. Cartier, G. Gagnon, S. Evans and J. Dolovich, Isolated late asthmatic reaction due to nickel sulfate without antibodies to nickel, Clin. allerg. 15 : 95-99 (1985).

3 - B. Gheysens, J. Auwerx, A. Van den Eeckhout and M. Demedts, Cobalt-induced bronchial asthma in diamond polishers, Chest 88 : 740-744 (1985).

4 - D. Choudat, P. Brochard, F.X. Lebas, J. Marsac and M. Philbert, Sarcoïdose ou pneumoconiose ? Coincidence ou relation ? Arch. Mal. Prof. 44 : 339-344 (1983).

5 - R.J. Knudson, R.C. Slatin, M.D. Lebowitz, B. Burrows, The maximal expiratory flow volume curve. Normal standards, variability, and effects of age, Am. Rev. Resp. Dis. 113 : 587-599 (1976).

6 - D. Choudat, A. Gaudichet, P. Brochard, F. Vallentin, T. Berriau, M. Philbert, J. Bignon and J. Proteau, L'exposition aux poussières de chrome chez les prothésistes dentaires. Arch. Mal. Prof. 46 : 318-320 (1985).

7 - W.N. Rom, J.E. Lockey, K.M. Rang and R. Johns, Pilot epidemiologic study of dental laboratory technicians. Am. Rev. Resp. Dis. 127 : 159 (1983).

8 - W.N. Rom, J.E. Lockey, J.S. Lee and Coll., Pneumoconiosis and exposures of dental laboratory technicians. Am. J. Public Health 74 : 1252-1257 (1984).

9 - H. Nomiyama, M. Yotoriyama and K. Nomiyama, Normal chromium levels in urine and blood of japanese subjects determined by direct flameless atomic absorption spectrophotometry, and valency of chromium in urine after exposure to hexavalent chromium. Am. ind. hyg. assoc. J. 41 : 98-102 (1980).

10- H. Welinder, M. Littorin, B. Gullberg and S. Skerfving, Elimination of chromium in urine after stainless steel welding. Scand. J. work environ. health 9 : 397-403 (1983).

CHAIRMEN SUMMARY: CLINICAL ASPECTS

B. Bloch and R. Söremark

There is insufficient benefit analysis in implant surgery. Retrospective surveys require team work and careful examination of medical records.
These may have inadequate detail as international coding of diseases and
procedures has only become more comprehensive since 1978 (I.C.D 8 of WHO).
Significant too is the need to survey the quality of life enjoyed by the
patient and to compare the incidences of complications between different
institutions and clinical results. An outline of such a survey in Sydney
metropolitan hospitals was given.

A new approach in designs and materials for a total hip replacement
prosthesis was presented. There is international collaboration in application and evaluation. The long term results are awaited with interest
(Laing).

The current activities of ISO/TC150 committee on Surgical Implants
was outlined. While full "fitness for purpose" standards are still difficult to achieve, significant progress has been made to specify compositional requirements and certain design specifications to satisfy international delegations of surgeons, scientists, manufacturers, and regulatory
health care authorities. The need for implant retrieval analyses was again
emphasized. The need for mandatory quality control particulary for "permanent" implants is desirable.

International audit or even a registry for such implants can be
achieved if the will is there. The meeting was reminded of long-term
problems in immunopathology and possible carcinogenicity making audit all
the more essential.

International standards can at least, if of quality, remove one uncertain factor in the implant field. However, we cannot specify surgical training and technique.

The problem of anchoring artificial joints in the bone and the failure implanted systems in bone is being studied worldwide.

The incidences of failure differ rather dramatically between different clinics. The surgical and fixation methods are very technique sensitive. The failure rate is presumably comparatively high in clinics where the treatment procedure is not often enough used. It can also be expected the failure rate increases when various treatment procedures are used simultaniously.

Interesting experiences with newer techniques involving the use of hip prostheses in Titanium screw-type avoiding the use of acrylic bone cement were presented. Long term experiences of the effects on host and implants remain to be seen, however.

It will seem, however, as the material and design of an artificial hip joint play less role in the rate of failure than the operative techniques, skill, and experience do. A correct management and a minimum of cold deformation and mechanical stresses of the hip prostheses and reconstruction plates reduce corrosion.

In edentolous patients various kinds of material have been tried for bone augmentation in the mandible and in the maxilla. Porous synthetic P and Ca containing bone substitutes were reported to give rather satisfactory results. Good results were shown when augmented bone areas were used for fixation of implants in the mandible and in the maxilla.

A variety of dental implants for the retention of bridge were shown. Some were designed for the distal part of corpus mandibulae, some for the maxillary sinus. Some showed brave surgical and implant procedures. Long-term effects will be of interest.

In the oral cavity, a mixture of dental alloys in permanent contact can enhance the corrosion process especially if the patient has a low salivary secretion rate and a low buffering capacity of the saliva. Certain oral microorganisms can also cause a pronounced corrosion in the oral cavity.

The interplay between corrosion enhancing and inhibiting substances in the saliva plays a dominant role for the development of the corrosion processes.

Cases were shown with intolerance reactions in the oral cavity and in the skin caused by corrosion products developed in the oral cavity. Corrosion products from dental alloys can hardly start a sensitization but merely cause a reaction or flare up in sensitized patients.

Dental technicians are exposed to dust of a number of metals. It was surprising to learn that there were none of the examined 58 technicians that had allergic reactions to nickel, chromium and cobalt.

The value of interdisciplinary collaboration as evidenced by this meeting was lauded.

QUALITY CONTROL

THE CORROSION OF DENTAL Ni - Cr ALLOYS

AN IN VITRO EVALUATION

Jean-Marc Meyer

School of Dentistry
University of Geneva
1211 Geneva 4, Switzerland

INTRODUCTION

Dental casting alloys based on the mickel-chromium system have been developped firstly as an economical alternative to gold alloys. They include a wide range of compositions, with nickel being between 69 and 81 weight percent, and chromium between 12 and 26 %. They may contain up to 12 % of cobalt, and/or up to 12 % of molybdenum. Minor alloying elements like Fe, Al, Mn, Ti, Ta, Ga, Nb, Si, B, Be, P and Sn may be added. It becomes therefore evident that the nickel-based dental casting alloys do not form a homogeneous group of alloys, and that the mechanical properties and the electrochemical behavior of a single alloy can not be representative of the whole family of Ni-Cr alloys. Furthermore, the variations in the oral environment (e.g. pH value) will not have the same effects on different alloys.

The purpose of the present study is therefore to demonstrate the wide range of responses elicited by Ni-Cr alloys of varying compositions, submitted to a simulated oral environment.

METHODS OF EVALUATING THE CORROSION OF DENTAL ALLOYS

Three main electrochemical characteristics can be observed to evaluate the corrosion behavior of dental alloys: the anodic polarization, the variation of the open-circuit potential with the time, and the polarization resistance. These parameters are obtained from samples immersed in a medium simulating the oral environment. A typical measuring system is shown in Figure 1. It consists of an electrochemical cell, and a series of instruments used to polarize the sample (potentiostat), and to measure the resulting changes in potential and current. These variations are recorded graphically and, in some instances, digitally for further mathematical analysis by a computer program.

The electrochemical cell contains three electrodes: the sample, or working electrode, a saturated calomel reference electrode (constant potential) and a platinum counter-electrode, whose surface is unaffected by

Table 1. Composition of the artificial saliva

Sodium chloride	(NaCl)	0.400	g
Potassium chloride	(KCl)	0.400	
Sodium dihydrogen phosphate	(NaH_2PO_4)	0.690	
Sodium sulfide	($Na_2S \cdot 9 H_2O$)	0.005	
Urea		1.000	
Calcium chloride	($CaCl_2 \cdot H_2O$)	0.795	

in 1000 ml of distilled water

the changes of potential and current produced in the system. The working electrode is made of a metallographically polished disk (diam. 10 mm) of the tested alloy.

The artificial saliva used to simulate the oral environment is maintained at 37 °C and saturated either with compressed air or with nitrogen to control the presence of oxygen. The composition of the artificial saliva (Meyer and Nally, 1974), derived from that proposed by Fusayama et al. (1963), is listed in Table 1.

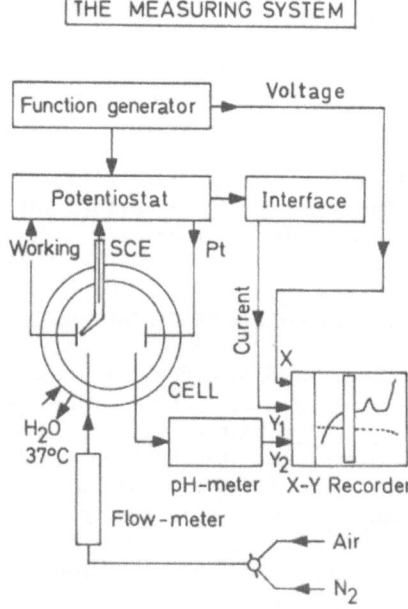

Fig. 1. Measuring system for evaluation of corrosion behavior of dental alloys in a simulated oral environment.

The anodic polarization of twelve dental alloys (Table 2) has been measured and the corresponding curves are shown in Figure 2. These alloys have been classified according to their composition.

Table 2. Classification of Ni-Cr dental alloys

Class I : 20 % Cr or more
        A Ducinox   (Ugine Carbone, France)
        B Howmedica III   (Howmedica Inc., U.S.A.)

Class II : 16 % Cr or more, more than 3 % Mo
        C Ticon   (Ticonium Co., U.S.A.)
        D Victory   (Unitek, U.S.A.)
        E Wiron S   (Bego, West Germany)

Class III : less than 16 % Cr, with Mo
        F Gemini II   (Kerr-Sybron Corp., U.S.A.)
        G Jelbon   (Jelenko Co., U.S.A.)
        H Microbond NP/2   (Howmedica Inc., U.S.A.)
        I Ultratek   (Metals for Modern Dentistry, U.S.A.)

Class IV : less than 16 % Cr, without Mo
        J Phenix   (Pouget-Nacache, France)
        K Ryco AR-1   (Courtin Ltd, England)

Class V : ternary base Ni-Co-Cr
        L DW 106   (Williams Gold Refining, U.S.A.)

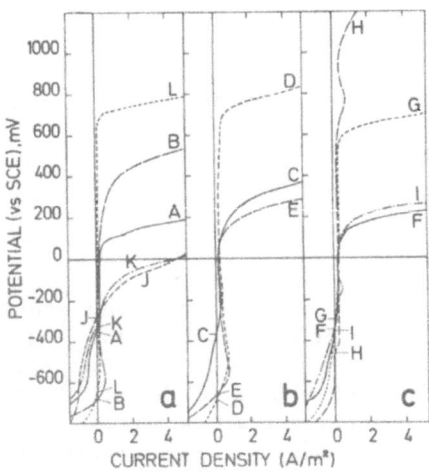

Fig. 2. Anodic polarization curves of 12 dental casting alloys (see Table 2).

It can be seen that the polarization curves differ markedly among the tested alloys. For example, alloys J and K (Class IV, less than 16 % Cr, without Mo) exhibit a sharp increase in the anodic current right after the corrosion potential (current density = 0), whereas alloys like D (Class II, 16 % Cr or more, with more than 3 % Mo), G and H (Class III, less than 16 % Cr, with Mo), and L (Class V, ternary base Ni-Co-Cr) display a wide range of very low current between the corrosion potential and the breakdown potential (sharp increase of the anodic current). This later behavior is characteristic of an alloy with a good resistance to corrosion.

The curves representing the variations of the open circuit potential with the time are shown in Figures 3 and 4. This information is very easy to obtain, since the samples are simply immersed in the testing medium, and the difference between their potential and that of a reference electrode is recorded during several hours. The slope of the resulting curves gives an information about the tendency of the tested material to resist corrosion : if the potential is increasing (going toward more positive values) with the time, the alloy is passivating. If the curve remains steady, the alloy is immune to corrosion, and if the curve is decreasing toward more negative values, the corrosion of the alloy is increasing with the time.

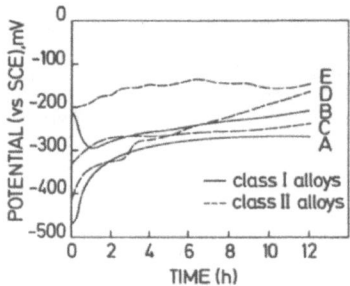

Fig. 3 Potential-time curves for alloys of Classes I and II.

The relative position of the curves within the potential scale gives also an indication of the nobility of the alloy : the higher the position, the more noble the alloy. It can thus be seen that alloys of Class III and V are more noble than those of Classes I, II and IV. Three alloys exhibit a decreasing value of the open circuit potential with the time: alloys I (Class III), J and K (Class IV). They are clearly corroding with the time. This behavior is characteristical for the Class IV alloys, since they do not contain any Mo, an element which markedly increases the resistance to corrosion. For alloy I, however, the tendency to corrosion may result from the presence of beryllium, which has been shown (Lee et al., 1985) to markedly reduce the resistance to corrosion.

The extent of corrosion is more precisely described by measuring the corrosion current, that is the anodic current, which flows at the corrosion potential. When displaying the polarization curve, the value recorded around the corrosion potential is the algebraic sum of the currents due to the simultaneous oxidation and reduction reactions taking place at that particular potential. When this sum is equal to zero, the potential is defined as the corrosion potential, and is very close to the open circuit potential. It thus corresponds to the situation observed in the mouth in absence of any possible polarization by other metallic restorations. At this corrosion potential, the anodic component of the resulting current is rarely null, since some oxidation is almost always taking place. If this oxidation could be measured, then the extent of corrosive degradation could be evaluated by the extent of this anodic current, called the corrosion current. Unfortunately, the corrosion current is impossible to measure directly from a polarization curve, and indirect methods are needed. The one which has been used in this laboratory is derived from the linear

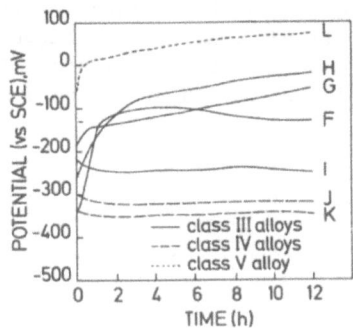

Fig 4. Potential-time curves for alloys of Classes III to V.

polarization technique described by Mansfeld (1973). It consists of a very narrow polarization range around the corrosion potential : from 30 mV below that potential and up to 30 mV above it. The slope of the tangent to the resulting curve, drawn at the corrosion potential, yields what is called the polarization resistance, a value inversely proportional to the corrosion current. The polarization resistance gives therefore a close evaluation of the tendency of the alloy to corrode in the particular conditions of the experiment : the higher the polarization resistance, the smaller the corrosion. The exact relation between the polarization resistance and the corrosion current is controlled by the values of the Tafel slopes, as shown in the equation developped by Stern and Geary (1957) :

$$R_p = \frac{1}{2.3} \cdot \frac{1}{i_{corr}} \cdot \frac{b_a \cdot b_c}{b_a + b_c}$$

where b_a and b_c are the Tafel slopes for the anodic and the cathodic reactions respectively.

To obtain the corrosion current from such an equation, it is necessary to know the values of the Tafel slopes for the tested alloy in the particular experimental conditions, usually through a preliminary experiment. The method developped by Mansfeld (1973) is useful, since it allows the simultaneous determination of the polarization resistance and the Tafel slopes, thus leading to the computation of the corrosion current in a single experiment. Basically, it consists of the comparison of the measured linear polarization curve with a theoretical curve computed by using the actual polarization resistance value and "guessed" values of the Tafel slopes. With a method of least squares, the theoretical curve is adjusted to the experimental curve by changing the values of the Tafel slopes, until the best fit is reached. The procedure is tedious when used with a trial and error method, but it has been optimized and automatized for computer processing : the guessing of the Tafel slope values is led to the optimal data by use of a simplex and of the recalculation of the fit (Wiskott, 1980). Such a method is very sensitive, and the state of the alloy surface has a strong influence on the results : the time elapsed between polishing and immersion in the artificial saliva, for example, may change the behavior of a given alloy from an active to a passive response. It should therefore be used with extreme care. The polarization resistance, however, is less sensitive to small variations in the surface condition of the sample, and its value determined from linear polarization curves are generally accurate enough to be used as a relatively safe indicator of the corrosion behavior of an alloy. Some data obtained on dental alloys immersed in the artificial saliva are given in Figure 5.

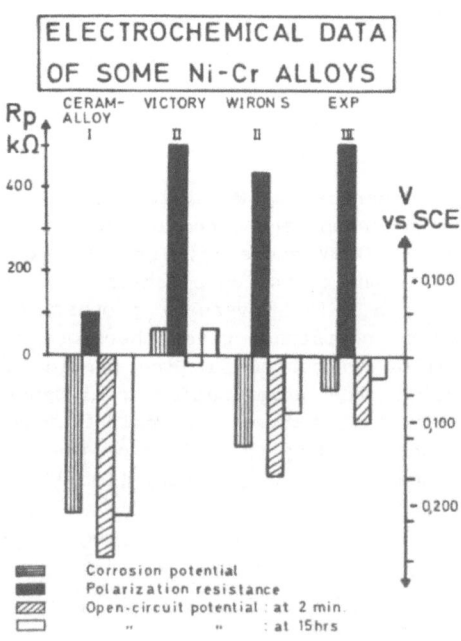

Fig. 5. Corrosion potentials and polarization resistances.
EXP refers to a Ni-V-Cr experimental alloy.

In order to understand the role of the individual components on the electrochemical behavior of alloys, some binary Ni-Cr alloys have been prepared and submitted to the usual corrosion tests already described. Their compositions are mentioned in Table 3.

Table 3. Composition of binary Ni-Cr alloys

Ni 91 – Cr 9 [*]	Ni 85 – Cr 15
Ni 89 – Cr 11	Ni 83 – Cr 17
Ni 87 – Cr 13	Ni 81 – Cr 19

[*] in weight percents

From the curves shown in Figure 6, it can be seen that the chromium content plays a clear role on the reduction of the anodic current level, and therefore on the extent of the corrosion damage resulting from the polarization. As indicated already in the classification appearing in Table 2, alloys with 16 wt. % of Cr or more are better protected against corrosion. It may be interesting to note that this classification (Meyer, 1977, Meyer et al., 1979) has been established well before the experimental values on Ni-Cr binary alloys presented here have been available (Meyer and Niney, 1985).

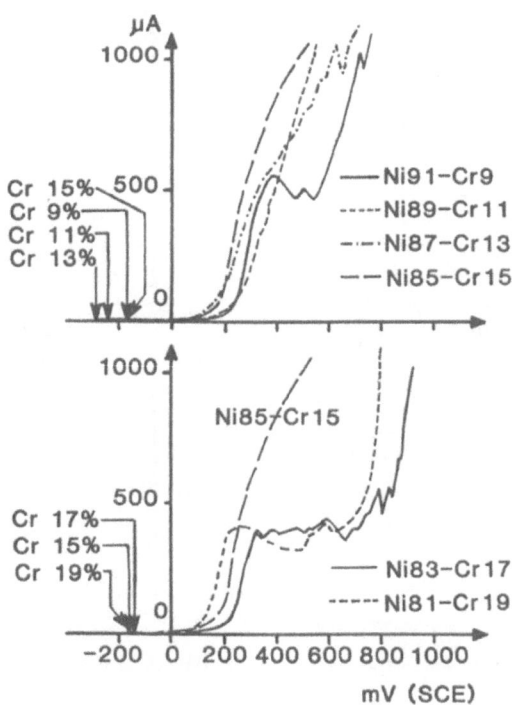

Fig. 6. Anodic polarization curves of Ni-Cr binary alloys.

Other elements are necessary, however, to reach a level of inalterability which is compatible with permanent use in the oral cavity. Among those favorable elements, molybdenum is of prime importance. It is present in all alloys of Classes II and III, with at least 3 wt. %. In some instances, alloys of these classes may display an excellent resistance to corrosion (see alloys D, G and H in Table 2 and Figure 2).

The technical requirements for a satisfactory processing in the dental laboratory have forced the manufacturers of the alloys to add such elements as aluminum, beryllium, boron, and silicon, among others, in minutes quantities (up to around 5 wt. %), to improve the mechanical properties or other important properties like castibility or precision of fit. The Table 4 gives some indication on the content of such elements in several commercial alloys, whereas their anodic polarization curves are shown in Figures 7, 8, and 9.

As a comparison, three base metal alloys with no nickel and no beryllium : Biocast, Dentitan (containing less than 10 wt. % Ti), and Neobond II, have been tested in the same conditions.

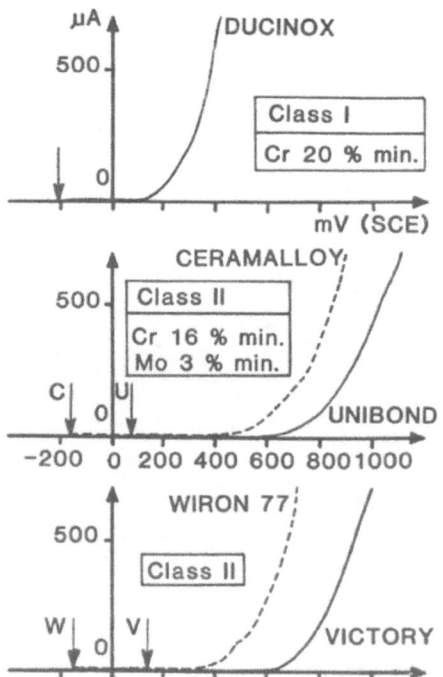

Fig. 7. Anodic polarization curves of Classes I and II alloys.

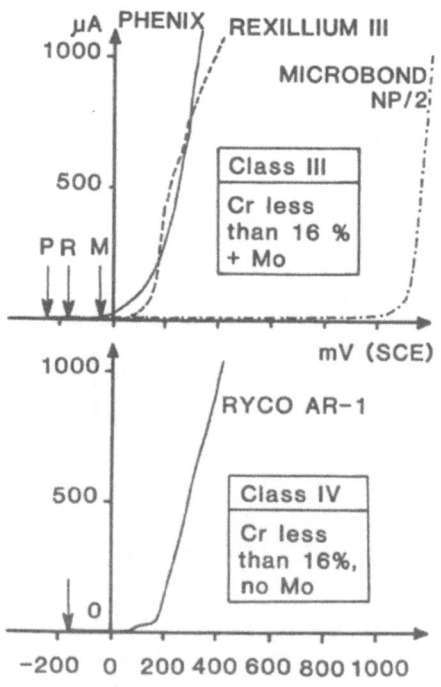

Fig. 8. Anodic polarization curves of Classes III and IV alloys.

Table 4. Additional elements in some dental alloys

Class	Alloy name	Cr	Mo	Al	Si	Be	B
I	Ducinox	22	--	--	2.3	--	0.24
II	Ceramalloy	17	4	0.02	4.3	--	0.35
II	Unibond	22	8	0.13	0.3	--	--
II	Victory	18	11	--	--	--	--
II	Wiron 77	16	5	4	0.8	--	--
III	Microbond NP/2	12	7	--	0.7	--	--
III	Phenix	15	4	0.02	5.1	--	0.45
III	Rexillium III	13	5	--	--	1.8	--
IV	Ryco AR-1	15	--	--	3.2	--	--

In weight percents.

In Figure 7, it appears clearly that the high Cr content is not sufficient to protect the alloy efficiently against corrosion. Alloys of Class II, however, are much better protected because of the presence of molybdenum, as indicated by the anodic curves having a breakdown potential higher than that of the Class I Ducinox, thus creating a relatively wide zone of security with very low anodic current. Furthermore, it should be noted that the four Class II alloys do not react similarly : two of them have their sharp increase in anodic current markedly displaced toward the higher polarization potentials (Unibond and Victory), as compared to the two others (Ceramalloy and Wiron 77). What is the cause of this behavior ? When looking at the compositions, it appears that both Ceramalloy and Wiron 77 have a higher silicon content, especially for Ceramalloy.

In Figure 8, the differences within the same class are even more important : alloys Phenix and Rexillium III have a breakdown potential very close to the corrosion potential, and they can therefore easily start to corrode if some polarization appears in the oral cavity. In the contrary, the alloy Microbond NP/2, which belongs to the same Class III, exhibits a very wide range between its corrosion potential and its breakdown potential. It is so very well protected against severe corrosive degradation. Here again, the cause of such a difference in the shape of the anodic polarization curves is to be found in the composition of the alloys. When referring to Table 4, the main differences among these three alloys is their silicon and beryllium content : Microbond NP/2 has only little silicon and no beryllium. The chromium and molybdenum contents are similar for the three alloys. It then appears that additional elements like silicon and beryllium have a strong negative effect on the corrosion resistance. In the second part of Figure 8, the Class IV alloy Ryco AR-1, which do not contain any molybdenum, but has a relatively high silicon content, is clearly exhibiting a poor corrosion resistance.

In Figure 9, the anodic polarization curves of three alloys with no nickel and no beryllium, based on the Co-Cr-Mo system, are presented as an example of products with a good resistance to corrosion. Their corrosion potential, as indicated by an arrow, is generally higher than that of Ni-Cr alloys.

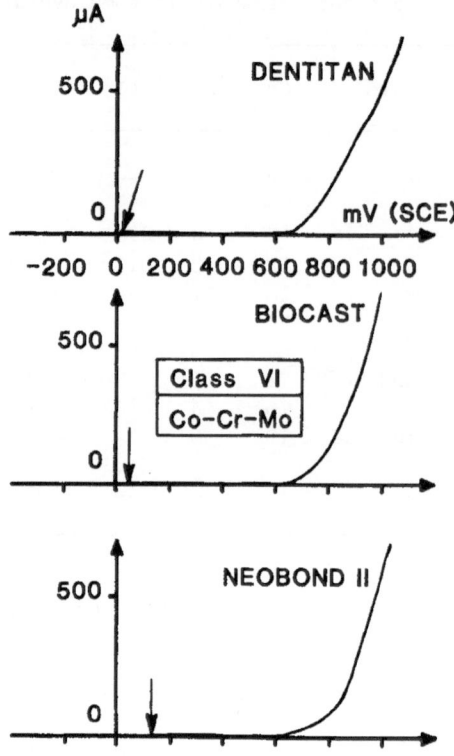

Fig. 9. Anodic polarization curves of Co-Cr-Mo dental alloys.

MICROSTRUCTURE AND CORROSION

It is easy to understand that a alloy with a single-phase structure will have a better resistance to corrosion than an alloy with a multi-phase structure. A good example of this is the case of alloys Ceramalloy (Class II, high silicon content) and Rexillium III (Class III, high beryllium content) : their microstructure is characterized by the extensive formation of a secondary phase, which has the aspect of an eutectic phase. Both alloys have shown a relatively poor resistance to corrosion. Even more, Rexillium III is currently used today as an alloy for the bonded bridge technique, because of its easy etching by chemical or electrochemical techniques, as it is generally the case with beryllium containing alloys. A recent study (Zidan, 1985) has clearly demonstrated this effect.

Alloys with no molybdenum are subject to pitting corrosion in chloride containing solutions, even if their microstructure appears to be single-phase.

RELEASE OF METALLIC IONS

One of the deepest concern about the use of nickel containing alloys is the risk of adverse reactions, like allergy, to such alloys. It is obvious that alloys, which easily corrode in the oral cavity, further enhance this risk, since the corrosion products released in the body are a source of metallic ions with potentially harmful effects. It was therefore interesting to evaluate the quantity of nickel ions released from Ni-Cr dental alloys immersed for 6 months in an artificial saliva. The results of the atomic absorption spectrophotometry are shown in Figure 10. Some alloys distinctly release more nickel ions than others : Jelbond (an alloy no more available today on the European market), e.g., has released 10 times more nickel ions than the other alloys. Products containing beryllium, like Ultratek and Lite Cast, released more Ni than corrosion-resistant products like Victory, for example. This is a further proof that the Ni-Cr dental alloys react differently, and that they should never be considered as a single, homogeneous family of alloys.

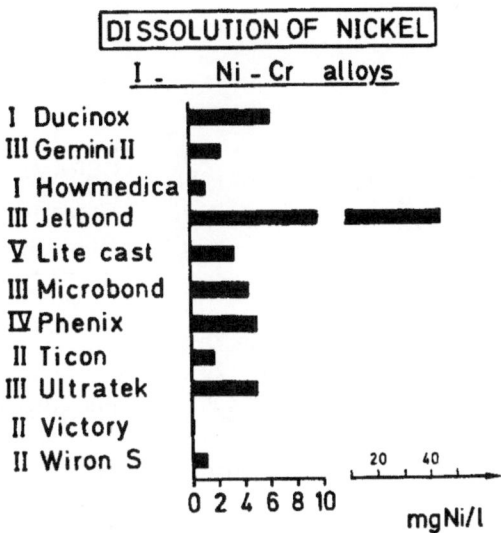

Fig. 10. Nickel ions released from Ni-Cr alloys immersed for
6 months in an artificial saliva.

THE INFLUENCE OF pH

It has been frequently hypothetized that pH should have a definite effect on the corrosion behavior of dental alloys. To evaluate the importance of such an effect, the artificial saliva used for the corrosion tests has been modified to obtain solutions with similar components, but with pH ranging from 3 to 9, instead of the 5.5 value of the standard artificial saliva. These modifications are shown in Table 5.

From the anodic polarization curves traced in these pH-modified artificial salivas, and shown in Figures 11 et 12, it clearly appears that the pH has a much minor effect on the corrosion behavior of the alloys than that of the composition, at least in the limits of pH values considered in this short study, and which more or less correspond to the physiological conditions. Only two alloys of the same class have been yet investigated : one with a high Si content and a low corrosion resistance, Ceramalloy, and one with a low Si content and a good corrosion resistance, Unibond. On both of these well contrasted alloys, the influence of the pH is minimal.

Table 5. Artificial saliva solutions with various pH

The desired pH is obtained by mixing standard artificial saliva with fixed amounts of acidic or basic solutions:

pH 3 Std art. saliva + Citric acid + Na_2HPO_4

pH 5 Std art. saliva + KH_2PO_4 + Na_2HPO_4

pH 7 Std art. saliva + KH_2PO_4 + Na_2HPO_4

pH 9 Std art. saliva + Borax + KH_2PO_4

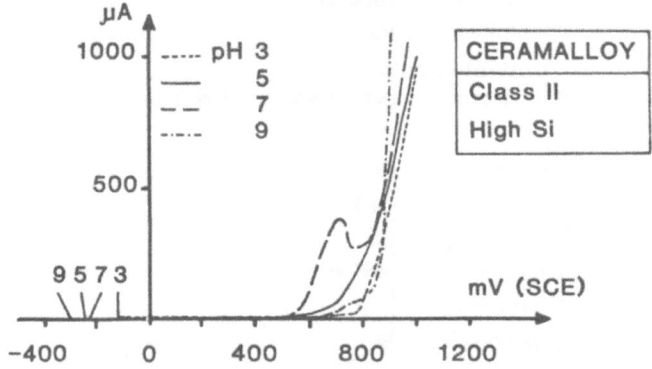

Fig. 11. Influence of the pH on the corrosion behavior of a Class II, high Si containing alloy.

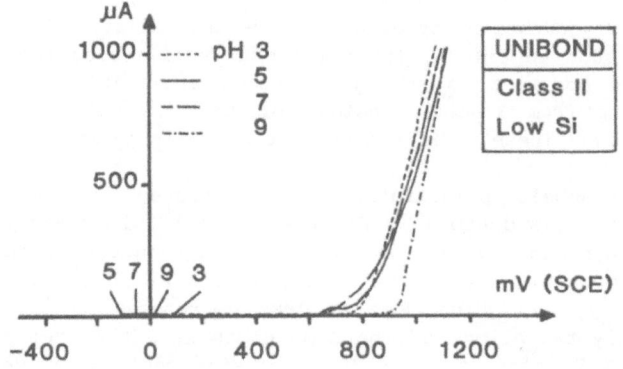

Fig 12. Influence of the pH on the corrosion behavior of a Class II, low Si containing alloy.

A list of open-circuit potentials measured in a given environment and sorted from the most noble to the least ones is called a galvanic series and is used to easily compare alloys in this environment. From all the potential-time curves obtained over the years for all kinds of dental alloys in this laboratory, the open-circuit potential values measured after 10 hours of immersion in the standard artificial saliva have been sorted and ranged to form a galvanic series, as shown in Figure 13. Such a series is useful to evaluate the potential risk of mixing alloys in the mouth.

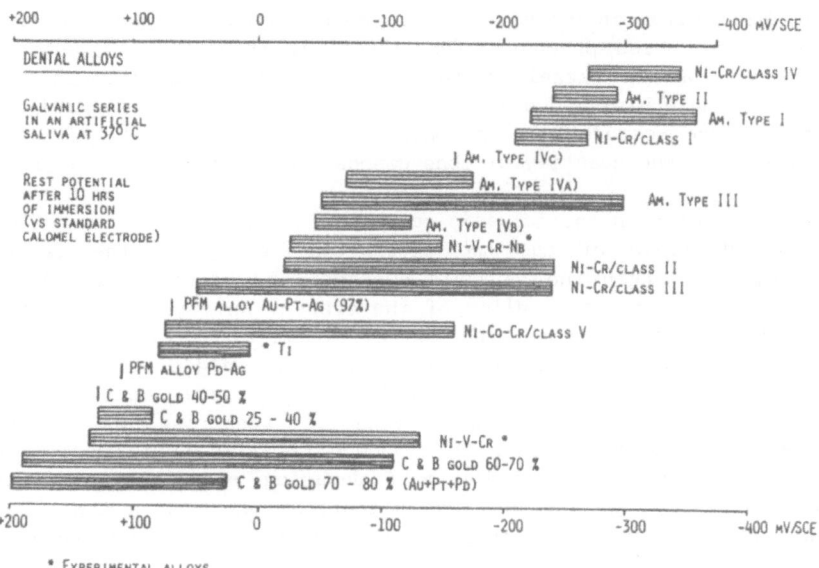

Fig. 13. A galvanic series for dental alloys, including
amalgams, precious and base-metal alloys, immersed
in an artificial saliva for 10 hours.

MIXING ALLOYS : THE GALVANIC EFFECT

In clinical practice, an alloy is seldomly unique in a patient's mouth: amalgams of various ages and types, precious alloys and base-metal alloys may co-exist, either in direct contact (permanently for restorations on adjacent teeth or repetitively for restorations on antagonist teeth), or well separated by several centimeters, in the same arch or in the opposite one. What is the effect of mixing the types of alloys in the same patient ? Could well separated metallic restorations interact with each other ? Which alloy shall most suffer from a permanent contact with an alloy of an other type ? Many questions often asked by the practitioner, or even by concerned patients themselves.

Basically, the corrosion behavior of an alloy is affected by a permanent contact with a more noble alloy. This effect is called galvanism, and can produce local currents that the patient can detect, more or less painfully, depending on his individual level of sensitivity. Complaints of galvanic pain frequently occur in the days following the placement of a new amalgam restoration, but usually pain shall gradually disappear with the aging of the amalgam restoration, corresponding to the completion of the slow reaction of mercury with the amalgam alloy. Galvanic effects may be even more dramatic than just pain: metallic elements can be progressively damaged, and cases have been reported (Wirz et al., 1980, Mazille and Brugirard, 1981) where stainless steel root posts, inside amalgam cores supporting gold restorations, have been completely destroyed by galvanic effect. The corrosion resistance of root pins, posts, and screws, vary widely, depending on the alloy composition: gold plated brass is the worst, followed closely by ferritic steels, austenitic stainless steels are better, but gold-platinum alloys, cobalt-based alloys, and titanium offer the best protection against corrosion (Meyer, 1985). The clinical consequences of such damages are most severe, since the whole restoration is ruined, with possible destruction of the root itself (necrosis due to heavy infiltration of metallic corrosion products, fragilization, fracture). Since stainless steel alloys contain nickel, nickel ions are released through galvanic corrosion of these pins and posts. But a similar effect is produced on Ni-Cr crown and bridge restorations in contact with more noble alloys: corrosion is enhanced, and the quantity of ions discharged in the body is increased.

One key factor in the galvanic corrosion is the difference between the corrosion potentials of the two interacting alloys : the larger the difference, the larger the galvanic current, and consequently the anodic oxidation of the less noble alloy of the system. An other important factor is the surface ratio between the two different alloys : the smaller the surface of the less noble alloy, the larger the anodic oxidation of this alloy. Since Ni-Cr alloys are frequently used in today dentistry for the porcelain-fused-to-metal technique (esthetic veneering of a metallic framework by dental porcelain), the exposed metallic surfaces are usually limited, and offer therefore the most damaging conditions for galvanic corrosion. Fortunately, the clinical situations where such small surfaces come into permanent contact with more noble alloys are not so frequent. The possibility of galvanic effects does exist, however, and it should never be overlooked by the practitioner, who shall always try to avoid to multiply the types of alloys, that he places in one patient's mouth. Referring to a galvanic series such as the one presented above may help him to choose at least among alloys with the smallest difference of their open-circuit potential, thus minimizing the effects of the galvanic couples, that he has formed.

The possible galvanic effect of a more noble alloy on a less corrosion resistant alloy, which is not placed in direct contact, but separated by a rather large distance (e.g. on the opposite side of the mouth), is not yet clearly demonstrated. Some authors (Phillips, 1973) declare that bones and tissues are conductors, and that currents may flow through them. It is still unclear, however, whether such conduction properties are sufficient to create an effective galvanic coupling between the two alloys, with the consecutive anodic oxidation and destruction of the less noble alloy. It is assumed at least that, if anything should happen in such a situation, the rate of the reactions shall be much slower than in the case of a direct and permanent contact, since the internal resistance of this in vivo circuit is much higher than that of a circuit with metallic conductors, as in the case of an alloy-to-alloy contact. In most instances, the effects shall be so slow that they could not be distinguished anymore from the general corrosion of the individual alloys. These interactions, whenever present, can therefore be considered as negligible when compared to the other forms of corrosion, and to the other clinical situations, which favor the anodic oxidation of nickel-based dental alloys.

CONCLUSION

The dental alloys containing nickel as the main component do not form a single group with uniform properties. On the contrary, large differences in compositions, including differences in the number and nature of the components and in their relative proportions, have led to the definition of classes of alloys, based both on the composition and on some electrochemical properties.

The main effect on the corrosion resistance of the Ni-Cr dental alloys is their composition : chromium associated with molybdenum provide a good resistance to corrosion, whereas additional elements like silicon and beryllium markedly decrease that resistance to corrosion. The release of Ni ions through anodic oxidation has been observed : it varies markedly from one alloy to the other, and it depends, like the general corrosion behavior of the alloy, mainly on the composition.

The influence of changes in pH of the artificial saliva is minimal, as observed on two alloys with a very contrasted corrosion resistance.

The electrochemical behavior of the alloys is closely linked with their microstructure. Alloys with secondary phases formed by additional elements like silicon and beryllium, have a much lower resistance to corrosion.

When mixing alloys in the oral cavity, the practitioner creates the conditions for galvanic corrosion. The extent of this form of corrosion varies according to the nature of the involved alloys, the clinical situations, the ratio of the alloy surfaces, and the individual factors associated with the patient. A general rule of safety is to avoid the placement of different types of alloys. A galvanic series of dental alloys in an artificial saliva is a helpful means in selecting alloys with the smallest difference in open circuit potential. When the clinical situation makes the presence of two different alloys unavoidable, the choice of the proper alloys could minimize the effects of the galvanic coupling.

Ni-Cr dental alloys could be used in place of precious alloys, but extreme care should be taken in the proper selection of products, whose properties may vary considerably. The dental technicians are not always aware of those differences in properties, and the selection is too often based on the economical factor only, thus leading to a possible unsatisfactory choice. Most of the precious alloys do not pose such a threat.

Acknowledgements

Thanks are expressed to Mr. Niney, from Comptoir Lyon-Allemand-Louyot, France, who provided the Ni-Cr binary alloys, and to all my collaborators, who have participated in the numerous investigations described here.

References

Fusayama, K., Katayori, T., Nomoto, S., 1963, Corrosion of Gold and Amalgam Placed in Contact with Each Other, J. Dent. Res., 42: 1183.

Lee, J., Lucas, L., O'Neal, J., Lacefield, W., and Lemons, J., 1985, In Vitro Corrosion Analyses of Ni-base Alloys, 63rd Gen. Sess. IADR, Las Vegas, J. Dent. Res., 64: 317, Abstract 1285.

Mansfeld, F., 1973, Tafel Slopes and Corrosion Rates from Polarization Resistance Measurements, Corrosion, 29: 397.

Mazille, H.M.J., Brugirard, J., 1981, Problèmes de corrosion de quelques métaux et alliages dentaires : comparaison de cas cliniques et d'essais de laboratoire, in : "Métallurgie Dentaire", P. Guiraldenq and J. Blanc-Benon, eds., Pyc Edition, Paris.

Meyer, J.M., 1977, Corrosion Resistance of Nickel-Chromium Dental Casting Alloys, Corrosion Science, 17: 971.

Meyer, J.M., 1985, Corrosion of Dental Root Posts, in : "Corrosion et Dégradation des Biomatériaux et leurs Incidences Cliniques (Premier Colloque Européen), Strasbourg, Mars 1984," Conseil de l'Europe, Strasbourg.

Meyer, J.M., Nally, J.N., 1974, Influence of Artificial Salivas on the Corrosion of Dental Alloys, 11th Annual Meeting, IADR, Continental European Division, Brussels.

Meyer, J.M., Niney, C., 1985, Unpublished results.

Meyer, J.M., Wirthner, J.M., Barraud, R., Susz, C.P., and Nally, J.N., 1979, Corrosion Studies on Nickel-Based Casting Alloys, in : "Corrosion and Degradation of Implant Materials, ASTM STP 684," B.C. Syrett and A. Acharya, Eds., American Society for Testing and Materials, Philadelphia.

Phillips, R.W., 1973, Tarnish and Corrosion, in : "Science of Dental Materials", W.B. Saunders Company, Philadelphia.

Stern, M., Geary, A.L., 1957, A Theoretical Analysis of the Shape of Polarization Curves, J. Electrochem. Soc. 104 : 56.

Wirz, J., Johner, M., Pohler, O., 1980, Korrosionsverhalten verschiedener Schrauben und Stifte in Wurzelkanal, Schweiz. Mschr. Zahnheilk., 90: 217.

Wiskott, A., 1980, Une approche de l'analyse des courbes de polarisation selon Mansfeld, Thèse No. 358, University of Geneva.

Zidan, O., 1985, Etched Base-Metal Alloys: Comparison of Relief Patterns, Bond Strengths and Fracture Modes, Dental Materials, 1: 209.

AUSTENITIC STAINLESS STEEL S350 FOR

SURGICAL IMPLANTS?

C. Gabbi[1], G. Melotti[2], E. Vita Finzi[3], and P. Tranquilli Leali[2]

Veterinary University, Parma (I)[1], Deltasider spa, Aosta and Verres[2], and Catholic University, (I) Roma (I)[3]

INTRODUCTION

The application of stainless steels as AISI 316 L + Mo and 316L Mo + N_2 as implantable materials is well known. Their biological response being established, the major stainless steel grades are now covered by international standards, i.e. ISO, ASTM, DIN, AFNOR and Italian UNI.

However, a limitation to the increasing use of these low cost materials which can be readily shaped lies in their mechanical properties.

Table 1. Mechanical properties

	Yield strength N/mm^2	Tensile strength N/mm^2	Elongation %
AISI 316	195	480	$\geq$ 45 min
AISI 316 + N_2	280	600	$\geq$ 45 min

PROPERTIES OF S350

The steel industry, both in Europe and North America, is currently interested in the development of super corrosion resistant stainless steels, having mechanical properties by far higher than those show above. In this field of research Deltasider has developed S350 a steel grade somewhat similar to other grades in the same class currently under study at other steel companies.

Table 2. Nominal composition of S350

C 0.04 Cr 21.0 Mn 5.0 Ni 13.0 Mo 2.25 N 0.3 V 0.2 Nb 0.25

As compared to the traditional grade 316L, Deltasider S350, an austenitic stainless steel, has higher mechanical properties and improved corrosion resistance.

Because of its austenitic structure, S350 cannot be hardened by heat treatment but only through cold reduction which, even with heavy deformation, will not modify the magnetic permeability of this material.

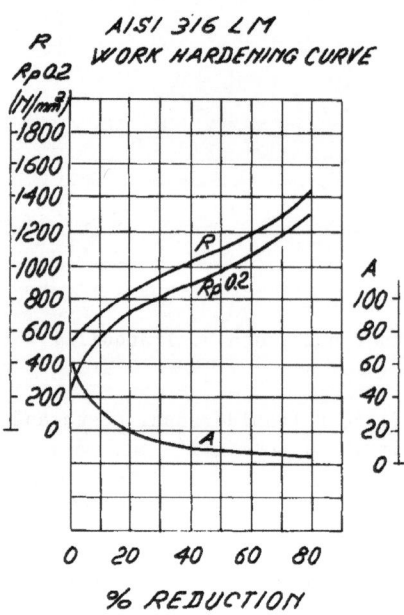 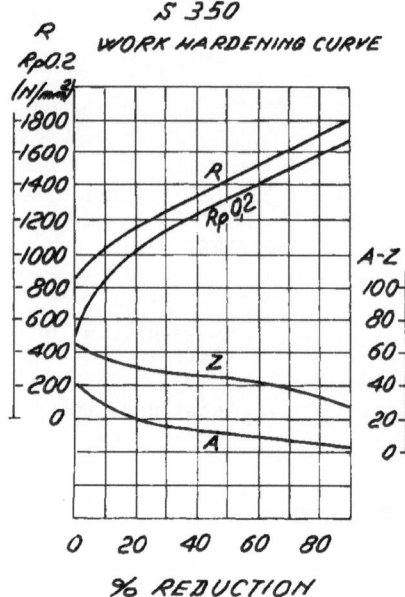

Figure 1.

Work hardening curves for both 316 and S350 are shown here for comparison.

Also mechanical properties appear to be far better than for AISI 316L Mo, as shown in the table hereunder.

Table 3. Mechanical properties of S350

	Yield strength RS N/mm^2	Tensile strength R N/mm^2	Elongation A %
S350	≥ 380 min	690 – 950	≥ 35 min

S350 exhibits best resistance to many corrosive media in the solution treated condition. Generally its corrosion resistance is better than 316L and 317L, as shown in the following table, where weight loss in mm/month is compared.

Table 4. Resistance to corrosion

Corrosive media	S350 mm/month	S316L mm/month
10% sulphuric acid, at 80° C	2.33	8.33
65% nitric acid, boiling	0.58	1.00
10% nitric + 1% hydrofluoric acid, at 80°C	5.75	36.83

Superiority of S350 over 316L has been confirmed by laboratory tests conducted to compare the resistance of both these grades to pitting corrosion and to crevice corrosion.

Pitting corrosion	↑	improved
Crevice corrosion	↑	improved
Stress corrosion	→	unchanged
Intergranular corrosion	→	unchanged

S350 can be fabricated with the same equipment used for 316L. Welding procedures are similar, provided that a filler metal of the same composition is used.

Machining is the only operation that has to be performed at significantly lower speeds than for 316L. Cutting speed on S350 has to be reduced by 50%, as in the case of modified 316L + N_2, this being the consequence of the higher work hardening rate of S350.

As a result, S350 has been approved for the manufacturing of mechanical components in several industries, e.g. pharmaceuticals, food. It has been used also for extremely heavy duty applications in the seacraft industry, e.g. connecting components for racing sailcrafts.

An extensive program of experimental tests has been carried out both in vitro and in vivo to verify the biological response to this grade also in the field of endoprotheses made by the investment casting process. Experimentation work was conducted by Prof. Tranquilli Reali, Catholic University, Rome and by Prof. Gabbi, University of Parma in view to determine the cell growth on the surface of small discs 5 to 8 mm diameter, as well as to check either the perfect adhesion of the bone tissue to metal specimens implanted in the long bones of rabbits or the reaction of tissues to discs placed within muscular tissues.

This program, now under way for over a year, with periodical checks after 3, 6, 9 and 12 mounths, is being conducted simultaneously on the same animals with similar implants made of AISI 316L Mo, AISI 316L Mo + N_2 and of the new S350 grade. So far the results have been satisfactory and bring to consider the possibility of extending the testing to finished components with the aim of replacing similar implants made of standard stainless steels.

Clearly, should the application tests results continue to prove satisfactory, the endoprotheses can be redesigned, e.g. the Thompson type hip implants that are made by the investment casting process, which are expected to have reduced size owing to the improved mechanical properties (up to 10 to 12% weight reduction).

As stated above, the costs of bringing S350 into final shape are somewhat higher, even if no particular difficulties are experienced as to:
- hot closed die forging
- working by abrasive wheels or tungsten carbide tools
- cold drawing into wire.

It will be possible therefore to make plates and fasteners, as is now the case for cobalt alloys, vitallium, steelites.

Here, tests are under way to verify the mutual behaviour of 316L Mo and S350 in the presence of organic liquids and corrosive agents, in order to check the electrochemical resistance of tightly pressed assemblies (plate-fastener).

CONSIDERATIONS ON PRODUCTION

The steel industry is higly interested in developing new stainless alloys.

A coordinated effort between surgery and engineering is essential in view of achieving reasonable optimization of such materials.

The S350 alloy now under testing represents an effort to promote a material having maximum reliability at a reasonable cost, particularly now that our Quality Assurance System is increasingly efficient. Under the Q.A. System the following goals are achieved:

- the guarantee that the finished material has enough potential to enable the end user in obtaining a product that is safely suitable for the intended application
- the assistance which is given to the user in the application of the material and the necessary cooperation to solve the problems involved in working and processing.
   This action is naturally supplemented by:
- the metallurgical designing of the alloy and of its manufacturing cycle
- the writing up of instructions and procedures to control the achievement of the new alloy and to make it repetitive
- the training of the personnel involved in production
- the inspection of all processing stages to insure full compliance with Q.A. procedures
- the identification of non conformance conditions and the deviation or scrapping of non conforming material.

So far this costly organization has been designed and implemented only for materials for use in the aircraft industry, for nuclear applications or for energy generation.

We feel that use of S350 for surgical implants will take full advantage of the "Quality Assurance System".

CONCLUSIONS

The results of the first tests, in vitro and in vivo, in view of extending the use of S350 to the field of surgical implants have been encouraging. Mechanical properties and corrosion resistance of S350 as a whole are a substantial improvement over the traditional AISI 316L and make it an interesting substitute where higher stresses are involved.

CORROSION AND BIOELECTROCHEMISTRY

G. Milazzo

President of the Bioelectrochemical Society
Director (retired) of the Chemistry Laboratories of the
(Italian National) Institute of Health
Piazza G. Verdi 9,   00198 Roma (Italy)

The problem of biocompatibility of materials, intended for use in prostheses to be implanted in human beings, can be split in the three subsequent branches :

1) Immunopathology of the implanted material

2) Its corrosion under the action of the contacting biological materials

3) Toxicity of the corrosion products, including carcinogenesis

The first, strictly biological point, will not be considered here, while the 2nd and the 3rd point, intimately connected together, will constitute the object of the present talk, limited to metallic materials.

Corrosion is an electrochemical process depending on the properties of both the metallic material utilized and the corroding liquid agent. To understand this process in every given case, the general rules governing the corrosion must be considered and the utilization of the potential-pH diagrams must be introduced in order to explore all corrosion possibilities (from the thermodynamic point of view) of a given material, intended for use in implants, for the discovery of passivity and of immunity domains of the investigated materials, which will identify the possibility of inhibiting or even excluding any corrosion.

In case a corrosion is found unavoidable, the toxicity of the corrosion-products must be investigated, taking into account their physico-chemical state (solid insoluble, solid partially soluble, complexes with chemical components of the contacting physiological liquids, free ions, etc...) and their relative toxicities.

If a human being needs a prosthesis because a broken or injured bone tissue must be repaired or replaced, the problems of suitability and biocompatibility immediately arise. It is obvious that under the general name <u>bone tissue</u> every hard tissue must be intended, which has to provide some mechanical founction and/or is submitted to some mechanical stress or wear.

The general problem of biocompatibility is a complex one and can be split in the following subquestions :

A) Mechanical resistence against mechanical stress and wear, the second point being extremely important in articulation and in dentistry

B) Tissue reponse

C) Resistence against chemical attack by contacting biological environment, i.e. corrosion

Corrosion in turn causes two other consequences :

C1) Progressive attack until possible destruction in the implanted prosthesis

C2) Appearance of toxic effects (including carcinogenesis) bacause of the toxicity of the products of the corrosion

It is, of course, impossible to consider all the above mentioned points in a short talk like this one. It is, therefore, necessary to restrict the objectives so that one point, at least, can be treated relatively exhaustively. This point of view is facilitated by the circumstance that points A and B are very different from point C. In fact, point A is purely mechanical and point B purely physiological. only point C can be investigated using electrochemical theories and techniques applied to events occuring in living bodies, i.e. bioelectrochemistry (1). Therefore, only point C will be considered in its origin, in its consequences and in the possibilities of avoiding any damage to the living organisms in which a prosthesis must be implanted.

Corrosion is the consequence of a typical electrochemical heterogeneous reaction (2). A very simple formulation in words of this electrochemical reaction states that "any element whose electrode potential under the given experimental conditions is more negative than the electrode potential, under the same experimental conditions of another element, displaces this latter from the solution in which it is present in ionic form". In biological systems, where the solvent is generally water, the investigation can be restricted to the behaviour of the metallic material under investigation, contacting an aqueous phase containing, of course, H^+ and OH^- ions.

The theoretical tool for this investigation is represented by the potential-pH diagrams describing the behaviour of the investigated material contacting an aqueous phase where H^+ ions and a simple non complexing anion are present.

To understand the construction and the use of these diagrams, nickel can be considered as a very clear example. The starting point for the construction of this diagram (and others of the same type for other metallic materials) is given by the reactions of Ni with H^+ and H_2O, their energetics and their electrode potentials at equilibrium . The equations collected in table 1 describing the reactions and the solubility equilibria of Ni-H$_2$O system are relevant for the potential-pH diagram of Ni (3).

On the basis of the equations giving the values of the potential U as a function of the concentration of the ionic species present in the solution (Ni^{2+}, HNiO$_2^-$ and H$^+$) for each one of the chemical equations (1) trough (10) and considering the solubility equations (11) and (12) the equilibrium diagram for the system Ni.-H$_2$O can be drawn at 25°C. It is given in fig. 1 for a concentration of Ni^{2+} ions equal to 10^{-6} M (1 pM).

Table 1 : Relevant reactions for the potential-pH diagram of the Ni-H$_2$O system at 25°C

N°	Reaction		Thermodynamic equation
1	Ni	Ni^{2+}	U = -0.250 + 0.0295 lg Ni^{2+}
2	Ni + 2H$_2$O	Ni(OH)$_2$ + 2H$^+$ + 2e$^-$	U = 0.110 - 0.0591 pH
3	Ni + 2H$_2$O	HNiO$_2^-$ + 3H$^+$ + 2e$^-$	U = 0.648 - 0.0886 pH + 0.0295 lg HNiO$_2^-$
4	3Ni^{2+}+4H$_2$O	Ni$_3$O$_4$ + 8H$^+$ + 2e$^-$	U = 1.977 - 0.2364 pH - 0.0886 lg Ni^{2+}
5	3NiO + H$_2$O	Ni$_3$O$_4$ + 2H$^+$ + 2e$^-$	U = 0.897 - 0.0591 pH
6	3HNiO$_2^-$ + H$^+$	Ni$_3$O$_4$ + 2H$_2$O + 2e$^-$	U = -0.718 + 0.02965 - 0.0886 lg HNiO$_2^-$
7	2Ni^{2+} + 3H$_2$O	Ni$_2$O$_3$ + 6H$^+$ + 2e$^-$	U = 1.753 - 0.1773 pH - 0.0591 lg Ni^{2+}
8	2Ni$_3$O$_4$ + H$_2$O	3Ni$_2$O$_3$ + 2H$^+$ + 2e$^-$	U = 1.305 - 0.0591 pH
9	Ni^{2+} + 2H$_2$O	NiO$_2$ + 4H$^+$ + 2e$^-$	U = 1.593 - 0.1182 pH - 0.0295 lg Ni^{2+}
10	Ni$_2$O$_3$ + H$_2$O	2NiO$_2$ + 2H$^+$ + 2e$^-$	U = 1.434 - 0.0591 pH

SOLUBILITY OF Ni-CONTAINING IONS IN THE PRESENCE OF SOLID Ni(OH)$_2$

11	Ni^{2+} + H$_2$O	Ni(OH)$_2$ + 2H$^+$	lg Ni^{2+} = 12.18 - 2 pH
12	NiO + H$_2$O	HNiO$_2^-$ + H$^+$	lg HNiO$_2^-$ = -18.22 + pH

(all numerical data are given for 25°C)

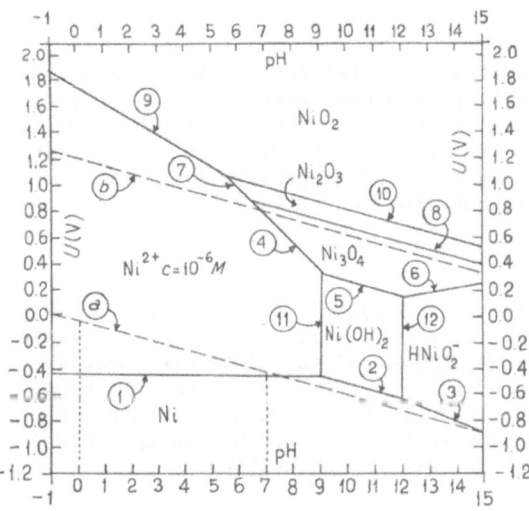

Fig. 1 : Potential-pH diagram of Nickel at 25°C

The diagram contains also the equilibrium lines of water _a_ and _b_. Line _a_ gives the theoretical equilibrium value of the potential of a hydrogen electrode, and line _b_ the one of oxygen electrode, respectively, under the pressure of 1 000 mBar (1 atmosphere) of gaseous H_2 and O_2 respectively, as a function of the pH.

Following Nernst's equation and considering the value of -0.250 V for the standard potential of the Ni/Ni^{2+} electrode, the potential of the metallic nickel contacting an aqueous solution of Ni^{2+} ions at 1 M concentration will be equal to -0.427 V (independant of pH), while the potential of the H_2/H^+ electrode varies from 0 (at pH = 0) to 0.427 V (at pH = 7.2). This means that Ni should be always corrodable within the limits of pH = -1 and 7.2, i.e. up to a pH just above the neutrality. It is not attackable in alkaline solutions ; in these solution, it is _immune_.

Further inspection and considerations of this diagram shows that a higher pH and/or more positive potentials solid hydroxides and oxides appear as reactions products, which under certain conditions may become so compact that any electrode reaction is completely inhibited : the metallic material has become _passive_.

From the diagram given in figure 1, a second diagram can be derived (figure 2) in which the domains where nickel can not be corroded and where it can not be attacked (because of immunity or passivity) are clearly evident.

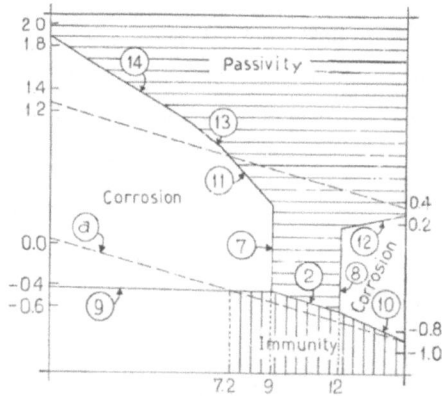

Fig. 2 : Immunity, corrosion and passivity domains of nickel

In the particular case of biological contacting liquids containing molecular species capable of producing complexes with Ni^{2+} ions (organic acids, aminoacids, etc...), thus altering the electrode potentials of Ni^{2+} ions (because of equilibrium reaction between Ni^{2+} ions, complexing agent and resulting complex) the potential-pH diagram will be modified by the energetic and electrochemical contributions originated by these complexing molecular species.

Thses considerations are purely thermodynamic and represent the necessary but possibly not yet sufficient basis for a reliable conclusion concerning the bahaviour and the suitability of materials when used in biological implants. Also kinetic effects must be considered, which can remarkably alter the potential-pH diagram. It is enough to consider the kinetic effects particularly relevant in the case of the hydrogen electrode reaction. Thses effects slow down the rate of the hydrogen electrode reaction on Ni, so that the (real) potential of the electrode on nickel becomes remarkably more negative than at real thermodynamic equilibrium.

To give a practical example, the overtension for the discharge of H^+ ions on a nickel electrode at pH = 0 at 25°C under a current density of 10^{-3} A/cm^2 is -0.33 V, i.e. the steady state potential of the H_2/H^+ electrode under these conditions is remarkably more negative than the equilibrium potential of the nickel electrode : nickel can no longer be corrosionally attacked under these real conditions, i.e. immunity domain has been extended toward lower pH values. Nickel could therefore be an acceptable material.

Similar potential-pH diagrams can be constructed for any other metallic material to investigate its corrosion properties, for example suitable alloys.

If the material to be implanted is a simple metal, which shows an immunity domain due to its positive standard potential, great care is necessary concerning its purity. If a metallic impurity is present in the form of a grain with a more positive potential than the one of the bulk metal, a galvanic element is formed as illustrated in fig. 3. This galvanic element shows an extremely low external resistance between metal bulk and impurity, so that a very small potential of this galvanic element can produce relatively high local current intensities, i.e. a strong local corrosion of the bulk metal. The bulk metal acts as negative pole of the above said local galvanic element and therefore passes into solution even if its standard potential is more positive than the one of the hydrogen electrode under the same experimental conditions.

This consideration is very important for alloys of the eutecticum type. Here a biphasic metallic material is present with grains of each one of the two metals of the eutecticum. Corrosion is in this case an unavoidable consequence.

In case the use of a particular alloy is desirable, bacause of other reasons, this alloy should be of the solid solution type, i.e. crystals with complete solubility of one metal in the other in the concentration domain utilized. In case this condition is still difficult to be fulfilled, conditions should be realized to improve the occurence of passivity to make corrosion impossible.

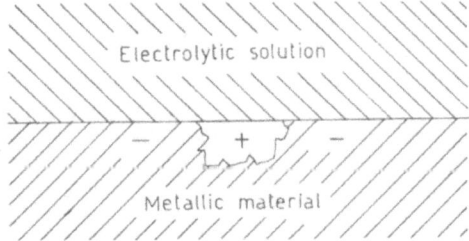

Fig. 3 : Imupurity grain in a bulk pure metal

All these investigations, realizing a special kind of quality control, are of course not the job of physicians, they are the job of bioelectro- chemists working in intimate cooperation with physicians, surgeons and biologists to realize the best possible starting material to produce prostheses for implants in living organism. But physicians, surgeons and physiologists should be aware of this basic knowledge in order to obtain from manufacturers biomaterials showing the above said properties required to avoid corrosion and its damage effects.

References

(1) For a somewhat more extensive definition of "Bioelectrochemistry" and its scientific content s. G. Milazzo, Bioelectrochemistry and Bioenergetics. An interdisciplinary survey in Bioelectrochemistry I, G. Milazzo and M. Blank (Editors), Plenum Press, New York and London (1983), p. 5-14.

(2) E. Deltombe, N. de Zoubov and M. Pourbaix, in Atlas d'Equilibres Electrochimiques, M. Pourbaix et al. (Editors), Gauthier-Villars, Paris (1963), p. 331-332.

(3) For a more extensive definition of the corrosion see G. Milazzo, Elektrochemie, Birkhäuser Verlag, Basel - Boston - Stuttgart (1981), p. 82-96.

MICRO-ANALYSIS OF THE COMPOSITION OF DENTAL IMPLANTS BEFORE AND AFTER

IMPLANTATION

Monique Simonoff[1], Bernard Berdeu[1], Yvan Llabador[1]
A. Garuet[2], P.F. Caitucoli[2], and F.X. Michelet[3]

1) Chimie Nucléaire, C.E.N.B.G. 33170 Gradignan
2) UER d'Odontologie 14 Cours de la Marne 33000 Bordeaux
3) Chirurgie maxillofaciale et Stomatologie CHR Bordeaux
   Place Amélie Rabat-Léon 33000 Bordeaux

SUMMARY

The metallic implants studied are of endo-osseous type. We determined the composition of two types of implants on identical samples, before and after implantation in the maxilla or the mandible, the implants gave rise to serious rejection phenomena. We used the PIXE method (particle induced X-ray emission) for these analyses.

The first type of implants corresponded to crutched implants, and the second type corresponded to blade implants. Although all the parts of the type I implants were supposed of identical composition we found the receptacle to be 64% Ni and the pegs (stem and pins) to be 70% Fe.

The implants removed showed evidence of migration of Ni and Fe from the receptacle and the pegs. An electrolytic cell using different parts of these implants showed a potential difference of $\sim$ 0.25V.

INTRODUCTION

Tissue intolerance to a metallic alloy depends on the nature of the implant itself. The alloys used may cause chemical, electrogalvanic or immunologic reactions with subjective or objective symptoms.

Certain types of metallic implants, Co-Cr-Ni alloys commonly used in odontology, have given rise to serious rejection phenomena. Due to these violent allergic or inflammatory phenomena, certain of these implants were removed some months after implantation.

Metallic migrations, although in trace amounts, have been measured in biopsies taken from the neighborhood of the implant. In this work, we determined by the PIXE method (particle induced X-ray emission) the composition of two types of implants on identical samples of a same lot, before and after implantation in the maxilla or the mandible.

MATERIAL AND METHODS

1. Implants

The metallic implants studied were endo-osseous types of two kinds (figure 1). The first type (Type I) corresponded to crutched implants, and consisted of an extra-tissular super-structure which emerges in the buccal cavity (a metallic receptacle to retain the dental prosthesis) and an endo-osseous substructure made of a central stem and four divergent pins of smaller diameter. The pins and central stem are driven by force into the maxilla or the mandible. The second type (Type II) corresponded to the blade implants. The lamellar infrastructure inserted in the bone supports a vertical blade (post). This post constitutes a stump on which a ceramo-metallic prosthesis is placed.

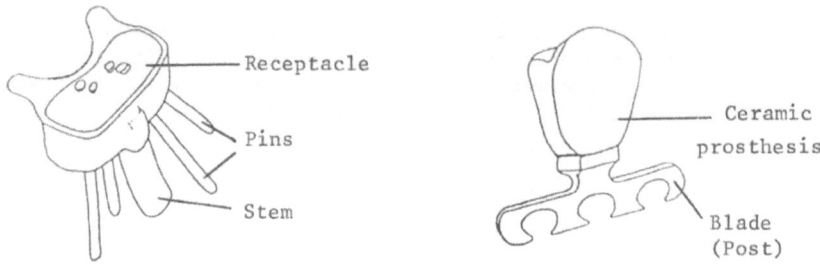

Fig. 1

(Type I) Crutched implant   (Type II) Blade implant

2. Preparation of samples

The implants were divided into sections by very fast rotating disks. The composition of these disks is essentially titanium (31.9%) and iron (62%), with some other elements (calcium, vanadium, manganese).

We have measured the trace element pollution that could arise from these disks and it was negligible.

After cleaning for ten minutes in an ultrasonic bath with triple distilled water the slices of implant were dried then weighed (range, 1 to 10 mg).

3. Dissolution of the samples

In our measurements we used the PIXE method with thin target (thickness < 1 mg/cm^2). The different samples were dissolved in 1 to 2 mL of concentrated hydrochloric and nitric acids and diluted to an equal volume with 10% hydrofluoric acid.

4. Targets

100 μL of the solution were diluted with 3 mL of triple-distilled water. Then 500 μL of a rubidium solution (RbCl) was added as internal standard for the quantitative measurement of the elements present.

One drop of the final solution was deposited on a thin film (2 μm) of polycarbonate in a plastic frame, forming the target for the PIXE analysis.

5. The PIXE system (Figure 2)

The PIXE system (1) as used in our laboratory has been described in reference.

A beam of 2.5 MeV protons from the 4 MV Van de Graaff accelerator at the "Centre d'Etudes Nucléaires" in Bordeaux is diffused by a 2 μm thick aluminium foil, situated 0.3 m from the target. A collimation system of carbon apertures defines the diffused beam onto the target, which is positioned at 45° to the incoming beam. This system provides a uniform intensity of irradiation over a constant target surface. The cross section of the beam is 10 mm^2. After passing through the thin targets, the beam impinges on a Faraday cup coupled with an electrometer, which is used to measure the beam current. The carbon collimator nearest to the target has triple function. It is the last beam diaphragm before the target, it is an electron suppressor when the current is directly measured on the target (thick targets) and it is a collimator for the detectors.

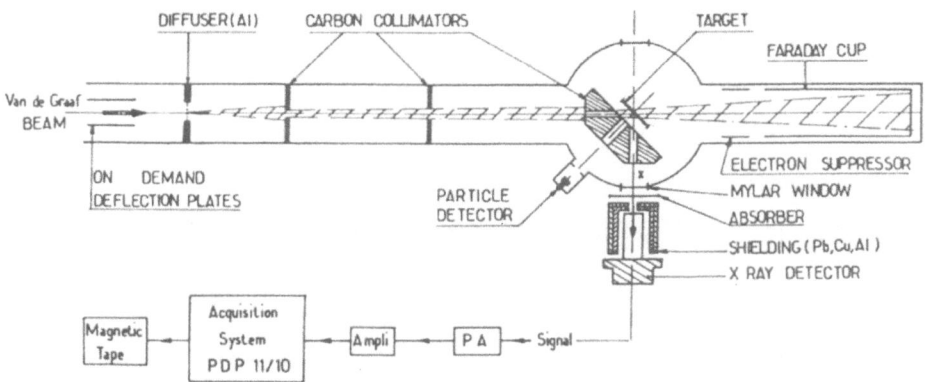

Fig. 2 : An overview of the PIXE chamber in Bordeaux.

X-rays emitted at 90° to the beam direction emerge from the vacuum chamber through a 25 μm aluminized mylar window and traverse an airspace before detection. Appropriate graphite X-ray absorbers are positioned in this space to reduce the intensity of the low-energy X-rays associated with biological samples. For chromium detection, a thickness of about 18–20 mg/cm^2 is sufficient to eliminate interference from the K, Ca X-rays.

The targets (ten on the holder) were irradiated in vacuum (10^{-5} – 10^{-6} mbar). The internal side of the chamber was entirely covered with plastic foil, covered with a very thin evaporated carbon film. By using an electrostatic "on-demand" beam deflection system for the X-ray detector system, pulse pile-up was kept low and no explicit dead time corrections had to be made. Photon spectroscopy was carried out by means of a 100 mm^2 intrinsic germanium X-ray detector, with a resolution of 180 eV (whm) for 5.9 keV Mn X-rays.

A surface barrier (Au-Si) detector inside the chamber was used for particle elastic scattering measurements and for monitoring purposes and was situated at 135° to the beam direction. The detector signals were fed through conventional amplifying and pulse processing circuits. The spectra were registered on magnetic tapes using PDP 11 minicomputer systems. Generally, 10 to 20 min irradiation with an integrated current of 5 to 10 μC was statistically sufficient for a chromium peak measurement. The sensitivity of the above experimental arrangement is of the order of 10^{-9} to 10^{-10} g for all of the elements of $Z > 14$.

6. Spectra

The X-ray spectra are treated by means of a computer program. Each peak is taken to be Gaussian and the background is subtracted after fitting to a polynomial or exponential equation of an appropriate degree. Figure 3 shows a typical fit obtained for a sample taken in the metallic part of the ceramo-metallic prosthesis (Blade implant).

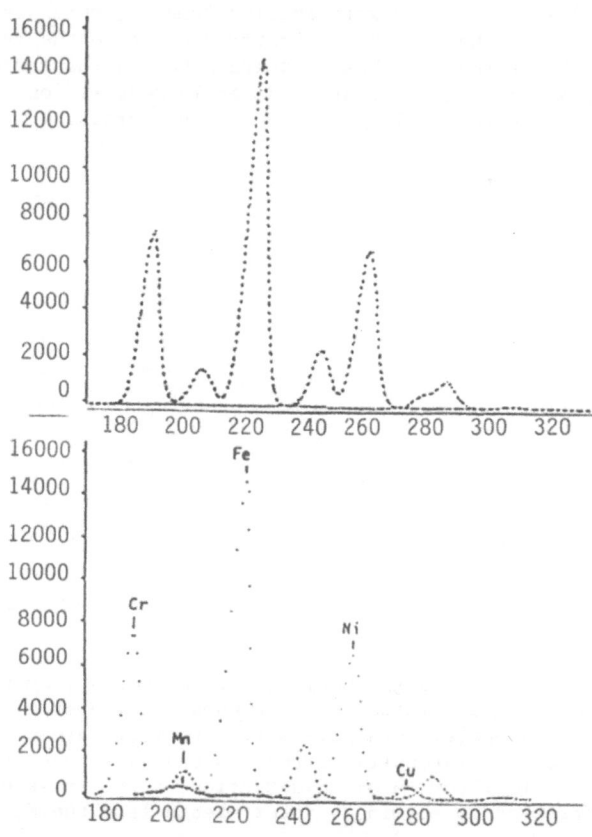

Fig. 3
a) A typical X-ray spectrum as measured in the PIXE chamber
b) The same spectrum treated by a computer program

RESULTS

The compositions cited are mean values obtained by analysing different sections of each part of the implant.

An identical unused implant is referred to as the reference implant.

Composition of the implants

a) Crutched implants

The crutched implants were from a non corrosive alloy whose chemical composition was stated to be : Co, 63% ; Cr, 27% ; Mo, 5,6% ; Ni, 2% ; Mn, 0.8% ; Fe, 0.6% with traces of Si, C, P, S as stated the manufacturer.

The table I gives the results in percentage obtained by PIXE in the different cut sections of receptacle, central stem and pins for chromium, manganese, iron and nickel.

Table I : Composition in percentage of crutched implants. Number of determinations ( ).

Element (in %)	Receptacle (18)	Central Stem (4)	Pins (23)
Cr	28.4 ± 1.9	20.0 ± 0.3	19.6 ± 1.7
Mn	1.5 ± 0.4	2.6 ± 0.2	1.8 ± 0.4
Fe	5.3 ± 0.6	70.0 ± 0.2	72.0 ± 1.9
Ni	64.8 ± 2.0	7.4 ± 0.1	6.6 ± 0.7

An example of a typical PIXE spectrum is given for each part of the crutched implant in Figures 4-5-6.

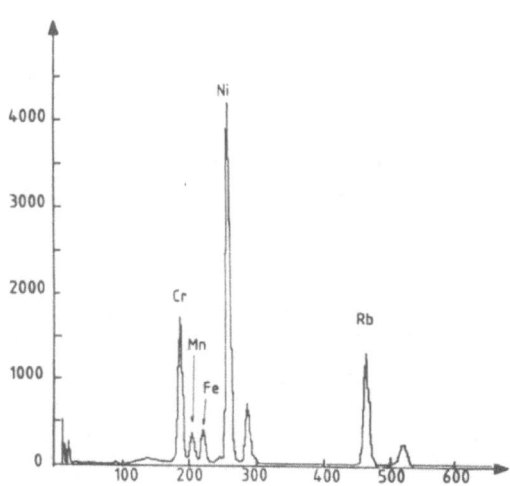

Fig. 4 : A PIXE spectrum of a crutched implant. The receptacle composed about two-thirds nickel and one-third chromium.

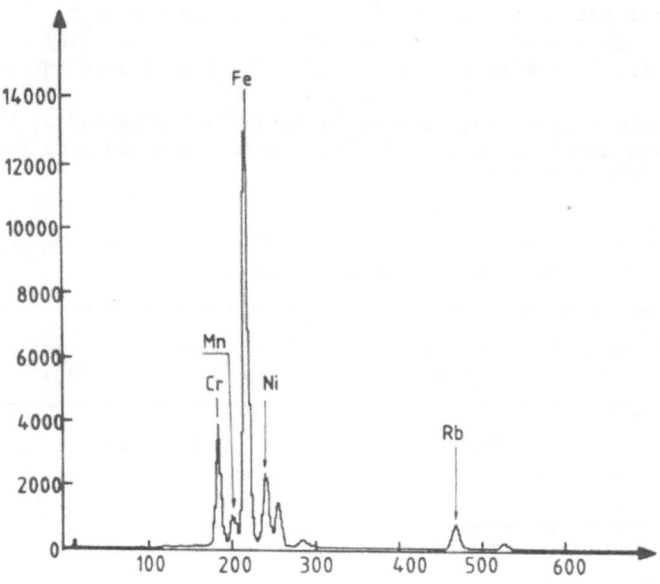

Fig. 5 : A PIXE spectrum of the central stem of a crutched
implant. The central stem contains essentially iron
(70%) and chromium (20%).

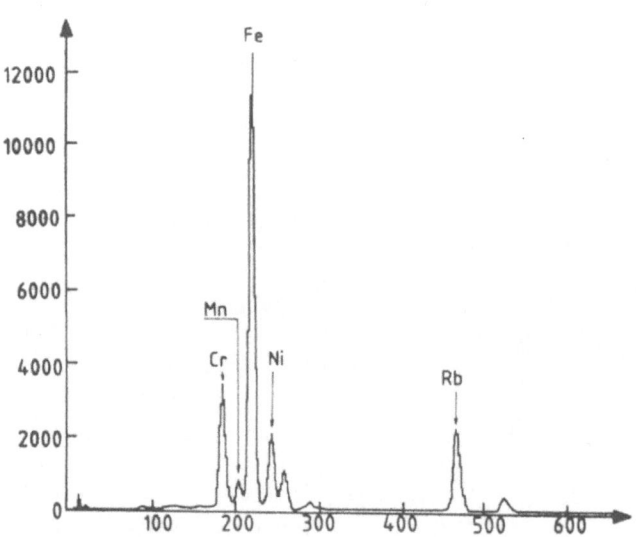

Fig. 6 : A PIXE spectrum of a pin of crutched implant. The
composition of the pins is similar to that of the
central stem.

336

b) Blade implants (Type II)

As shown in table II for blade implants, the post was pure titanium and the metallic part of the ceramo-metallic prosthesis was a mixture of iron and nickel.

Table II : The composition of the two parts of blade implants.
( ) Number of determinations

Element (in %)	Metal of the ceramo metallic prosthesis (9)	Blade (9)
Ti		100
Cr	23.9 ± 0.2	
Mn	1.6 ± 0.6	
Fe	45.8 ± 0.9	
Ni	25.6 ± 0.5	
Cu	3.1 ± 0.8	

Comparison between extracted and reference implants

The next figures (7a-b-c) show the evolution for Cr, Mn, Fe, Ni, after extraction from the mandible or maxilla (dot line), compared to reference implants which have not been used (dashed line).

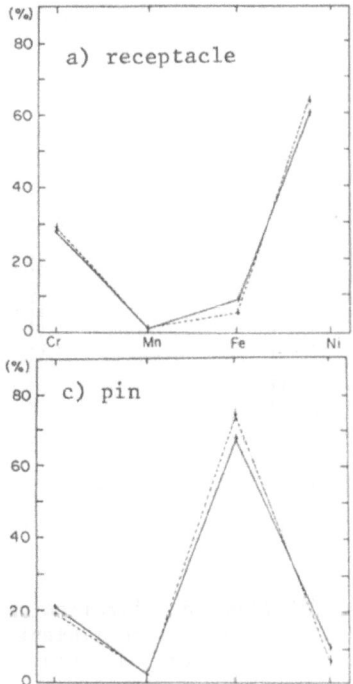
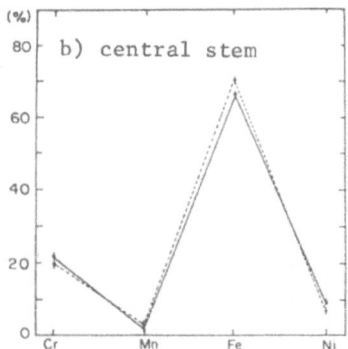

Figure 7

The receptacle of the extracted crutched implant contains less Fe, but more Ni, than does the receptacle of the reference implant. The inverse is found for the central stem and the pins, which contain more Fe and less Ni.

The number of blade implants analysed was insufficient to permit their discussion at this stage.

The variations of composition observed for the Type I implants are small, but not due to experimental error. We believe them to arise during the period of implantation, in situ, as a result of electrochemical corrosion phenomena and perhaps a cross migration of Ni and Fe between the receptacle and the pegs.

Electrolytic cell (fig. 8)

An electrolytic cell was set up the receptacle alloy as one electrode and the alloy of the stem (pins) as the other. Saliva or plasma served as electrolyte. The potential difference was found to be 0.24 V.

The cell using the two parts of the Type II implant as electrodes gave a potential difference of 0.27 V.

After a 6 month period, the electrodes of the two cells were cut into sections and analyzed by the PIXE method, the differences in composition were not significant. Analysis of the electrolyte fluid showed striking differences after incubation (Fig. 8a, b). We observe the migration of Cr, Mn, Fe and Ni. The nickel migration is the most significant and corresponds to 76% of the total metallic migration in the electrolyte.

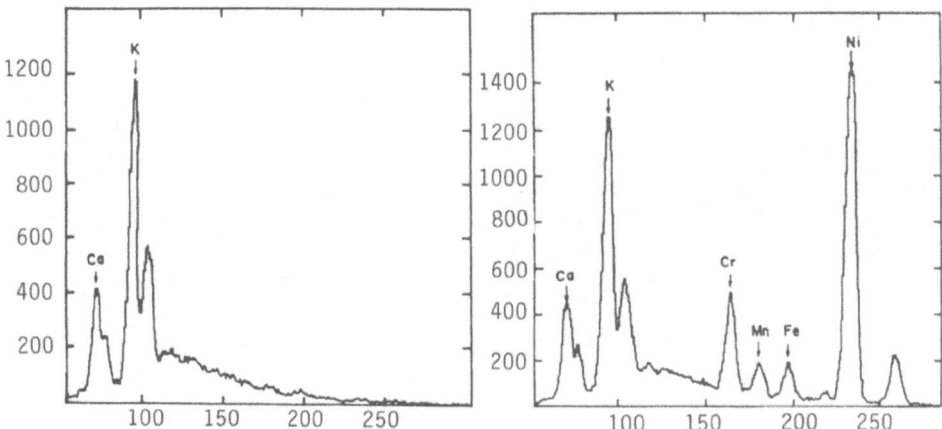

Fig. 8(a): Composition of electrolyte fluid (saliva) before incubation.

Fig. 8(b): Showing migration of metals from implant into electrolyte fluid.

DISCUSSION

PIXE analysis is a useful tool for the investigation of metallosis, an inflammatory tissue reaction in the vicinity of metal implants. Tissue sampling was undertaken when an implant had to be removed for medical reasons.

The samples were always taken from the capsule of connective tissue formed around the implant. For comparison, additional samples were taken at distances 1 to 4 cm from the implant on the same mandible or maxilla ; these so-called unaffected tissue was taken from patients during the extraction of the implant after large reactive and inflammatory damage.

These measurements and the results will be described in detail elsewere (2). They show an increasing concentration of nickel in the tissue around the implant. The mechanistic model of the implant corrosion and the metabolism of the corrosion products in the metallosis time has been investigated in details with animals (3,4). During the first 2 or 3 months after implantation there is strong corrosion of the implant. In the clinical cases this corrosion would occur by a splitting off of the individual crystallites (5 ,6). In the tissue these crystallites would be degraded into microcrystallites. It is in this first impregnation phase that the concentrations of the different implant components would greatly increase. After this period a capsule of connective tissue will form around the implant, corresponding to the cessation of the increase of the concentrations and a strong decrease in corrosion due to a shielding of the implant by the capsule.

But, because the concentrations of the alloy constituents in the contact tissue are not correlated to the concentration of the alloy,itself the effect cannot be explained by difference in the transport velocity and biological control mechanisms alone (7, 8).

Chemical, metallographic and micro analytical investigations indicate the need for further investigation because of the varying interpretation of the different findings and it is not possible to draw final conclusions solely on the basis of these analyses (4).

The highest accumulation in the contact tissue appears a short time after implantation. Thull (9) observed large variations of the electrical potential by investigating the electro-chemical behavior of metallic implants under condition similar to those in the tissue.

We observe in electrolytic cell in natural or artificial saliva potential difference of 0.24 V with using the receptacle alloy (Cr, 28% - Fe, 5% - Ni, 65%) as one electrode and the alloy of the pins (Cr, 20% - Fe, 72% - Ni, 7%) for the other.

After 6 months in the cell, we found essentially in the saliva an important migration of nickel. In the tissue of all biopsies taken in the neighborhood of such typical implants we observed (2) essentially an increase in nickel a factor : (2.5 ± 1.1). The results for chromium were within in the error limit between biopsies and control tissue.

Our investigation by microanalysis of all the different parts of the implants (Type I) show, after some months implantation there is a loss of nickel of 2-3% in the central stem and the pins inserted in the bone with an increase in the receptacle and the surrounding tissue. Co-Cr-Ni alloys are frequently used for orthopaedic implants in spite of the increasing number of reports dealing with reactions of intolerance to these metals. In addition clinical and scientific research suggest a relation between the use of the alloys and the increase of allergic reaction to the metal (10, 11).

CONCLUSION

Our measurements permit the following conclusions to be drawn :

The composition of commercially supplied implants may be very different from that specified for the reference and made of several kinds of alloys.

Different metals or alloys in the same implant give rise to electro-chemical phenomena and should be avoided. In a typical (Cr, 20% ; Fe, 70% ; Ni, 10%) the loss of nickel from the stem and pins can be observed after some months implantation with an increase in nickel in the neighborhood of the implant.

References

1) M. Simonoff, Y. Llabador, G.N. Simonoff, P. Besse, and C. Conri, Cieangiographically determined coronary artery disease and plasma chromium level for 150 subjects, Nucl. Inst. and Meth. 231 :368 (1984)

2) A. Garuet, M. Simonoff, B. Berdeu, Y. Llabador, F.X. Michelet, and P.F. Caitucoli, Measurement of nickel and chromium at the site of metallic dental implants, NATO-OTAN Advanced Research Workshop on Biological incidences of Co-Cr-Ni alloys used in Orthopeadic Surgery and Stomatology, Bishenberg, France, 30Sept-4 Oct., 1985

3) F. Lux, J. Schuster, and E. Zeisler, A mechanistic model for the metabolism of corrosion products and of biological trace elements in metallosis tissue based on results obtained by activation analysis, J. of Radioanal. Chem. 32 : 229 (1976)

4) E. Dielert, and W. Winter, Biochemical and histotological investigations in the ambient tissue of metallic implants, Trace Element - Analytical Chemistry in Medicine and Biology, Vol. 2, 1983, ed. Walter de Gruyter & Co., Berlin-New York - Printed in Germany

5) F. Lux, and R. Zeisler, J. Radioanal. Chem. 19 : 289 (1974)

6) R. Zeisler, Thesis, Technische Universität München, 1974

7) R. Michel, and J. Zilkens, Untersuchungen zum Verhalten von Metallspuren im umgebenden Gewebe von AO-Winkelplatten mit Hilfe der Neutronnenaktivierungsanalyse. Z. Orthip. 116 : 666 (1978)

8) R. Michel, J. Hofmann, R. Holm, and J. Zilkens, Zum übertritt von Kororionsprodukten aus Stahlimplantaten in das Kontaktgewebe. Untersuchungen der Implantatoberfläche mit ESCA und instrumentelle Neutronenaktivierungsanalyse des Kontaktgewebes. Z. Ortop. 118 : 793 (1980)

9) R. Thull, Implantatwerkstoffe für die Endoprothetik, Schiele u. Schön, Berlin 1978

10) H.F. Hildebrand, B. Roumazeille, J. Decoulx, M.C. Herlant-Peers, P. Ostapczuk, M. Stoeppler, and J.F. Mercier, Proc. 3rd. Int. Conf. on Nickel Metabolism and Toxicology, Paris Sept. 1984

11) P. Ostapczuk, M. Froning, and M. Stoeppler, Proc. 3rd. Int. Conf. on Nickel Metabolism and Toxicology, Paris, Sept. 1984

SURFACE TREATMENT OF METALLIC BIOMATERIALS

TO IMPROVE BIOLOGICAL AND ECONOMICAL COMPATIBILITY

D.Muster*, M.Champy*, S.Szmukler*, C.Baltzinger**, and
C.Burggraf***

*Laboratoire de Stomatologie, Univ. Strasbourg I , France
**Laboratoire de Métallurgie Structurale, Univ. Metz,
***Laboratoire de Cristallographie, Univ. Strasbourg I

INTRODUCTION

The future of a surgical implant depends upon the bone-material interface, both for the tolerance and function in the physiological environment (Muster and Champy, 1978).

Improvement of metallic biomaterials designed for internal fixation or endoprosthesis requires a more precise knowledge of the structure and stability of the bone metal interface. We applied in this purpose surface physics methods for surface characterization at high resolution : SEM (scanning electron microscopy), TEM (transmission electron microscopy), ESCA (electron spectrometry for chemical analyses) or XPS (X-Ray photoelectron spectroscopy), AES (Auger electron spectroscopy : Carrière et al.,1985, Sundgren et al.,1985). These methods provide information about structure and binding mechanisms at the level of 10-30 A (Bouzouita, 1980; Muster et al.,1982), enabling thus to value the first stages of corrosion processes (Szmukler, 1982; Muster et al.,1982; Muster et al.,1983). We were among the first to apply these methods for this area of concern.

CORROSION OF IMPLANTS

We used small maxillo-facial Champy osteosynthesis plates made of stainless steel 316L (67% Fe, 18% Cr, 12% Ni, 3% Mo), provided by Martin (Germany), either at a virgin state or after 1.5 month to 2 years of implantation. It must be noticed that more than one-third of the patients don't come again for the removal of the plates after consolidation : the temporary implant becomes thus permanent.

An in vitro study has been conducted on plates after 30 months immersion times in standard physiological solutions (Earle type). Very little corrosion occured; in the remaining solution analyzed by spectroscopic atomic absorption, the presence of iron and nickel has been detected; chromium and molybdenum have not been detected (Szmukler et al., 1984). More representative physiological solutions are used actually.

Corrosion increases with time of implantation (Muster et al., 1983). We noticed however that implantation of 1.5 month may lead to a higher rate of corrosion than implantation of 9 months. But the studies have been made on different persons. We may conclude that the physiological liquid is not always of the same reactivity from one individual to the other one and depends also on the surgical handling (Comte, 1984; Sutow and Pollack, 1981; Williams, 1981; Williams et al., 1985). As the evolution of corrosion in vivo with time cannot be studied on the same person, such a law may only be deduced from statistical results.

The corrosion seems to be intergranular, starting at the grain boundaries wich are always high reactive centers and thus brittle areas for a material. An approximate evaluation of the amount of metal eliminated from the alloy has been done in estimating the volume of defects on the 2 years implant. About 0.1 mg of alloy has been dispersed in the body with a 4 holes plate of 72 mm^2 external surface.

Surface methods, in particular ESCA, have been used to characterize the surface of the implant itself.

In this method, the sample is placed under ultra-high vacuum and irradiated with soft X-rays of known energy hν ($\approx$1200 to 1500 eV). Electrons are ejected from different electronic levels of the atoms; their kinetic energy E_C depends on the energy of the X-rays and on the binding energy E_B of the electron ejected : $E_C = h\nu - E_B$. The measurement of E_C gives the different values of E_B. If the chemical environment is modified, the electrons of the atoms will have slightly different binding energies; this is called "chemical shift".

Spectra obtained on a virgin plate and on an osteosynthesis plate extracted after 9 months implantation show interesting features (Fig.1).

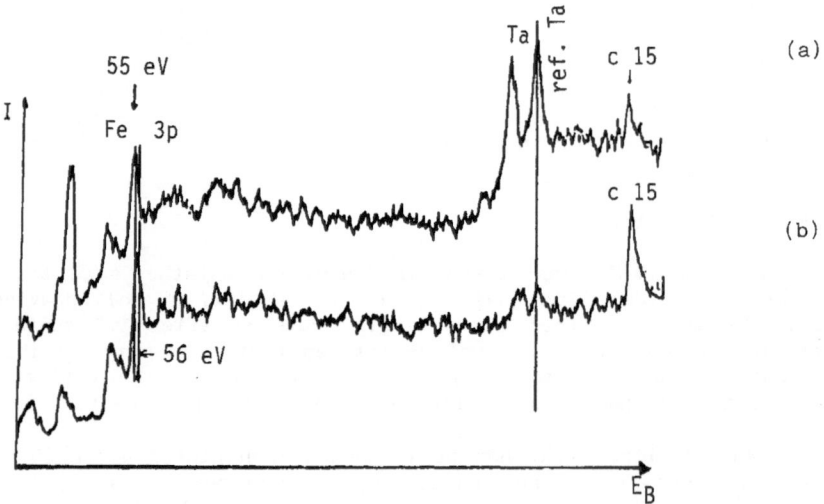

Fig.1.   ESCA spectra (number of photoelectrons versus binding energy E_B) obtained on (a) virgin plates and on (b) 9 months implants (Ta is used as reference).

We see the characteristic carbon peak 1s and a peak due to iron. A tantalum sheet has been analysed simultaneously in order to get a reference in energy. We notice that the peak of iron does not correspond to the same energy in the case of the virgin plate and after 9 months of implantation. We thus see that the oxidation of iron goes from Fe   to Fe   favouring oxides formation of the hematite type which are eleminated from the alloy; the structure of the alloy is thus destroyed and the elements will be eliminated from the plate.

SURFACE TREATMENT OF IMPLANTS

Faced with the reality of corrosion, we thus need appropriate means to avoid it. We may use a corrosion resistant high-technological material such as titanium and its alloys, tantalum, zirconium or platinum, but they are expensive. Furthermore, these metals show good biocompatibility but have poor mechanical properties.

Our actual aim is therefore to use traditional materials, less expensive, with well confirmed mechanical behaviour and to improve the biocompatibility of the materials used through appropriate surface treatments.

Our research work is directed in two ways : the formation of protective coatings on the implant and the formation of out of equilibrium compounds through ion implantation.

Different methods of coating (Holland, 1970) may be used. Vacuum evaporation, sputtering, galvanoplasty, chemical vapour deposition (Pauleau, 1985) are the most utilized. However, it is difficult to obtain with them continuous and regular films on complex shaped samples.

We have retained the use of thermal decomposition of metal carbonyls (Hieber, 1961; Calderazzo et al.,1966; Syrkin et al.,1975; Cotton, 1976; Nutt and Wawner, 1981; Vogt, 1982; Connor, 1983). A typical metal carbonyl molecule consists of a metal ion surrounded by six carbon monoxide groups. The dissociation of metal carbonyls in vapour phase can be represented in general as a series of elementary steps: vaporisation (sublimation) of the carbonyl, directing vapours toward the sample, chemical reactions in the vapour phase and on the sample (thermal dissociation of the carbonyl and other reactions), adsorption-desorption processes on the sample, nucleation and coating crystals growing. In practice, each step of the process may play a non neglectable role and requires a specific study. The reactions are :

$$M(CO)_6 \rightarrow M + 6(CO) \quad \text{and} \quad 2(CO) \rightarrow C + CO_2$$

One of the advantages of this process is the relatively low temperature at which these reactions occur. Deposition of the carbonyl vapour is achieved in the 200-600°C range, whereas most halide compounds require temperatures above 1000°C. Therefore, there is less risk of surface damage during deposition. Other advantages include easy availability of the starting products, high speed, ease of metallisation of samples of complicated shapes, possible automatisation of the process and high quality of the metallic layers obtained. All these advantages are making this method particularly promising.

The carbonyls are heated to a temperature T_1 which is the temperature of sublimation of the compound. Vapours of the metal carbonyl compound fill the whole space maintained under vacuum or in an argon atmosphere, under temperature T_1. The material to be coated, of complex geometrical form, with holes and humps, is placed in this atmosphere and the whole is brought to the temperature T_2 ($T_2 > T_1$); at this temperature T_2 the vapour is decomposed, a metallic film precipitates. That decomposition occurs at the surface of the material which has the convenient temperature; in case of no turbulence the precipitate is uniform.

We give the example of the reactions occuring with iron carbonyl

$$2Fe_2(CO)_9 \rightarrow 2Fe(CO)_5 + 2Fe + 8CO \qquad \text{at } 120°C$$

$$Fe(CO)_5 \rightarrow Fe + 5CO \qquad \qquad \text{at } 300°C$$

Our first experiments have been done under vacuum in sealed glass tubes; the metallic films obtained are very pure and well crystallized. Films of a few microns in thickness are obtained easily. Their crystallographic structure has been controlled with X-ray diffraction; in the case of iron we get the structure Im3m. The adherence on glass is of medium quality, but it is improved by ion plating with inert gas ions. General descriptions are given by Picraux et al.,1984 and by Brenier et al.,1985.

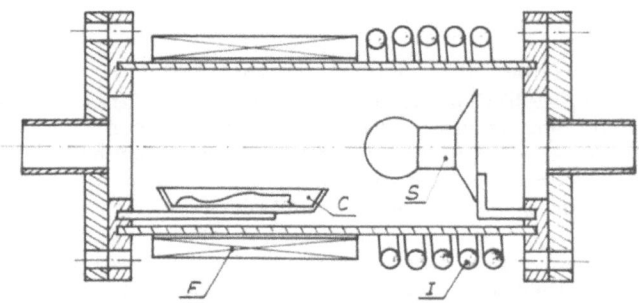

Fig.2.   Schematic view of the reactor used to produce uniform coatings after decomposition of metal carbonyls.
S = sample to treat; C = metal carbonyl powder;
F = furnace giving the temperature T_1 in the reactor
I = inductor to heat the sample at the temperature T_2

We have built a reactor in which those decompositions are now done (fig.2). This reactor is using a system of two heating coils; the furnace gives a temperature T_1, the inductor (high frequency) gives the temperature T_2 to the piece that has to be coated.

The reaction may be done either in an inert gas atmosphere or at low static vacuum (10^{-2} torr). The surface which has been coated in this way with a layer that is uniform is then bombarded with inert gas ions (Ar^+, He^+, Kr^+, Ne^+). This treatment is done either in a plasma chamber with a residual gas pressure of 10^{-2} mm Hg, or with an ion gun which has been adapted to this work.

On fig.3 is shown such an ion gun. Fig.4 shows a general view of an ion bombardment plant (Froger, 1981).

(a)

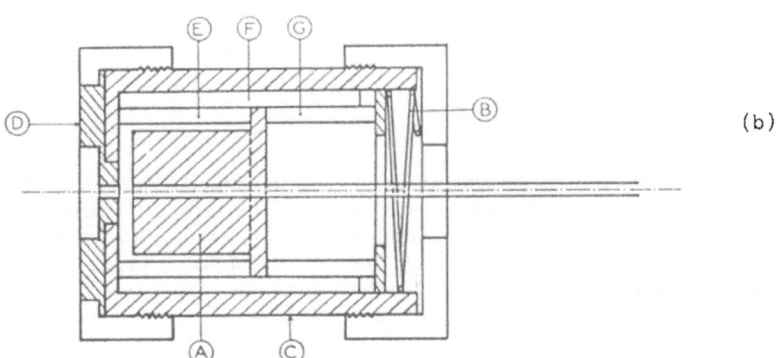

(b)

Fig.3. Ion gun for treatment of surfaces through
        ion implantation a) General view, b) Diagram
        (A) anode, (B) spring, (C) cathode),(D) dia-
        phragm, (E),(F),(G) insulating tubes (silica glass)

Fig.4. General view of the installation used for ion bombardment.

This gun alloys us also to get reactive ion bombardment.

Actually we get very good layers of iron. Experiments are done on Cr, Mo, Co, Mn, Ir and W. The coatings we produce by now are constituted of only one kind of atoms (Ir or W). Studies are underway to stabilize at the surface ternary compounds by means of ion plating, in order to get good biocompatibility with a substrate of molybdenum alloys. For example, we are trying to synthetize a bidimensional ternary compound containing the main elements of a very performant stainless steel like S350 described by Melotti and Vita Finzi (1982), Gabbi et al., 1985.

Adherence tests of the coatings we have realized have been done with the pulling-off method. An important development of this method has been performed by Szmukler, 1984.

CONCLUSION

In this work we have studied the in vivo corrosion of Champy type implants (osteosynthesis plates). By the use of scanning electron microscopy and of X-ray photoelectron spectrometry, we have shown that the mechanism of corrosion is of the intergranular type; an estimation of the amount of metallic ions leached into the organism has been done; our aim is to get a lower dissolved quantily.

In the second part of this work, we suggest to use classical materials having good mechanical properties and to apply to them surface treatments. We thus propose to coat the plates with metallic layers; stabilizing them by means of ion plating we may obtain bidimensional compounds on the surface, out of equilibrium, but very stable and having a good adherence to the substrate. Tests of adherence have been performed and an ion gun has been built for the surface treatments.

Firstly we propose to use at the surface films of iridium or tungsten stabilized by ion implantation. Secondly we want to stabilize by means of ion plating layers of three elements having a good adherence.

REFERENCES

Bouzouita M., 1974, Etude de l'interface os-métal par diffraction et spectrométrie électronique,
Thèse Doct. Chir. Dent., Strasbourg

Brenier R., Perez A., Thevenard P., Treilleux M., Capra T., 1985, Ion beam mixing of iron-titanium multilayers,
Materials, Science and Engineering, 69 : 83-88

Calderazzo F., Ercoli R., Natta G., 1968, Metal carbonyls : preparation structure and properties,
in, "Organic Syntheses via Metal Carbonyls", I.Wender and P. Pino, ed., Interscience Wiley Publ., New-York, 1-189

Carrière B., Deville J.P., Humbert P., 1985, Les informations chimiques obtenues par spectroscopie Auger,
J. Microsc. Electron., 10 : 29-61

Champy M., Lodde J.P., Schmitt R., Jaeger J.H., Muster D., 1978, Mandibular osteosynthesis by miniaturized screwed plates via a buccal approach;
J. of maxillo-facial Surgery, 5 : 14-21

Comte P., 1984, Metallurgical observations of biomaterials,
in, "Contemporary Biomaterials", J.W. Boretos, M. Eden, ed., Noyes Publ. Park Ridge, 66-91

Connor J.A., 1983, Metal carbonyls,
Organomet. Chem., 11 : 164-177

Cotton F.A., 1976, Basic inorganic chemistry
J. Wiley & Sons, 473-495

Froger M., 1981, Etude et mise au point d'un dispositif d'amincissement par bombardement ionique. Application à l'étude de quelques verres synthétiques et naturels au microscope électronique.
Thèse de 3è cycle, Université Strasbourg I.

Gabbi C., Melotti G., Tranquilli-Leali P., Vita Finzi E., 1985 S350 en orthopédie (Priv. Comm.)

Hieber W., Beck W., Zeitler G., 1961,
Neuere Anschauungen über Reaktionsweisen der Metallcarbonyle, insbesondere der Mangancarbonyls.
Angw. Chem., 73, 11 : 364-368

Holland L., 1970, Vacuum deposition of thin films,
Chapmann & Hall, London

Melotti G., Vita Finzi E., 1982, L'acciaio inossidable in ortopedia
Min. Ortop., 33 : 768-773

Muster D., Champy M., 1978 , Le problème d'interface os-biomatériaux. Acta odontostomatol , 121 : 109-124

Muster D., Jaeger J.H., Bouzouita M., Burggraf C., Baltzinger C. 1982, Application of Physical surface methods to the study of the stability and structure of bone-metal interfaces.
Anat. Clin. 4 : 183-188

Muster D., Szmukler S., Baltzinger C., Burggraf C., 1982 Fundamentals of bone-metal biocompatibility,
J. Dent. Assoc. South Africa, 37 : 845-847

Nutt S.R., Wawner F.E., 1981, CVD Coatings from metal carbonyl on SiC filaments, Ceram. Eng. Sci. Proc., 81, 2 : 840-848

Pauleau Y., 1985, Les revêtements métallurgiques de tungstène
   déposés par réaction chimique en phase gazeuse,
   Bull. Soc. Chim. de France, 4 : 583-593
Picraux S.T., Pope L.E., 1984, Tailored surface modification
   by ion implantation and laser treatment,
   Science, 226, 4675 : 615-622
Sundgren J.E., Bodö P., Lundström I., Berggren A., Hellem S.,
   1985, Auger Electron spectroscopic studies of stainless-
   steel implants,
   J. of Biomed. Mat. Res., 19 : 663-671
Sutow E.J., Pollack S.R., 1981, The biocompatibility of certain
   stainless steels, in :"Biocompatibility of clinical implant
   materials", vol.1, DF. Williams, ed.,
   CRC Press, Boca Raton, 45-98
Syrkin V.G., Prokhorov V.N., Ramanova L.N., 1975
   Preparation of iron coatings with prescribed properties
   by a carbonyl method,
   Zh. Prikl. Khim. (Leningrad), 48 : 2487-2490
Szmukler S., 1982 , Approche de la stabilité des interfaces os-
   métal par les méthodes de la physique des surfaces : étude
   de la corrosion de plaques d'ostéosynthèse,
   Thèse de Doct. Chir. Dent., Strasbourg.
Szmukler S., 1984, L'adhésion des couches minces
   Rapport de DEA, Chimie Appliquée, option matériaux,
   Université de Strasbourg.
Szmukler S., Muster D., Champy M., Baltzinger C., Burggraf C.,
   1985, Comportement des interfaces os-métal - Approche par
   les méthodes de la physique des surfaces, in : "Actes du
   Colloque Corrosion et dégradation des biomatériaux et leurs
   incidences cliniques", D. Muster, ed.,
   Conseil de l'Europe, Strasbourg, 145-152
Vogt G.J., 1982, Low-temperature chemical vapor deposition of
   tungsten from tungsten hexacarbonyl,
   J. Vac. Sci. Technol., 20 (4) : 1336-1340
Williams D.F., 1981, The properties and clinical uses of cobalt
   chromium alloys ,in : "Biocompatibility of clinical implant
   materials", vol.1, D.F. Williams, ed.,
   CRC Press, Boca Raton , 99-27
Williams D.F., Askill I.N., Smith R., 1985, Protein adsorption
   and desorption phenomena on clean metal surfaces,
   J. Biomed. Mat. Res., 19 : 313-320

CORROSION AND PROTECTION OF SURGICAL AND DENTAL METALLIC IMPLANTS

C. Sella*, J.C. Martin*,
J. Lecoeur**, J.P. Bellier**,
and J.P. Davidas***

*   Lab. de Physique des Matériaux, C.N.R.S., Meudon 92195
**   Lab. Electrochimie Interfaciale, C.N.R.S., Meudon 92195
*** Univ. Paris VII, Soc. Fr. des Biomatériaux et Systèmes
       Implantables, France

SUMMARY

Corrosion resistance of metals and alloys can be enhanced by protective coatings or by special alloying additives. In this paper the protection efficiency of Al_2O_3 coatings, deposited by RF sputtering onto NiCr and CoCr alloys, is tested. An intermedialy layer of Al_2O_3 cermet is deposited and this enables the stresses at the interface metal-coating to be reduced. Electrochemical studies, in different biological media (synthetical blood plasma and artifical saliva) are performed. The potentiodynamic current potential curves analysis is carried out. Using the faradic currents amplitude, the corrosion resistance of different metals and also the efficiency of the protective coating can be rapidly and simply tested.

INTRODUCTION

Most of metals used for orthopaedic and stomatologic implants and prostheses belong to the families of nickel base and cobalt base super-alloys elaborated for the industries of advanced technology (space - aeronautic - nuclear, ...) and have been experimented as biomaterials.

Metal corrosion in biological electrolytes, such as saliva and blood plasma, is an electrochemical process resulting from several oxidation reactions occuring simultaneously. All anodic surface reactions are not always corrosion reactions, a partial oxidation of some products of the solution can occur, so that the measured current corresponds to the addition of both an oxidation and a corrosion current.

EXPERIMENTAL TECHNIQUE

The corrosion resistance is evaluated from potentiodynamic anodic polarization curves giving the current density dependence on the electrode potential. A standard three electrode system is used (fig. 1). The working electrode is the specimen, the reference electrode, with a constant potential, is a saturated calomel electrode (sce) and the counter electrode is made of vitreous carbon. Potential is imposed on the working electrode versus the reference electrode and the current flowing between the specimen and the counter electrode is recorded continuously

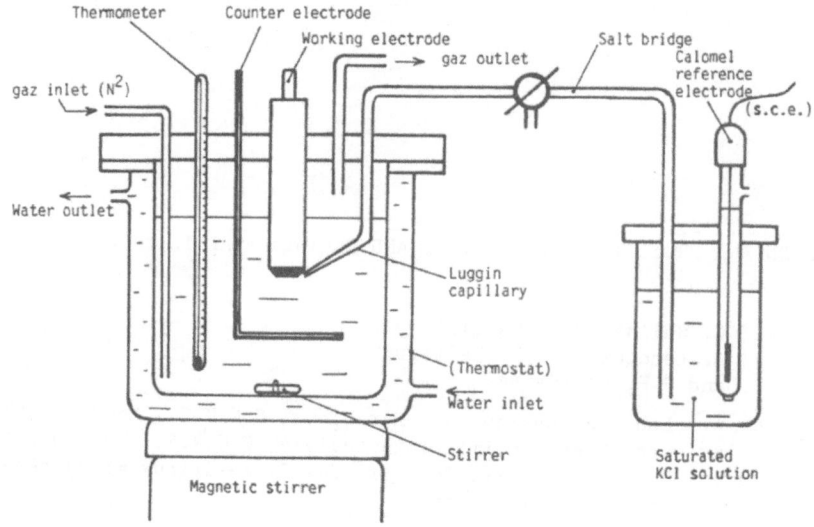

Fig. 1 : Electrochemical cell

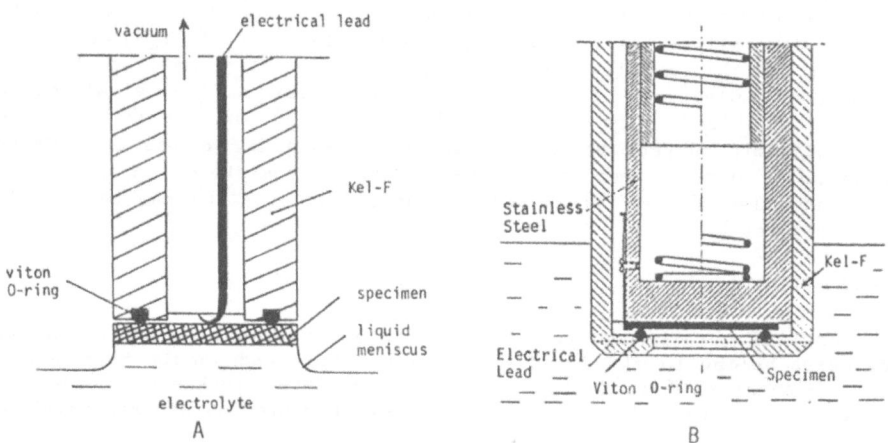

Fig. 2 : Specimen holder

on a linear scale. During the testing period the voltage alters at the sweep rate of 20 mV per second, between - 1,2 and + 1,5 V/sce.

Two types of specimen holder have been used. The first one (fig. 2A) is used in the case of an uncoated sample. The dipping technique is applied, so that only the surface is in contact with the solution. The second one (fig. 2B) is used with coated specimens, the O-ring means that only the surface is in contact with the electrolyte.

This technique permits the study of a lot of experimental conditions : - metals : chemical composition, casting, heat treatment, internal stresses, machining, grinding, sandblasting, surface quality, roughness, superficial contamination, coating defects ...,

- solutions : chemical composition, temperature, pH, aeration, stirring, organic or biological substances adsorption...

TABLE 1

Artificial saliva		Synthetic blood plasma	
Na_2HPO_4	0.260 g	NaCl	6.800 g
NaCl	0.700 g	$CaCl_2$	0.200 g
KSCN	0.330 g	KCl	0.400 g
KCl	1.200 g	$MgSO_4$	0.100 g
KH_2PO_4	0.200 g	$NaHCO_3$	2.200 g
$NaHCO_3$	1.500 g	Na_2HPO_4	0.126 g
		NaH_2PO_4	0.026 g
		Glucose	1.000 g
Ultra purified water to 1000 g		Ultra purified water to 1000 g	

Test specimens consist of disks with a thickness of 1 to 3 mm and a diameter of 12 mm. For some of them the surface was rough-cast, for others the surface was polished to a mirror finish using standard metallographic techniques, with fine diamond plastes dow to 1 um. The specimens are studied using a scanning electron microscope, both before and after corrosion tests. A compositional analysis is made for each specimen using an electron probe microanalyser.

The in vitro experimentation required the use of artifical saliva and synthetic blood plasma. The composition of these solutions is shown in table 1.

The artificial saliva was buffered with lactic acid to a PH of 6.7 and maintained at 37°C. These solutions were either oxygenated by air bubbling or oxygen depleted by nitrogen bubbling.

CORROSION BEHAVIOUR OF NiCr AND CoCr ALLOYS USED IN ORTHOPAEDIC SURGERY
AND STOMATOLOGY

Composition

Table II chemical shows the composition of the Nickel base and
Cobalt base implant alloys tested. This composition was determined by
electron probe microanalyser.

TABLE II

Type	Ref.	Composition (in weight %)
NiCr	Platinel	74 Ni - 16 Cr - 8.5 Fe - 0.2 Ti - 1 Si
without Mo	SP 20	66 Ni - 22 Cr - 9 Fe - 3 Si
With Mo + W	BiO 21	64 Ni - 18 Cr - 6.2 Fe - 8.5 Mo - 1.3 W - 1.2 Si
With Mo	AD - 1	72 Ni - 20 Cr - 6 Mo - 1.2 Si
		low B, C, Mn content for all specimens
CoCrMo	C 60	66 Co - 28 Cr - 5.5 Mo
	SP 2000	60 Co - 30 Cr - 1.5 Fe - 4 Ni - 1.5 Mo - 3 Si - 0.1 Pt
	AD - 2	67 Co - 27 Cr - 3 Mo - 1 Si
		low Fe, Ni, Si, Mn, C, P content for all specimens

Polarization curves

The potentiodynamic polarization curves for different alloys are
shown in fig. 3 for artificial saliva solutions and in fig. 4 for
artificial plasma. For corresponding alloys the shape of the curves are
very similar, demonstrating that in a given alloy corrosion is not very
different whether it occurs in artificial saliva or in synthetic plasma.
Furthermore, the corrosion in a given solution, depends strongly on the
alloy studied. For example, fig. 3 shows the curves for different NiCr
and CoCr alloys. In the case of Ni based alloys, including a wide variety
of compositions having a Nickel content ranging between 64 and 74 weight
% and a chromium content of between 16 and 22 %, the shape of the curve
depend strongly on the composition.

These alloys may contain up to 10-12 % of molybdenum and minor
alloying elements like Fe, Si may be added. The passivation domain is
reduced for alloys with less than 16 % Cr and without Mo and an important
hysteresis is seen (fig. 3A). This hysteresis could be due to the
kinetics of the formation of corrosion products. The addition of Mo
increases the passivation domain (fig. 3B) and a larger passivation
domain is obtained by the addition of Mo plus W (fig. 3C).

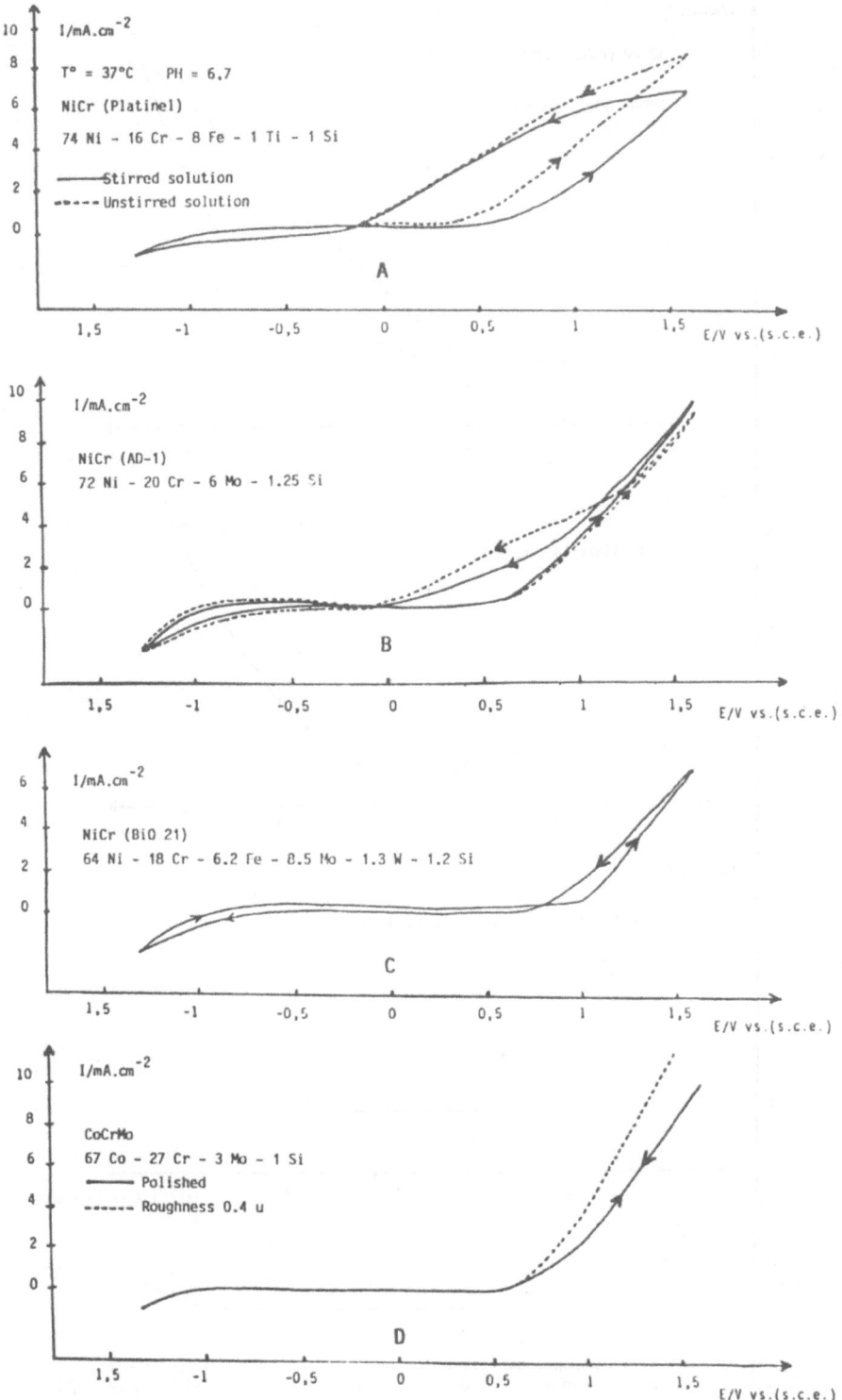

Figure 3   ARTIFICIAL SALIVA

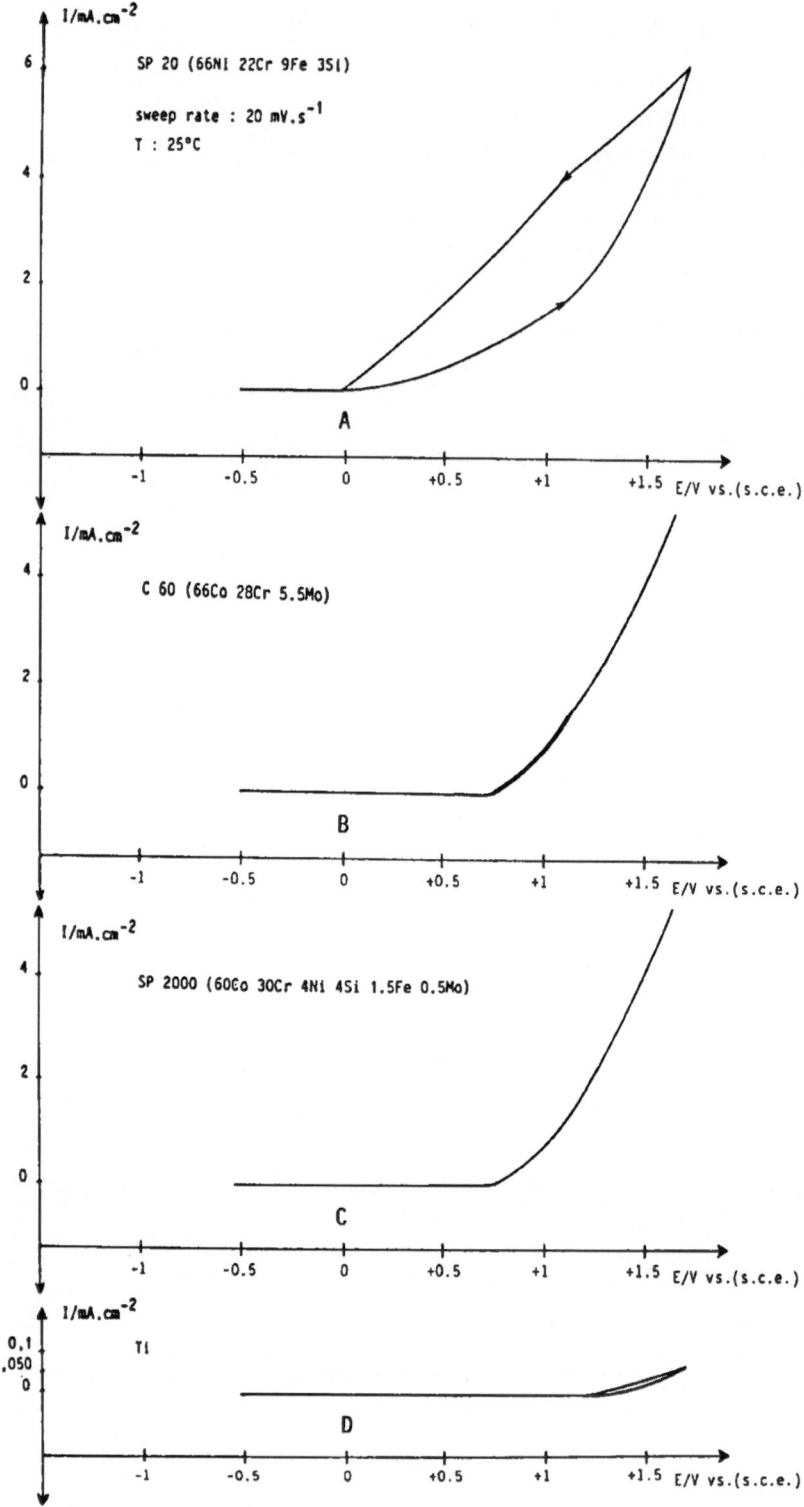

Figure 4   ARTIFICIAL PLASMA

In this last case, the hysteresis is greatly reduced.

CoCrMo stellites form a more homogeneous group : their composition and microstructure are vely close (table II) so that all polarization curves are very similar for all of the alloys studied. (fig. 3D, 4B, 4C). No hysteresis is observed and the passivation domain is roughly the same for all of them . The alloying elements form a homogeneous solution, showing in the as-cast condition a fully austenitic structure, with interdendritically precipitated $M_{23}C_6$ carbides (fig. 11). The corrosion resistance of these alloys can be attributed to this homogeneous structure and the cumulative effect of chromium plus molybdenum.

It can be noticed that, in comparison with the NiCr and CoCr alloys, titanium (fig. 4D and 17D) exhibits a superior corrosion resistance as it has a very large passive zone and also very low current densities (50 times lower).

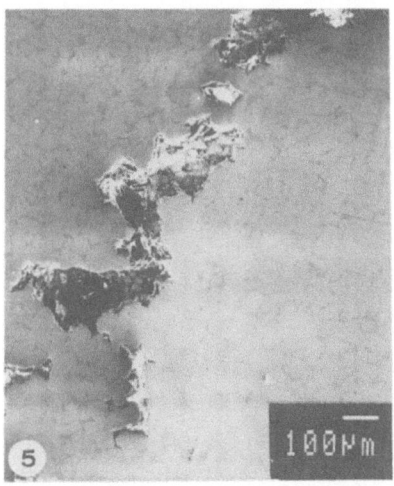 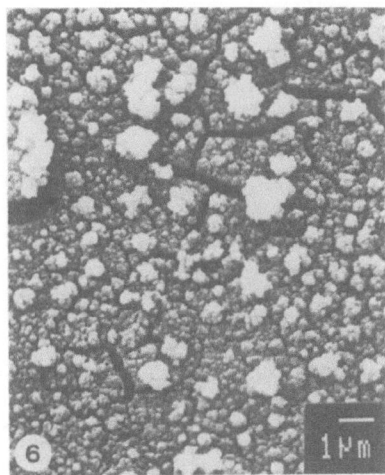

Fig. 5 : Scanning electron micrograph before corrosion test of a NiCr (SP 20) polished surface with cast defects

Fig. 6 : Corrosion products layer on NiCr (SP 20) after in vitro corrosion in artificial saliva

Scanning electron microscope studies of the alloys surface in corrosion

Fig. 5 shows a NiCr alloy before the corrosion test ; a large cast defect can be observed. After electrochemical corrosion the surface is covered by the corrosion products (fig. 6). After removing this layer, corrosion pits are observed in the case of Platinel (fig. 7 and 8) and SP 20 (fig. 9 and 10) alloys which have a low chromium content and no molybdenum.

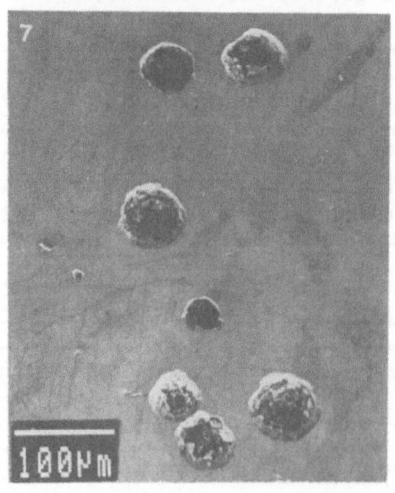
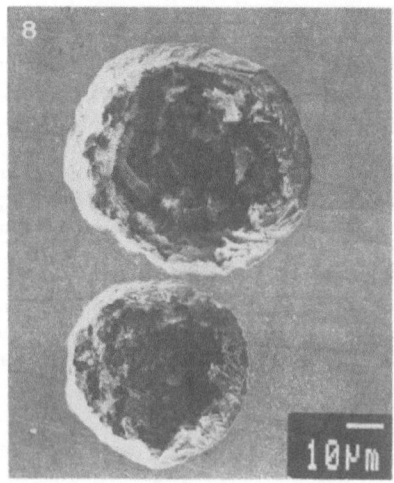

Fig. 7-8 : Ni - 74 Ni - 16 Cr - 9 Fe (Platinel) in saliva

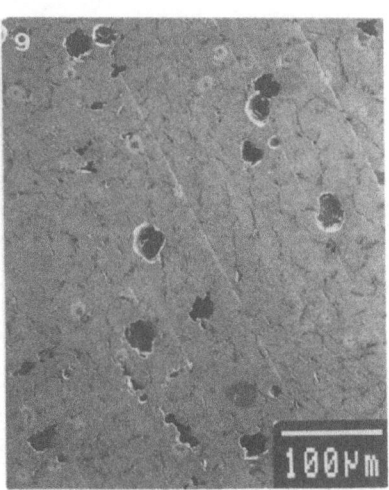
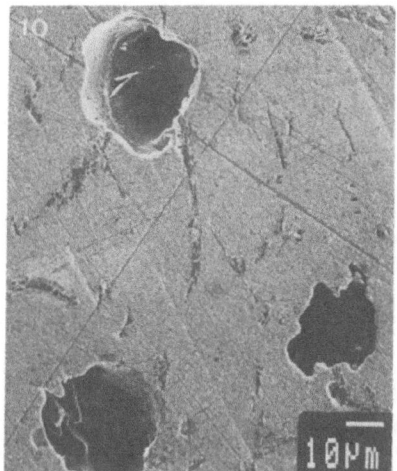

Fig. 9-10 : 66 Ni - 22 Cr - 9 Fe (SP 20) in plasma

Fig. 7-8-9-10 : Corrosion pits observed after removing the corrosion
products on the two NiCr alloys (without Mo)

The addition of molybdenum or Mo plus tungsten and an increase in
the chromium content greatly enhance the resistance to pitting and
crevice corrosion : no pits are observed after electrochemical corrosion
of such alloys (fig. 11 and 12). However intergranular Mo rich carbide

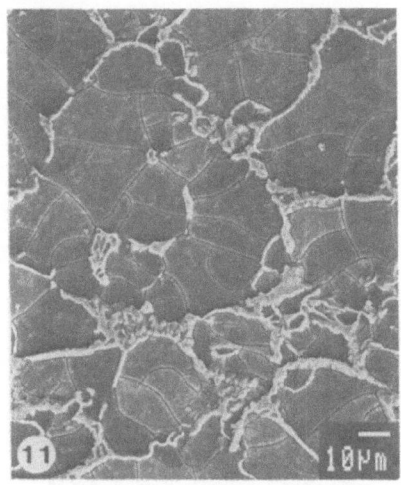 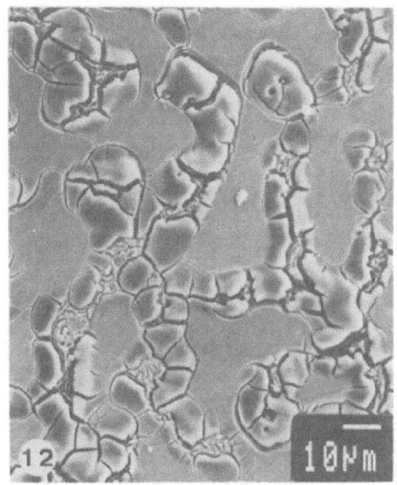

Fig. 11-12 : The addition of Mo or Mo + W enhances the resistance to pitting and crevice corrosion :

Fig. 11 : 67 Ni - 27 Cr - 6 Mo (AD-1 alloy in saliva)

Fig. 12 : 64 Ni - 18 Cr - 8.5 Mo - 1.3 W (BIO 21 in saliva)

precipitates are observed in relief since the corrosion rate is lower within these carbides than it is on the metal grains (fig. 11).

Similar surface aspects are observed in the case of CoCrMo before corrosion (fig. 13) and after a few polarization cycles in artificial saliva a thick layer, due to corrosion in the transpassive zone, appears and after drying gets the appearance of earthenware (fig. 14).

After removing this layer the precipitated $M_{23}C_6$ carbides are more visible using an optical microscope (fig. 15) or a scanning electron microscope (fig. 16).

PROTECTIVE COATINGS

Corrosion resistance of metals and alloys can be enhanced by protective coatings or by special alloying additives (i.e : Al added in hot corrosion resistant superalloys provides the constant presence of an Al_2O_3 protective surface layer).

In this paper the protection efficiency of Al_2O_3 coatings deposited by RF sputtering into NiCr and CoCr alloys is tested. The protective efficiency of the coatings is evaluated rapidly and simply from the potentiodynamic current-potential curves (fig. 17).

In the case of a simple Al_2O_3 coating, the passivation domain increases and the corrosion current is reduced a lot but not completely.

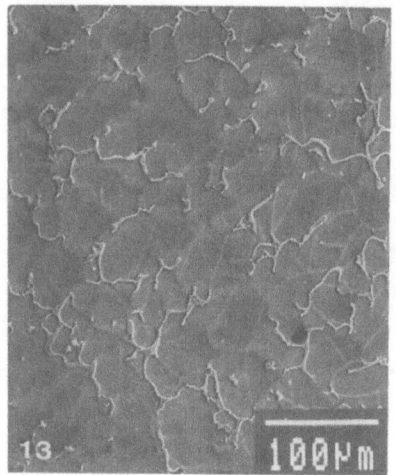 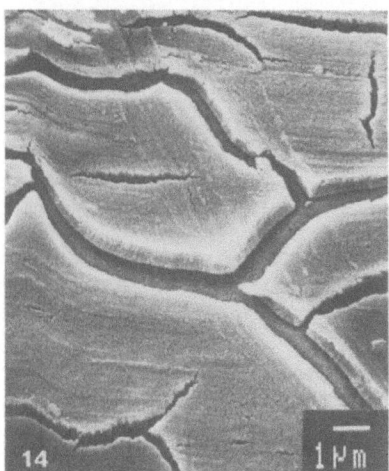

Fig. 13 : CoCrMo stellite surface before corrosion test

Fig. 14 : Corrosion products after many polarization cycles

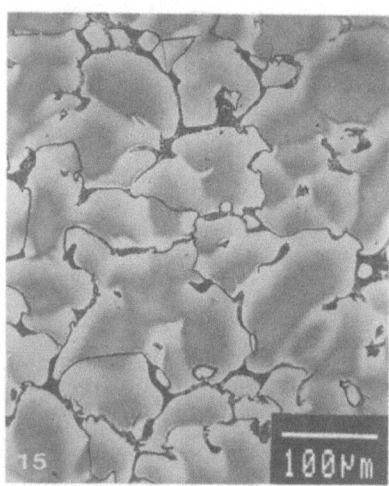 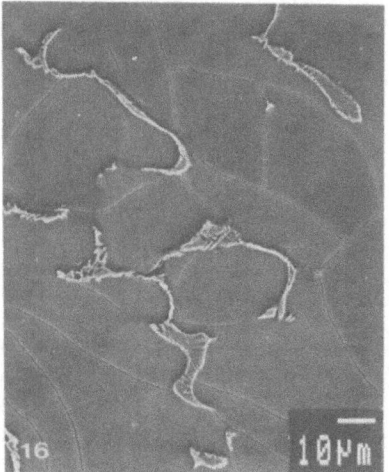

Fig. 15-16 : Surface aspect of the same CoCrMo specimen after removing the corrosion products layer. The corrosion rate within Mo ride carbides (in relief) is lower than it is on the metal grains.

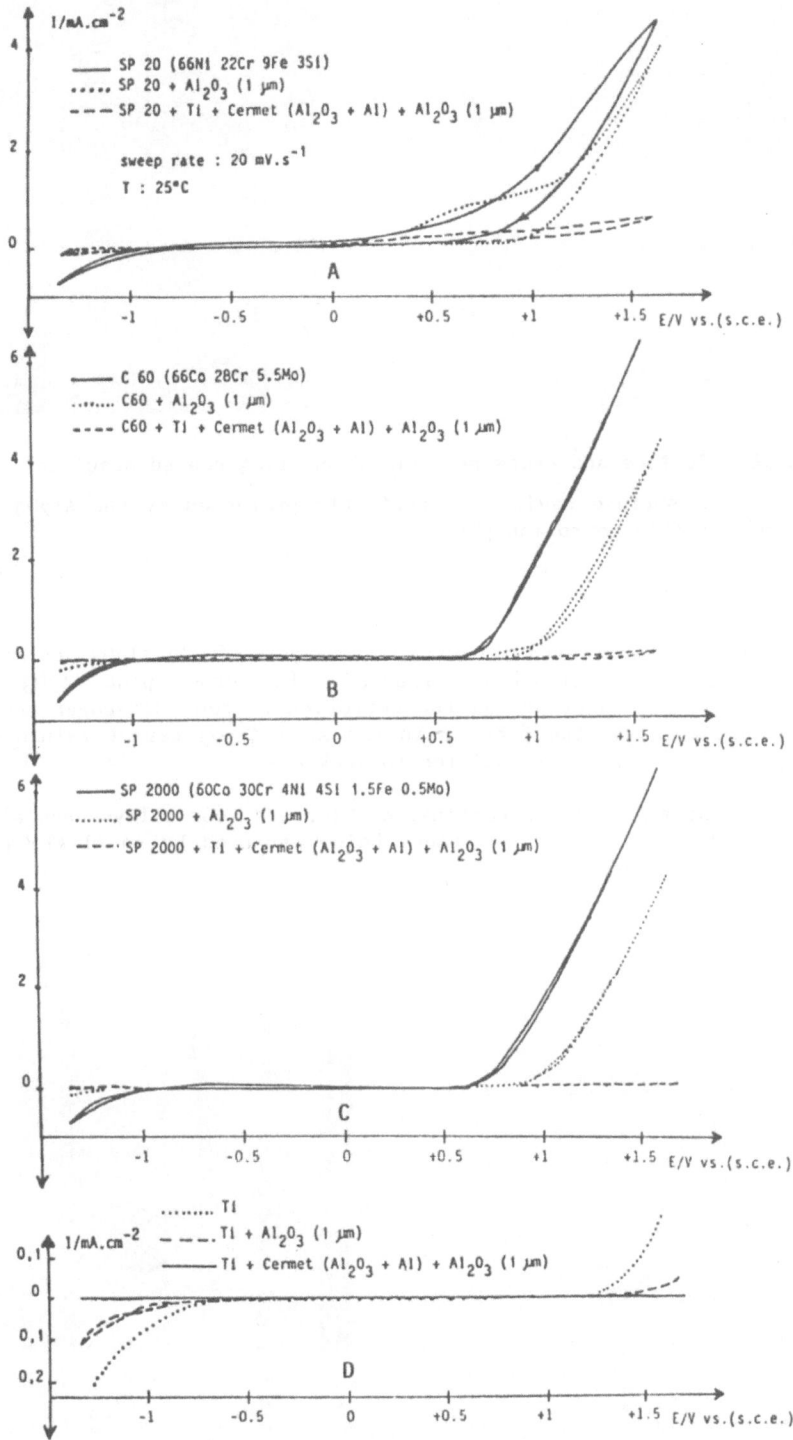

Figure 17   ARTIFICIAL SALIVA

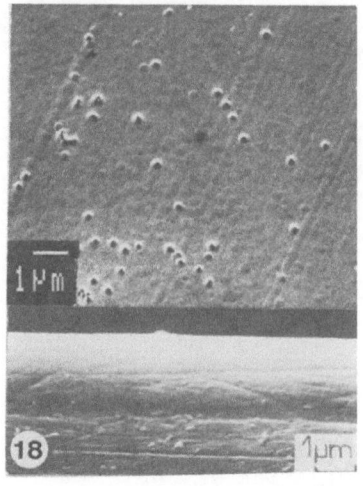
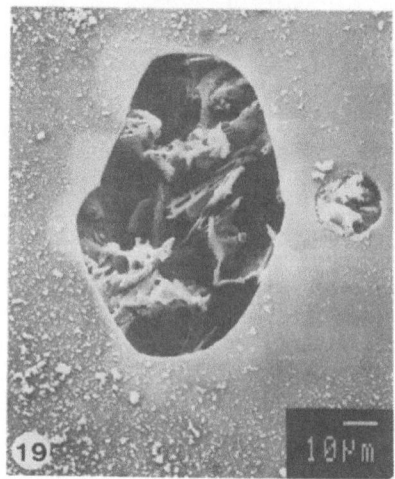

Fig. 18 : Surface and cross section of an Al₂O₃ coated specimen

Fig. 19 : A surface cavity not perfectly protected by the Al₂O₃ coating has induced this corrosion pit.

Using a sputtered intermediary layer of Al-Al₂O₃ cermet, the stresses at the interface are reduced, the cermet plus Al₂O₃ coating becomes strongly adherent and the corrosion current disappear completely (fig. 17). Moreover the free Al in the Al + Al₂O₃ cermet reinforces the Al₂O₃ layer in parts susceptibles to weakness.

Titanium protective coatings on NiCr and CoCr alloys have also been tested, their corrosion resistance being enhanced by an Al-Al₂O₃ cermet plus Al₂O₃ coating.

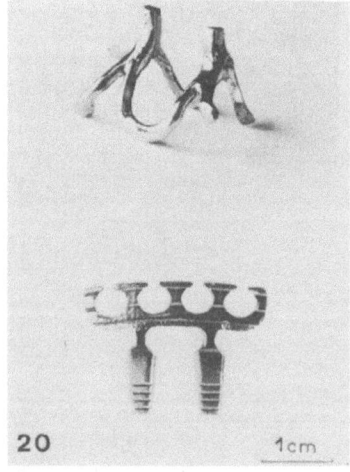
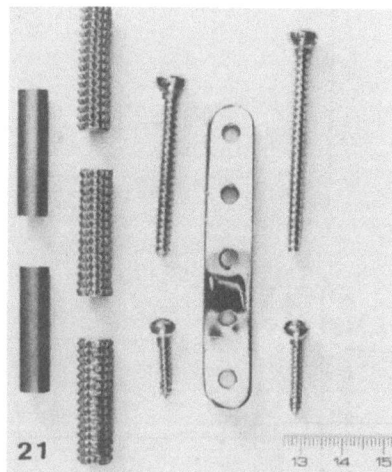

Fig. 20-21 : Surgical and dental metallic implants protected by an Al₂O₃ coating.

Fig. 18 show the surface aspect and the cross section of a specimen protected by a cermet plus Al2O3 coating. No corrosion is observed after many polarization cycles in artificial saliva or plasma.

The two main problems in the protective efficiency are due to the adherence of the coating and the surface quality and porosity of the metal, as shown in fig. 19 where a surface cavity has not been perfectly protected by the coating and thus a corrosion pit has been induced. Surgical and dental metallic implants protected by a multilayer coating are shown on fig. 20 and 21.

DISCUSSION

New deposition techniques are now experimented, such as ion plating and ion vapour deposition, in which the adhesion of the film is improved by high energy ion bombardment of the growing film.

New multilayer coatings such as TiN plus Al2O3 or TiC plus Al2O3 turn to be a better solution because their adhesion to the NiCr or CoCr substrate seems to be better than that of a cermet plus Al2O3 coating.

These in vitro corrosion test results are confirmed by cell culture as well as by clinical experience. Signs of corrosion have not been observed on any of the Al2O3 coated dental implants removed several years after implantation.

The electrochemical corrosion tests allow us to identify and quantify (by atomic absorption) all release materials which are susceptible to be diffused from metals implanted in vivo, both in situ and in the general circulation. Such release products might induce cytotoxic effects, allergic effects, ...

Work is being performed in collaboration with M.F. Harmand (Bordeaux), M.F. Sigot (Compiègne) and X. Ronot (Paris) using human cell culture models in order to evaluate the biological effects of both release products and surface corrosion layers.

Acknowledgment

We are pleased to acknowledge M. Akoun J. Andro, C. Baherze, A. Boutry, B. Cohen, C. Kohler, H. Oussalem, J.P. Plaut for their useful contributions to this work.

CHAIRMAN SUMMARY: QUALITY CONTROL

G. Milazzo

Let us start with the introductory talk about corrosion in general.
This talk attempted to summarize the majority, if not all, of the fundamental problems arising from the implantation of metallic prostheses in
living bodies. A few other papers in this session were devoted to particular
aspects of corrosion and its mechanisms, and one pointed to the possible
inhibition of corrosion using surface layers of Al/Al_2O_3, i.e. utilizing
the passivation properties of aluminium.

Other papers presented in previous sessions also dealt with corrosion,
e.g. those on tissue response, on *in vivo* corrosion, on biocompatibility,
etc... For details, see the individual summaries.

It is clear that damage of any kind would accur without solubilization by intimate contact of the metallic implanted material with the biological environment. The metallic material constituting the implant, can be
solubilized or interact with biological environment only if corrosion
occurs. This statement immediately points toward the need for electrochemical research to identify passitivity and immunity regions in the appropriate corrosion diagrams.

These diagrams are in any case necessary, sometimes, however, not
sufficient because corrosion is often accelerated by the presence of
particular organic compounds capable of reacting to form complexes with
the free ions present. The occurrence of complexes reduces the concentration of free metallic ions in the medium contacting the implant, therefore
displacing the position of the electrochemical equilibrium and as a consequence increasing the chemical attack on the implant.

A dramatic example which can be mentioned is that of stainless steel, i.e. Cr-Ni-Fe alloys, which are extremely resistant to attack by strong acids (hydrochloric, nitric, sulfuric) but are easily attacked by very weak acids like lactic and other similar organic acids.

Efforts must therefore be made to find new materials which can be used under immunity conditions rather than under passivity conditions. Indeed the passivity can be destroyed by formation of complexes. The previously mentioned example of stainless steel is quite convincing.

Considering the impossibility of reproducing *in vitro* the identical conditions which exist in biological tissues contacting metallic implants the conclusion is again reached that preference should be given to immune materials over passivated materials.

This means that before any new material is introduced in clinical practice, or even proposed, its corrosion properties must be thoroughly investigated, first under simple, reproductible physico-chemical conditions. Only those materials passing this first test (that is, showing immunity and/or passivity domains) should be investigated furhter under conditions simulating, in so far as possible, true biological conditions. For this second task, intimate cooperation between biologists, physicians and surgeons on one side and physico-chemists, electrochemists and biophysicists on the other side is absolutely necessary and unavoidable.

To realize and facilitate this cooperation, it is perhaps worthwhile to mention that a scientific Society exists which was founded several years ago, the "Bioelectrochemical Society". The aim of this society is to bring together scientists with different basic education (biological-physicochemical) to discuss complex and difficult biological problems which cannot be solved neither by biologists, nor by electrochemists or physicochemists, alone. It is obvious that biomedical problems are among the most important ones requiring cooperation as already discussed. Information on this society can be obtained from its Secretary General, Prof. H. Metzner (Universität Tübingen, Institut für Chemische Pflanzenphysiologie, Corrensstrasse 41, D-7400 TUBINGEN, B.R.D).

On the other hand, patients cannot wait until the rigorous thermodymic and kinetic research produces the best possible material. Therefore series of meaningful empirically thought experiments approaching one or another parameter suitable to give information on the behaviour of the tested materials must also be accepted. This point was illustrated by another paper presented in this session. Materials producing the best results can then be provisionally accepted and/or submitted to further basic investigation.

CONCLUSIONS
AND
RECOMMENDATIONS

CONCLUSIONS AND RECOMMENDATIONS

The participants of this interdisciplinary workshop have studied during five days the relationship *Implant-Patient* under its different aspects: fondamental, biological, clinical, immunopathological and technological as well.

During a special session at the last day, the participants have elaborated recommendations and preventive measures to be addressed to patients, dental technicians, clinicians, scientifics, manufacturers and public authorities. One of the most important recommendation is the propagation of information among the concerned groups, this means to inquire and to inform.

Recommendations to Patients

1. Think to make remove a temporary implant when it has acomplished its therapeutical function (after 1 to 2 years)

2. Draw the clinician's attention to possible allergic antecedants to metals.

3. When a patient state signs of intolerance, he should consult his surgeon (orthopaedist or dentist), since intolerance is not consequently the clinician's error. The clinician, however, is the first and the best to judge the causes of an intolerance.

Recommendations to Clinicians

1. Inquire himself and inform the patient about possible risks.

2. Withdraw ostesynthesis materials more systematically and more rapidly, and in any case where exist a risk for the patient's health. - (Except when the patient runs a greater risk by a second surgical intervention: anesthesia and others)

3. Before implanting osteosynthesis material, the surgeon or dentist should inquire into allergic antecedants of the patient and require an allergic test to metals. - Sometimes, however, prosthetic material is implanted in an emergency (maxillo-facial surgery, osteosynthesis in orthopaedic surgery) Thus it is very difficult for the surgeon to make respective investigations.

4. Respect the indications peculiar for each material, e.g. not to modify the form of an implant if it is not scheduled for.

5. Use strictly identical materials for multiple implatations: 2 identical plates, plates and screws made of identical material, etc...

6. Use nickel-containing alloys only when there is no other possibility.

7. Use only alloys of best quality.

Recommendations to Manufacturers

1. Produce only best quality with the best advanced technology.

2. Increase investment for research specific for biomaterials and encourage development of products designed for biological use. Indeed, many materials were designed for a different use than biomaterials (e.g. aerospace industries).

3. Avoid completely the presence of silicium and beryllium as alloy components.

4. Abandon alloys with high nickel content.

5. A regulation is needed concerning the threshold of nickel release from an alloy.

6. All non-precious alloys must undergo an electrochemical control preferably in biological milieu.

7. Perform corrosion tests in biological milieu.

8. Manufacture preformed implants in order to avoid any torsion during fixation (in particular for maxillo-facial and backbone surgery).

Recommendations to Scientists

1. Perform corrosion and degradation tests under biological conditions.

2. Assess the biocompatibility of a material by *in vitro* and *in vivo* tests.

3. Enhance research on the metabolization of corrosion products.

4. Undertake investigations on molecular basis and in the field of electrochemistry.

5. Carry out epidemiological studies.

Recommendations to Dental Technicians

1. Follow strictly the instructions for use given by the manufacturer in order not to decrease the quality of an alloy or of other materials.

2. Practise own quality control tests.

3. For health purposes work in as good conditions as possible and avoid inhalation of airborne particles, which may induce pathologies of respiratory tract and interstitial tissue.

1. Stimulate information.

2. Envisage certification and specification of products with indications of limit of use. Establish check-lists with alloy composition and performed analytical and biological tests. This may stimulate industrialists to propose only products of most possible quality.

3. The outcome should be international standardization.

It is pointed out that the mainly encountered problems with Cr-Co-Ni alloys are essentially induced by nickel. The other metals – chromium and cobalt - may also lead to unwanted biological effects. It must be underlined that the nickel release from an alloy should be much more considered than its content in the alloy.

Furthermore, the high cobalt content in recently elaborated alloys may have biological consequences which we cannot yet measure correctly at present.

The most frequently encountered effects are osteolysis, tissular sclerosis and especially immunological reactions. No clinical evidence is given for a disquieting relationship between these alloys and the advent of neoplasms in human organism.

CONTRIBUTORS

Toxicity, 21-27, 37, 41, 78,
        101, 107, 124, 159,
        161, 207, 325
Toxicology, 183, 222
Transferrin, 38
Tribromosalicylanilide,190
Tumor, 11-17, 25, 90-92, 125
Tumorigenesis, 89
  malignant, 11, 161
Tungsten, 84, 323, 347, 357
Turbulence, 344
TXRF-47, 53

Ulceration, 202, 270
Ultratec, 315
Ultratrace, 59-71
Unibond, 313-315
Urin, 31-42, 46-53, 69-71, 91,
        108-111, 134-150, 296
Urticaria, 279-280

Vanadium, 332
Vascularization, 256
Vasoconstriction, 125
Vasodilitation, 213

Victory(as alloy), 313-315
Vitallium, 13, 63, 83-85, 101,
        145-149, 267, 323
Vitamin, 37
  B12, 37, 150
  C, 280
Volatilization, electrothermal,
        34
Voltametry,
  adsorption differential
        pulse, 46-53
  adsorption square wave,
        134-150

Welding, 323
WHO, 6, 242, 299
Wiron-77, 313
Wiron-88, 175-181

Yolk sac, 277

Zeeman effect, 34, 48
Zinc, 36, 170, 190
Zirconium, 101-103, 343